KU-660-044

THE WORLD OF BUTTERFLIES

THE WORLD OF BUTTERFLIES

VALERIO SBORDONI
SAVERIO FORESTIERO

GUILD PUBLISHING LONDON

Colour illustrations
Walter Aquenza, Milan: 1, 18A, 18B, 24, 26, 27, 64, 69, 102B
Giambattista Bertelli, Brescia: 10, 11, 16, 17, 19, 32, 70, 76, 82, 83, 91, 93, 99, 100, 111
Luciano Corbella, Milan: 68, 101, 102A
Lorenzo Orlandi, Milan: 22, 75
Gabriele Pozzi, Milan: 5, 6, 7, 8, 9, 12, 13, 15, 20A, 20B, 21, 23, 28, 29, 30, 31, 33, 34, 35, 36, 37, 38, 39, 40, 41, 42, 43, 44, 45, 46, 47, 48, 49, 50, 51, 52, 53, 54, 55, 56, 57, 58, 59, 60, 61, 65, 66, 67, 71, 72, 73, 74, 79, 80, 84, 85, 88, 89, 90, 92, 94, 95, 96, 97, 103, 104, 105, 106, 107, 108, 109, 110
Aldo Ripamonti, Milan: 2, 3, 4, 25, 77, 78, 86, 87, 98, 112
Romano Rizzato, Milan: 62, 63, 81
Lino Simeoni, Verona: 14

Black and white drawings
Vittorio Salarolo, Verona
Giorgio Seppi, Trento
Lino Simeoni, Verona

Translated from the Italian by Neil Stratton, Hugh Young, and Bruce Penman

This edition published 1985 by Book Club Associates
By arrangement with Arnoldo Mondadori Editore S.p.A. and Blandford Press

Typeset by
Rowland Phototypesetting (London) Limited
Printed and bound in Italy by
Officine Grafiche di Arnoldo Mondadori Editore, Verona

Contents

Foreword

As far as the general public is concerned, butterflies and moths are the best known and possibly the best loved of insects. The splendid colours that often decorate their wings are without doubt the basis for this popularity. The phenomenon of metamorphosis is another reason for their fame since in this group the change can be so spectacular as to become a symbol of transformation itself (the "worm" changes into an "angelic butterfly"). Lepidoptera have also been the subject of thousands of scientific papers, and they continually offer experts new areas for research. That more than 3,000 scientific articles were published on them in 1978 gives us some idea of the interest they generate. In addition to this extensive scientific literature there is an equally rich popular one, which emphasizes the attractiveness of Lepidoptera both to the layman and to the expert. However, if one were to ask the man in the street what he actually knows about these animals, the answer would be brief – a few names, some general facts about their life cycles, and a few curious features of the behaviour of a small number of species: the cabbage white, the swallowtails, the processionary moths, the saturnids, and a few others. Consequently this book by Valerio Sbordoni and Saverio Forestiero, which is not just a list of species enlivened by beautiful colour plates but an attempt to open a window on a world unknown to most people, is most welcome. It does not confine itself to displaying multicoloured butterflies and a list of names; instead it explains the biological significance of their coloration, the mechanisms of their evolution and its intimate connection with that of the higher plants, and the causes of the diversity found in animals and discusses the distribution of members of the Lepidoptera. The meaning of the exceptional mimicry displayed by these insects is explained, as are certain unusual behavioural features, their migrations, and how they relate to a wide variety of environments, these apparently fragile animals having colonized virtually every corner of the dry land. The result is a true picture of a group of animals that has always been closely associated with human beings but about which, as this book will show the reader, little has been known until now.

From the approach chosen by Sbordoni and Forestiero in their discussion of butterflies and moths it will become clear how much can be learned from the study of animals when they are investigated over the whole complex of phenomena that underlie their evolution and govern their life. A book like this provides us with an incentive to explore the world around us and to look more closely at the countryside, which all too often we view only carelessly and superficially. We shall thus realize that the countryside is also a habitat and hope to stimulate among people that feeling for nature that ought to be the inheritance of a true civilization.

Sandro Ruffo

Preface

When we were asked some years ago to write a popular book on butterflies and moths, we accepted, confident that there would be plenty of available information to offer the reader on these popular, captivating insects and in part that we should be able to clarify the less obvious processes and mechanisms of natural history. Our plan was to provide thorough answers to the questions anyone might pose on behaviour, ecological habits, and metamorphosis, on the reasons for the differences between the sexes and between individuals and species, and on the evolution of butterflies and moths. This apparently simple objective turned out to be very difficult to achieve because it became clear that, to answer these questions, it would be necessary to explain quite a bit of biology, minimizing as far as possible the number of terms and theoretical content.

Our book attempts to offer a different way of looking at butterflies and moths, and we must record our debt to two famous pioneers of this approach, E. B. Ford and A. B. Klots, whose writings moulded a whole generation of lepidopterists. Popular literature on butterflies is certainly not scant; in recent years natural-history publishing has seen the appearance of a number of field guides as well as large illustrated catalogues. A feature common to most of these books is the use of photographs, made possible by the recent wide distribution of technologically advanced equipment. The form of illustration chosen for this book deliberately runs counter to this trend as we felt that drawings would be more suitable both as a way of illustrating concepts and phenomena and as a link with the great natural-history illustration of the last century. In the book we have attempted to show the diversity of butterflies and moths and their adaptations, trying wherever possible to explain these in terms of evolution. We have taken particular care over the choice of examples by consulting hundreds of scientific works, especially original research and reviews published in the last fifteen to twenty years. Other examples have been taken directly from our personal experience and from that of our fellow lepidopterists. The illustrations are based on material in both museums and private collections as well as on specimens photographed in the wild.

The arrangement of the text and the plates needs some further explanation; for we have tried, as far as possible, to interrelate the questions treated in the text and plates: the text generally deals with concepts and, where appropriate, draws on individual species as examples; in the plates and the captions relating to them, on the other hand, the material receives greater attention. Thus, for example, in the chapter on adaptations against predation the plates show some typical adaptations of adults and caterpillars, whereas within a theoretical and systematic framework the text deals with these same adaptations independently of the developmental stage that displays the adaptation.

The nomenclature of butterflies and moths is highly complex, and no one who is not an expert can be familiar with it in detail. We have therefore tried to use the most widely adopted names of genera and species in the recent literature. Given the popular nature of the

work, it did not seem appropriate to conform to the increasingly common tendency to place the name of the author after a taxon.
We should like to thank numerous people for their help in the production of this book: Marina Cobolli Sbordoni continually helped us at the various stages of the project. Donatella Cesaroni and Tommaso Racheli have collaborated in the editing of particular sections, the former having contributed to the discussion of speciation and the latter to that of the families of the butterflies. Many people have offered advice or material for the illustrations: Daniele Baiocchi, Emilio Balletto, Luciano Bullini, Carlo Prola, Romolo Prota, Piero Provera, Tommaso Racheli, and Vincenzo Vomero. Charles Remington has been of particular assistance not only in providing detailed information on the research undertaken by himself and his fellow workers on North American forms but also in critically scrutinizing plates. A similar revision of the plates has been undertaken with enthusiasm and care by Tommaso Racheli. Carmen De Carlo provided efficient assistance in the final stages of the preparation for the work. Raimondo Cardona has offered us the benefit of his ethnolinguistic skill, opening up a quite new cultural perspective for naturalists. Much of the illustrated material is in the collections at the University of Rome and the Museo Civico di Storia Naturale in Verona, and we wish to thank those in charge of the collections, Augusto Vigna Taglianti and Sandro Ruffo, respectively. We should also like to express our sincere acknowledgment to the latter for having stimulated and encouraged us from the outset.
We thank Mondadori, the publishers of the original edition, for having believed in a book of this type on butterflies and moths, and in particular we wish to express our warmest appreciation to Giuseppe Agostini and Giuseppe Parisi for their editing of the manuscript.
During the long gestation of this book we have learned a great deal about butterflies and moths, and we have been completely absorbed. We hope that the reader may find the same pleasure in the discovery of these fascinating insects.

Valerio Sbordoni
Saverio Forestiero

Structure, Origin, and Relationships of Butterflies and Moths

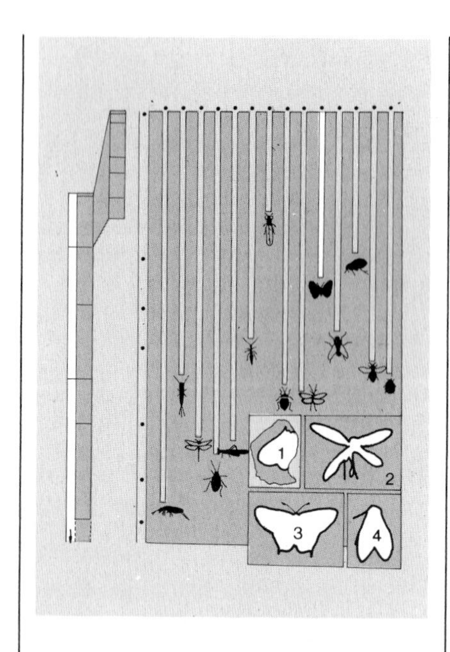

Plate 1 **The age of the Lepidoptera and their fossils**

A relatively recent insect order, fossil finds date the appearance of the butterflies and moths from the beginning of the Cretaceous (about 130 million years ago). The main diagram illustrates the origin of the Lepidoptera in relation to thirteen other common insect groups. The insert shows four fossil species: (1) *Doritites bosniaski* (female, Papilionidae, in Miocene marls from Monte Gabbro, Pisa, Italy); (2) *Undopterix sukatshevae* (Micropterigidae, lower Cretaceous, Siberia); (3) *Prodryas persephone* (Nymphalidae, Oligocene, from Florissant, Colorado); (4) *Zygaena miocaenica* (Zygaenidae, in bituminous rocks from the Miocene, Württemberg, Germany).

Butterflies and moths are a fairly homogeneous group of insects that forms the order Lepidoptera. The name is derived from the Greek words λεπις, "scale," and πτερόν, "wing," referring to the scales covering the surfaces of the wings; the structure and pigments of these scales are responsible for the extraordinary variety of wing colours. The insect order Lepidoptera is one of the largest and most important of orders: About 165,000 species have been recognized and described. This figure will inevitably rise in the future as hitherto unexplored regions, where the fauna is still virtually unknown, are investigated and specimens in museums and other public and private collections are reexamined according to new classification criteria.

Like all other insects, adult Lepidoptera have a chitinous external skeleton (*exoskeleton*) and a body divided into three regions, head, thorax, and abdomen. A pair of legs articulates with each of the three thoracic segments — the pro-, meso-, and metathoraces — and two pairs of wings are inserted on the meso- and metathoraces. In common with other *holometabolous* insects lepidopterans develop indirectly and pass through several very different stages, beginning with the egg and followed by the usually *eruciform* (caterpillarlike) larva, the pupa, and finally the imago, or adult. Consequently they undergo complete metamorphosis. The following is a brief diagnosis listing the characteristics that, taken together, distinguish this order from all other insects:
Winged, with complete life cycle, terrestrial, only rarely aquatic, insects, small, medium, or large in size [from 1 to 100 mm (0.04 to 4 in) in length with a wingspread of from 2 to 270 mm (0.08 to 10.6 in)], two pairs of membranous wings more or less densely covered by scales; suctorial mouthparts or, more rarely, chewing mouthparts in the adult stage; larvae eruciform with mouthparts typically chewing.

Origin of Moths and Butterflies and Fossil Forms

Despite their extraordinary evolutionary diversity and richness — in the animal kingdom only the order Coleoptera has more described members — the Lepidoptera are, in comparison with other insects, a relatively young group. The Collembola, which many now regard as a separate class, were present as long ago as the Devonian, 350 million years B.P. (before the present), and the forests of the Carboniferous, dating from 300 million years B.P., were the home of numerous insects, such as dragonflies, including giant ones, cockroaches, and grasshopper like forms, that were similar to the modern insects.

Until a few years ago it was not possible to date the origin of butterflies and moths with any degree of accuracy because of the paucity of fossil remains (there were no more than 200 of these in all, some of which were mere fragments or individual scales, which were difficult to classify), and almost all came from the Tertiary rocks or amber of the Eocene, Oligocene, and Miocene. The most ancient of these remains were at most 50 million years old; and that some of them were of butterflies and moths similar to modern families thought to have been highly evolved suggested that the origins of the group went back further than had been first thought. On the other hand certain finds attributed to the Triassic (about 200 million years ago), which were initially described as Lepidoptera, have subsequently been recognized as belonging to other insect orders.

However, a series of recent discoveries has shed more light on the evolutionary history of the butterflies and moths: Fossil species belonging to two primitive lepidopteran families (the Micropterigidae and the Incurvariidae) have been found in amber from Lebanon and in sedimentary rocks in Siberia. All date from the lower Cretaceous, some 100 to 130 million years B.P. Two micropterigid fossils in particular, the

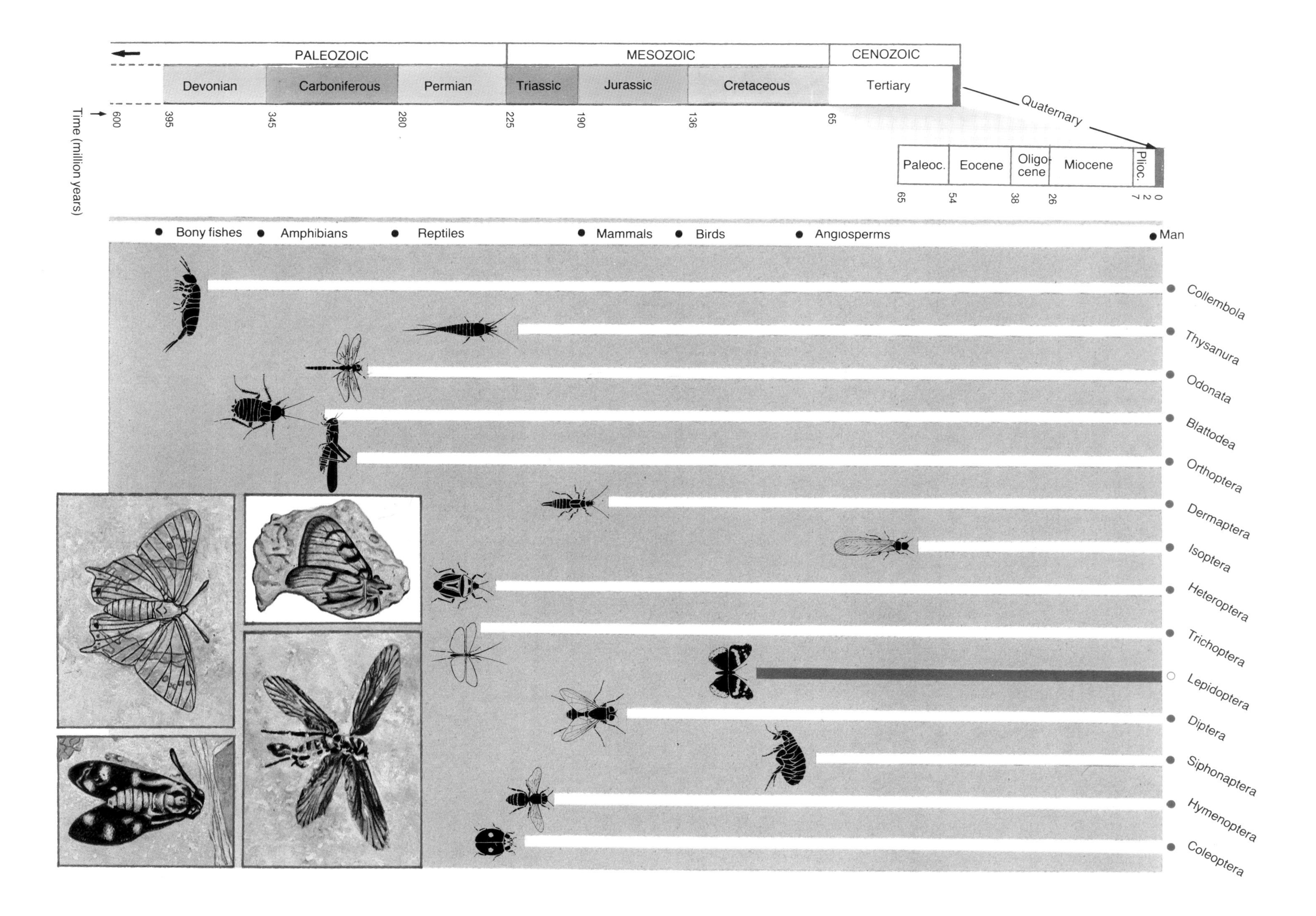
PALEOZOIC
MESOZOIC
CENOZOIC
Devonian
Carboniferous
Permian
Triassic
Jurassic
Cretaceous
Tertiary
Quaternary
Paleoc.
Eocene
Oligo-cene
Miocene
Plioc.
600
395
345
280
225
190
136
65
54
38
26
7
2
0
Time (million years)
Bony fishes
Amphibians
Reptiles
Mammals
Birds
Angiosperms
Man
Collembola
Thysanura
Odonata
Blattodea
Orthoptera
Dermaptera
Isoptera
Heteroptera
Trichoptera
Lepidoptera
Diptera
Siphonaptera
Hymenoptera
Coleoptera

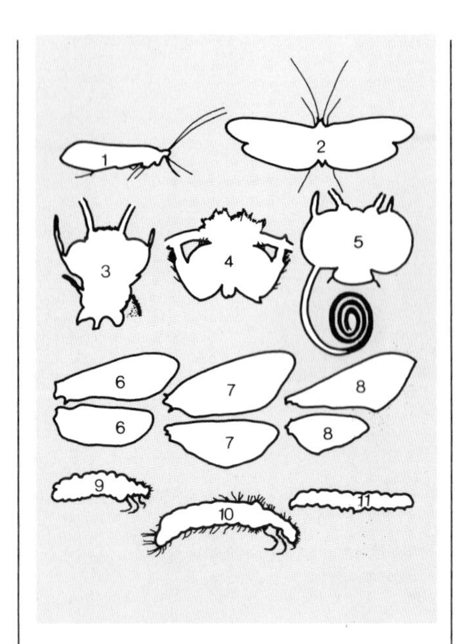

Plate 2 **Lepidoptera and their nearest relations, the Trichoptera**

This plate compares certain homologous structures of the Lepidoptera and the Trichoptera (caddis flies). Like most moths caddis flies at rest hold their wings over their abdomens rather like a roof (1). The wing surface of the Trichoptera is characterized by numerous hairs (2). The structure of the head and mouthparts shows certain affinities between the Trichoptera and the Lepidoptera in the extreme reduction of the mandibles and in the development of the galeae, which is limited in the Trichoptera (3) and certain families of Lepidoptera, such as the Tineidae (4), although it is very marked in other families (5). The wing venation of the Trichoptera (6, *Stenopsychodes*) is similar to, but much more complex than, that of the Lepidoptera, which, for example, almost always lack the fourth median vein (M_4). Furthermore, in the Trichoptera and the primitive Lepidoptera the wing venation of both wings is similar (7, *Oncopera*, Hepialidae), whereas it is somewhat different in the Ditrysia (8, *Hippotion*, Sphingidae). Caddis-fly larvae are aquatic and often construct a protective case (9) and breathe through tracheal gills (10, the larva removed from its protective case). Lepidopteran larvae are terrestrial and characterized by having abdominal prolegs (11, larva of *Anetus*, Hepialidae).

Siberian *Undopterix sukatshevae* (see Plate 1) and the Lebanese *Parasabatinca aftimacrai*, are very well preserved and leave no doubt as to their exact relationships. From these finds we are now certain that butterflies and moths were present at least by the early Cretaceous. The first fossils of flowering plants, the angiosperms, with whose evolution that of the Lepidoptera is certainly bound up, are also from the early Cretaceous (see the chapter Ecological Relationships).

However, that the first known angiosperms were already clearly differentiated into forms similar to modern ones (magnolias) suggests that their evolution had actually begun considerably earlier, perhaps in the Jurassic or even the Triassic. The same argument would apply to butterflies and moths, whose ancestors should possess characters markedly different from those of modern forms; these characters should appear intermediate to the caddis flies (order Trichoptera), which are generally recognized as the Lepidoptera's closest relatives.

It is highly likely that the Lepidoptera and the Trichoptera shared a common ancestor. Some features of the two orders are compared in Plate 2. The Trichoptera are distinguished principally by their wing venation, by having wings almost entirely covered by hairs (only rarely by scales), and by having a wing-coupling apparatus very different from that of the Lepidoptera. Trichoptera larvae are aquatic and may be either eruciform, as in the Lepidoptera (but protected by special movable cases secreted or built by themselves), or *campodeiform*, in which case they generally lack a movable case. The Trichoptera have *suctorial*, filtering mouthparts as adults and chewing mouthparts as larvae. However, although the Micropterigidae are certainly highly primitive lepidopterans, they cannot be regarded as the actual transitional form between the Trichoptera and the Lepidoptera. The first known trichopteran fossil dates from the lower Permian, more than 250 million years B.P. and more than 100 million years before the first find that can definitely be assigned to the Lepidoptera. It was possibly during this period that from a group of insects with terrestrial and aquatic eruciform larvae and with adults with membranous wings the Lepidoptera began to develop along a separate evolutionary line, adapting in the caterpillar stage to specialized phytophagous diets. According to this hypothesis, the evolution of the caterpillars would have occurred before that of the adult stage. Butterflies and moths would subsequently have gradually acquired their specific characters by evolving special structures for feeding on nectar as the angiosperms and their flowers took their places in the ecosystems. This hypothesis is supported by the recent discovery (published in 1983) in a Cretaceous deposit in Massachusetts of a fossil egg belonging to the Noctuidae, which shows that a "modern" form of moth was present as early as 75 million years B.P.

General Morphological Features

Following is a more detailed description of the principal morphological features of butterflies and moths.

Head

The head is small, subglobular, with two large compound eyes and one or more pairs of simple eyes (*ocelli*) — not always present. The antennae are composed of several segments and vary considerably, in both form and length, not only from one family to another but also between males and females of the same species. They may be *filiform*, *clavate*, or comblike on one side only (*pectinate*) or on both sides (*bipectinate*). Numerous sensilla of different types are located on the antennae and these receive olfactory,

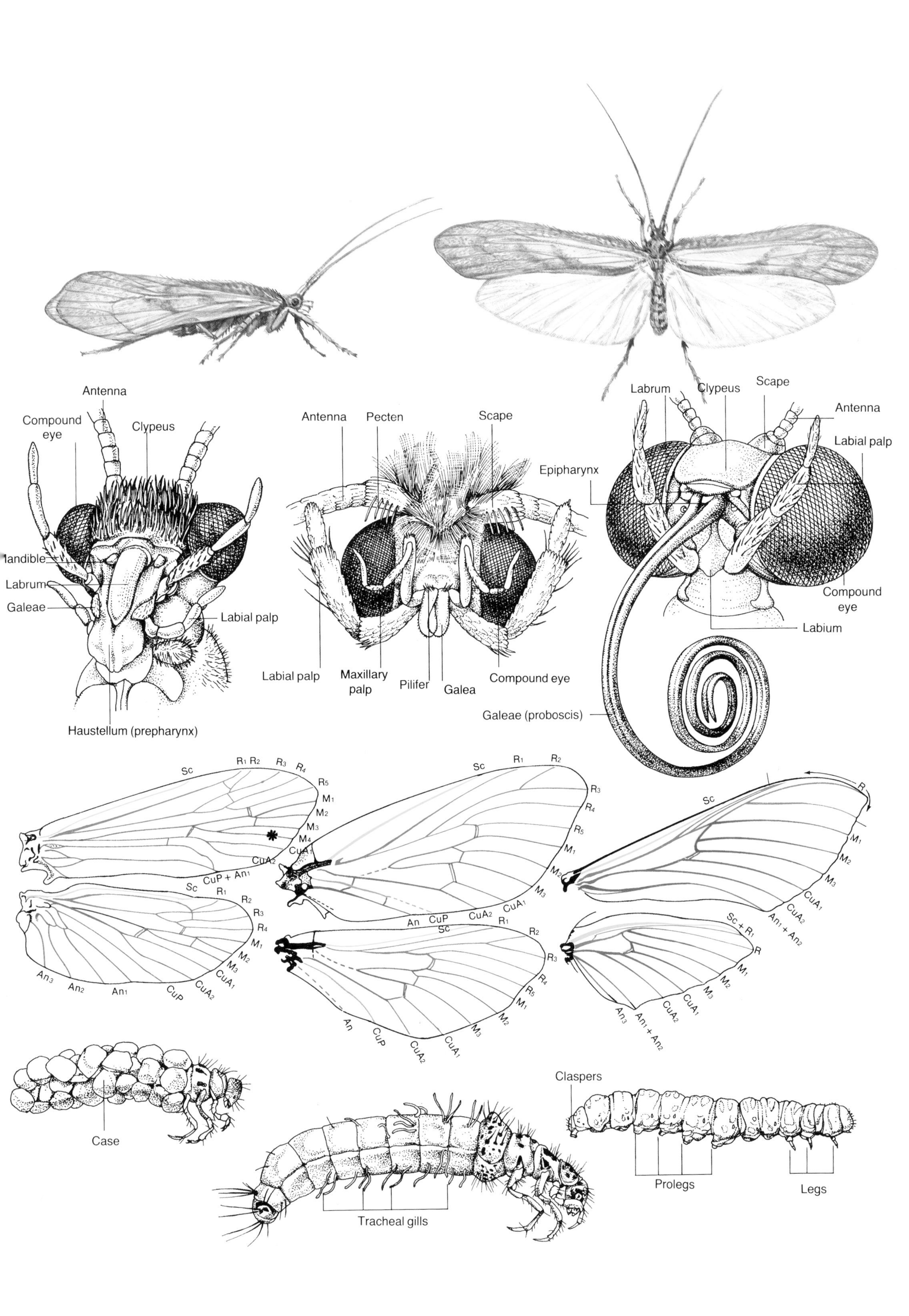

Antenna
Compound eye
Clypeus
Mandible
Labrum
Galeae
Labial palp
Haustellum (prepharynx)
Antenna
Pecten
Scape
Labial palp
Maxillary palp
Pilifer
Galea
Compound eye
Labrum
Clypeus
Scape
Antenna
Labial palp
Epipharynx
Compound eye
Labium
Galeae (proboscis)
Sc
R1 R2
R3
R4
R5
M1
M2
M3
M4
CuA1
CuA2
CuP + An1
Sc
R1
R2
R3
R4
M1
M2
M3
CuA1
CuA2
CuP
An1
An2
An3
Sc
R1
R2
R3
R4
R5
M1
M2
M3
CuA1
CuA2
CuP
An
Sc
R1
R2
R3
R4
R5
M1
M2
M3
CuA1
CuA2
CuP
An
Sc
R
M1
M2
M3
CuA1
CuA2
An1 + An2
Sc + R1
R
M1
M2
M3
CuA1
CuA2
An1 + An2
An3
Case
Tracheal gills
Claspers
Prolegs
Legs

tactile, and other stimuli. In most cases the mouthparts are highly unusual, being of a suctorial type derived from chewing mouthparts present in certain fossil forms and in the primitive Micropterigidae.

Usually the *mandibles* are greatly reduced or completely absent; the upper lip (*labrum*) is also reduced but projects laterally as two bristle-bearing lobes. In some cases, however, the two outer lobes of the mandibular maxillae (*galeae*) are developed to an extraordinary degree and enclose a channel, forming a sort of tube known as the *proboscis*. The structure varies in length and in certain sphingids may reach almost 30 cm (1 ft). At rest the proboscis is coiled into a more or less tight spiral beneath the head. The liquids on which the butterfly or moth feeds are siphoned through this structure by means of a pump formed by modified muscles and sclerites in the pharyngeal region. Apart from the galeae the other parts of the maxillae, such as the *cardo*, the *stipes*, the *lacinia*, and the *maxillary palps*, which may be composed of from one to five segments, are reduced or rudimentary. The lower lip (*labium*) is also poorly differentiated, and at times it is virtually reduced to a membrane. On the other hand the *labial palps*, which are composed of between one and three segments, are well developed and sometimes highly elongated.

The Micropterigidae, as already mentioned, differ markedly from other Lepidoptera in that they retain chewing mouthparts, which are regarded as primitive and from which the suctorial type just described is believed to be derived. Chewing mouthparts that are modified to a greater or lesser degree are present in many other insects (Orthoptera, Coleoptera, Neuroptera, Trichoptera, and so on) as well as in caterpillars. In the Micropterigidae there are a labrum with *pilifers* (a well-developed pair of mandibles toothed at their ends), maxillae with distinct cardo and stipes and with an elongated maxillary palp with five segments, and finally the labium with a fairly well-developed ligula and with labial palps consisting of three segments.

Intermediate forms exist between these two types of mouthparts, and in many families they are incompletely developed, reduced, or atrophied.

Thorax

The lepidopteran thorax is rather compact and consists of three closely joined segments. Each thoracic — and indeed each abdominal — segment consists of a *tergite* (dorsal), a *sternite* (ventral), and a pair of *pleurites* (lateral) that are divided into an *epimeron* and an *episternum*. The prothorax is normally reduced, and only in primitive forms is it developed to any degree. It generally bears two wing-like expansions known as *patagia*, which are regarded as rudimentary prothoracic wing processes. The mesothorax is the most highly developed segment, and in addition to the first pair of wings it bears a pair of movable processes, the *tegulae*, which cover the wing bases. The metathorax is considerably smaller, especially dorsally. In addition to the second pair of wings the metathorax of certain families, such as the Pyralidae and Noctuidae, bears a pair of tympanal organs for the reception of sound vibrations.

Typically the slender legs are composed of five parts: *coxa*, *trochanter*, *femur*, *tibia*, and *tarsus*. The front, prothoracic, leg has an elongated, partially movable coxa, and the tibia frequently bears an articulated and more or less elongated spur known as the *epiphysis*. This takes various forms and is used in cleaning the antennae and possibly at times even the proboscis. The middle, mesothoracic, and hind, metathoracic, legs differ from the front ones in generally having thicker coxae, which are always fused to the thorax and are each divided into an anterior *eucoxa* and a posterior *meron*. The middle tibiae bear a pair of terminal spurs at their distal end, while the hind tibiae have two pairs of spurs, one distal and the other median. On all legs the tarsus, or terminal portion, is divided into

Structure, Origin, and Relationships of Butterflies and Moths

five segments, or *tarsomeres*. The first of these is always longer than the others, and the final tarsal segment ends in two simple or forked claws (the *pretarsus*) and a structure known as the *empodium*, which consists of a fleshy median portion, the *pulvillus*, and two lateral lobes, the *paronychia*.

In certain families the legs have tended to become rudimentary to a more or less marked degree. The prolegs are reduced in many butterflies, in particular in lycaenid, libytheid, and riodinid males, while they are extremely rudimentary throughout the Nymphalidae. The hind legs are reduced in the males of certain geometrids and hepialids. Finally some female psychids have such underdeveloped legs that they are virtually *apodal* (limbless) as well as *apterous* (wingless).

The wings of the Lepidoptera consist of a double membrane supported by tubular structures known as *veins*. Their basic shape is roughly triangular with more or less rounded angles, it being possible to distinguish a *humeral* angle, an *apical* angle (*apex*), and an inner angle (*tornus*) and three margins: the costal (or *costa*), the inner, and the outer (or apical) margins. In the hind wing the inner angle and margin are known as the *anal* angle and margin. The forewings are generally larger; the hind wings display a variety of features. Sometimes a groove is present along the anal margin, sometimes they are folded like a fan, and quite often they have tails of varying length. In the Pterophoridae the four wings are distinctively fringed.

The two wing surfaces are more or less densely covered with scales and in some primitive Lepidoptera with small, unjointed, spinelike processes known as *microtrichia* or *aculei*. The scales are very highly modified, flattened, platelike bristles (*macrotrichia*), with a very short pedicel inserted into sockets in the cuticle, each of which is associated with a *trichogen* cell. They usually overlap — sometimes in a double layer — so as to cover the whole of the wing surface.

The scales house the various pigments responsible for chemical coloration. *Melanin* is the most common of these pigments, producing blacks and greys as well as contributing to red-browns and yellows. *Pteridines*, which produce brilliant reds, oranges, yellows, and some whites, are less common, whereas *flavones*, *ommochromes*, and *carotenoids* are responsible for most yellows and browns. Many of the metallic and other iridescent colours displayed by butterflies and some moths have a physical or structural origin. These apparent colours stem from the particular microstructures of the scales, which diffract and diffuse the light and give rise to interference phenomena. When a lepidopteran scale is examined under a scanning electron microscope, a highly regular series of parallel vertical ridges, or *striae*, can be seen on its outer cuticle (Plate 4). The scales responsible for physical coloration display numerous horizontal grooves (or thin lamellae) on both sides of the vertical ridges. The upper and lower sides of these grooves are only a fraction of a micrometer (a micrometer is itself a thousandth of a millimeter) apart. Interference colours arise when the wavelengths of the light reflected by the horizontal grooves or lamellae are proportional to the distance between their sides. The iridescent effect is increased by the presence on these scales of a dark pigment (melanin) that absorbs a large part of the transmitted light and thereby enhances the reflected component. The greens and blues of butterflies and in particular the ever-changing colours of the upper wing surfaces of morpho butterflies (*Morpho* spp., Plate 41), are determined by this mechanism, whereas in the same butterflies the coloration of the under wing surfaces is chemical.

Scales may vary in form in relation to their function, not only between different species or families but also on the same individual, and in practice virtually every transitional stage between a cylindrical hair and a well-defined scale can be found. The males of many species possess

Structure, Origin, and Relationships of Butterflies and Moths

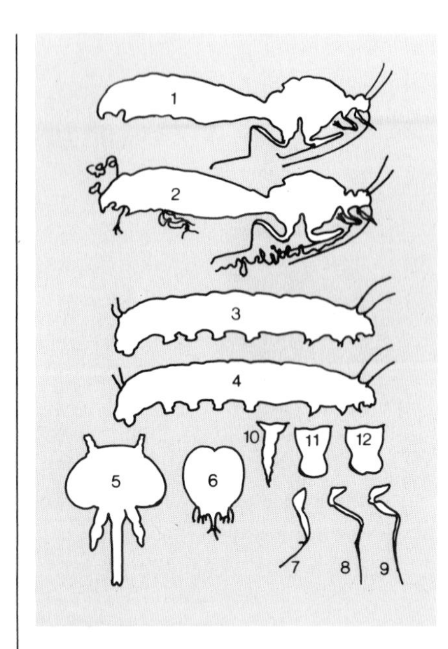

Plate 3 **The morphology and anatomy of the adult monarch and its caterpillar**

The external (1) and internal (2) structure of the adult and the caterpillar (3, 4) of the monarch (*Danaus plexippus*). The monarch in Figure 1 has had its wings and its covering of scales and hairs removed to show the articulation of the thorax, the abdomen, and the appendages. Structure of the head of (5) an adult butterfly and (6) a caterpillar. Note the differences in the mouthparts, which are suctorial in the adult but chewing in the caterpillar, the absence of compound eyes in the latter, and the presence of the spinneret, whose duct opens between the labial palps. (7, 8, 9) First, second, and third thoracic legs of the adult. (10) Thoracic leg of a caterpillar; (11, 12) two different kinds of proleg.

specialized glandular scales, known as *androconia* or *plumules*, which are situated both on the wings and on other body parts or appendages, such as the antennae, mouthparts, thorax, legs, abdomen, and genitalia. Androconia may be scattered between normal scales or clustered together in patches, tufts, or fringes. Sometimes they may be concealed in wing pockets or folds, and they may be protected by rows of scales or bristles. Androconia are connected to glands that secrete *pheromones*, special substances which act as chemical messengers between different individuals and which are particularly involved in sexual attraction.

Wing venation follows fixed patterns and is used as an important systematic character. Various systems of nomenclature have been proposed for the classification of the veins; the one used here, that of Henry Comstock, is widely employed in the recent literature. Generally the following veins are present on both wings: the *subcostal* (which supports the costal margin), the *radial*, the *median*, the *cubital*, and the *anal* (or *vannal*). The venation of butterflies and moths resembles that of the Trichoptera although it differs in several respects, such as the lower number of cross veins, the lack of the fourth median vein, and the absence of the *thyridium*, a smooth and transparent spot close to the fork of the median vein. There are two principal patterns of venation: The first, which is nearer to that of the Trichoptera, occurs in the Monotrysia, where the venation of the two pairs of wings is very similar; the second occurs in the Ditrysia, where the venation of the hind wings is reduced and clearly different from that of the forewings. In addition at the center of each hind wing there is an elongated area known as the *discal cell*, which is completely bounded by veins. The forewings are normally linked to the hind ones by special wing-coupling mechanisms, which vary from group to group. The Micropterigoidea and the Eriocranioidea have a *jugo-frenate* apparatus, consisting of a *jugum*, a projecting lobe on the inner edge of the forewing, and a *frenulum*, a small group of bristles on the costa of the hind wing. The coupling is achieved by the hooking of the frenulum to the jugum, when the frenulum lies under the forewing (in the Micropterigidae and Mnesarchaeidae), or through the downward pressure exerted by the jugum on the hind wing and by the upward pressure of the frenulum on the ventral surface of the forewing (in the Eriocraniidae). In the Hapialidae the jugum is a narrow bristle- or scale-bearing lobe strengthened by a branch of the anal vein while the frenulum is completely absent; coupling is brought about by the fitting of the costa of the hind wing between the jugum and the inner edge of the forewing (the *jugate* type). In many Ditrysia the jugum is absent and the frenulum consists of a row of long bristles situated at the base of the hind wing. The frenulum engages a structure on the forewing known as the *retinaculum*, which normally displays sexual dimorphism. In males the retinaculum is a chitinous lobe situated close to the wing base, whereas in females it consists of a group of hairs or strong scales angled forwards and located close to the hind edge of the discal cell. This type of coupling apparatus is known as *frenate*. In other Ditrysia and in particular in the Papilionoidea, Hesperioidea, and Bombycoidea no special structures exist, but coupling is effectuated by the marked expansion of the hind-wing base supported by one or more veins. This is the *amplexiform* type. Various intermediate forms occur between these four kinds of wing-coupling apparatus.

The flight of butterflies and moths, like that of other insects, is governed by complex mechanisms involving the thoracic muscles, divided into indirect (vertical and longitudinal) and direct, the fulcrum (a process of the pleuron), and the axillary sclerites that articulate with it. The powerful indirect muscles, which run vertically from the inner surface of the tergum to that of the sternum, contract and depress the thoracic box. The lowering of the tergum automatically raises the wings, and the latter hinge on the

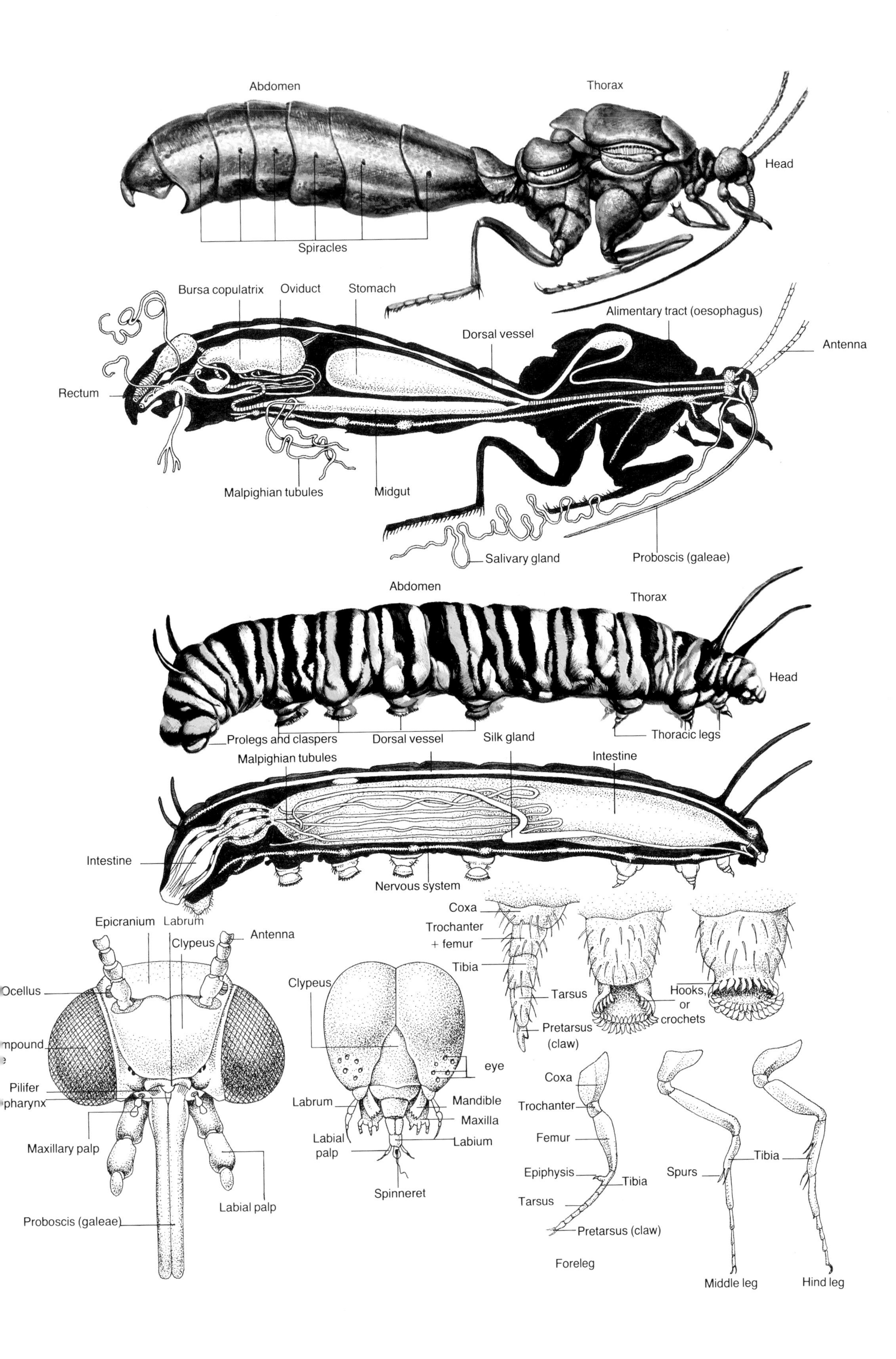

Abdomen
Thorax
Head
Spiracles
Bursa copulatrix
Oviduct
Stomach
Alimentary tract (oesophagus)
Dorsal vessel
Antenna
Rectum
Malpighian tubules
Midgut
Salivary gland
Proboscis (galeae)
Abdomen
Thorax
Head
Prolegs and claspers
Dorsal vessel
Silk gland
Thoracic legs
Malpighian tubules
Intestine
Intestine
Nervous system
Epicranium
Labrum
Clypeus
Antenna
Ocellus
mpound
Pilifer
pharynx
Maxillary palp
Labial palp
Proboscis (galeae)
Clypeus
eye
Labrum
Mandible
Maxilla
Labial palp
Labium
Spinneret
Coxa
Trochanter + femur
Tibia
Tarsus
Pretarsus (claw)
Hooks, or crochets
Coxa
Trochanter
Femur
Epiphysis
Tibia
Tarsus
Pretarsus (claw)
Foreleg
Spurs
Tibia
Middle leg
Hind leg

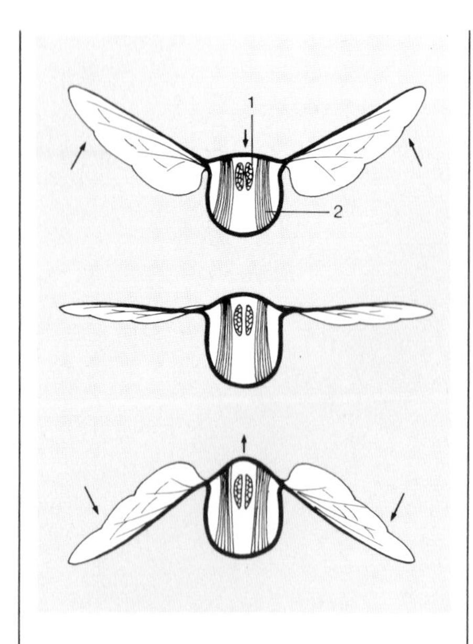

Diagram of the mechanism of flight in the Lepidoptera and other insects [(1) indirect longitudinal muscles; (2) indirect vertical muscles]. A section through the thorax and the role of the vertical and longitudinal indirect muscles are shown at three stages in flight. The wings are raised by contracting the vertical muscles, which lower the thoracic box, while the wings are lowered by the longitudinal muscles extending the thoracic box. Note the changing angle of the wings at different phases; this angle is principally governed by direct muscles. (Adapted from Snodgrass.)

fulcrum by means of the axillary sclerites. The thorax is then restored to its initial shape, and hence the wings are lowered through contraction of longitudinal muscles. The direct muscles, which are inserted on the sclerites at the base of the wing, actively produce forward, backward, and twisting movements of the wings. The speed and power of the flight depend upon the form and surface area of the wings and on the rate at which they beat — normally between two and forty times a second. Temperature has a great effect, and in many cases there is a process of preheating, in which the wings are rapidly vibrated on the ground in order to reach the threshold temperature, which is for many species in the region of 30°C. The sphingids are the fastest flyers and can exceed 50 km/hr (30 mi/hr). At rest the wings may be held rooflike over the abdomen or open and in the same plane as the body. In many moths the rear wings are folded up under the forewings when at rest. The wings may be rudimentary (*brachyptery*) or completely atrophied to the point of disappearing altogether (aptery). In most cases these phenomena involve only females (see the chapter Diversity and Evolution of Butterflies and Moths), and they are frequently associated with the adaptation to harsh conditions, occurring, for example, in species living at high altitudes where the adults are active only during the winter, and in desert species. In some island and sand-dune moths both sexes are flightless.

Abdomen

The abdomen of the Lepidoptera is cylindrical and fusiform, is more or less elongate, and consists of ten segments. The first eight of these are free, while the last two are more or less fused together and modified to form the external genitalia. Typically each segment consists of a dorsal tergite and a ventral sternite (the first two sternites are fused) and two membranous lateral pleurites. The spiracles, through which air enters the tracheae of the respiratory system, open on the pleurites; for the first six to eight segments there is one pair of spiracles per segment. There are also two pairs of reduced spiracles on the thorax. In some families there are a variety of *tympanal organs* on the first, second, and third abdominal segments. The abdomen also bears odoriferous organs in various locations. These are present in both sexes but vary considerably from one family to another, and they are sometimes structurally complex in the males.

As has already been mentioned, the male and female genitalia are derived from the transformation of the ninth and tenth abdominal segments. The nomenclature for the various parts that make up the apparatus is rather complex; only a very general outline will be given here. In the male there is a roughly triangular or trapezoidal dorsal plate known as the *tegumen*. This is derived from the ninth tergite, and it extends backwards as a single or double hooked process known as the *uncus*, derived from the tenth tergite. Ventrally the ninth sternite has been transformed into an archlike *vinculum*, which may extend forwards as a *median process*, or *saccus*, and laterally as a pair of platelike processes, the *valvae*, which are highly variable in form. The tenth sternite is horseshoe or jaw shaped — hence known as the *gnathos* — and with the uncus it forms a sclerotized ring within which the anal tube is situated. The penis, or *aedeagus*, lies in the center of the structure, between the valvae. It is more or less mobile and has a part, the *vesica*, that can be extruded. It is supported by sclerotizations of the intersegmental membrane between the ninth and tenth segments, a dorsal *fultura superiore* and a ventral *juxta*.

In females the structures are simpler. Depending upon the group, there is a genital aperture for oviposition known as the *oviduct*, which may or may not be separate from the genital aperture for copulation, the *ostium*

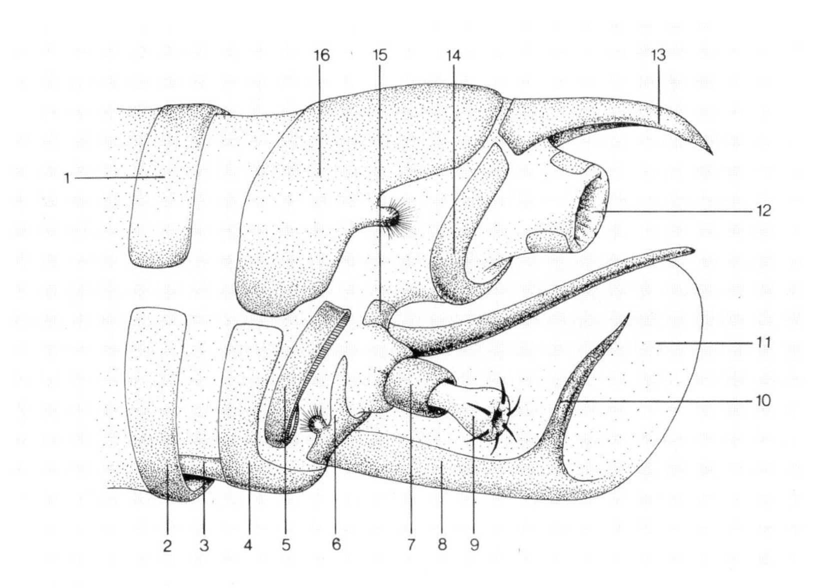

Diagram of the external genitalia of a male butterfly, viewed in profile. (1) Eighth abdominal tergite; (2) eighth sternite; (3) saccus; (4) vinculum; (5) section of left valva near base; (6) fultura inferiore (juxta); (7) penis (aedeagus); (8) lower process of the valva; (9) vesica, partly evaginated, with sclerotizations; (10) harpes; (11) valvula; (12) anus; (13) uncus; (14) gnathos; (15) fultura superiore; (16) tegumen (ninth tergite). (Adapted from Bourgogne.)

bursae, which leads via a copulatory duct, the *ductus bursae*, to the *bursa copulatrix*. Three principal structural models (see the chapter Systematic Review of Butterfly and Moth Families) are distinguished by the arrangement of the genital openings and the anus and the role of the last three abdominal segments. The *monotrysia* type has a single genital aperture for both copulation and oviposition, which may open into the terminal region of the rectum (cloacal aperture) or may be independent of the anus; the *exopore* type has two genital openings and an anus that are all separate although all situated on the ninth segment; finally the *ditrysia* type has separate genital openings situated on the seventh or eighth segment (ostium bursae) and the ninth (oviduct), the latter being more or less closely associated with the anus. Genital plates and other more or less sclerotized structures are also associated with female genitalia, and certain families have an ovipositor, which may be capable of piercing objects and is more or less retractile. The morphology of the male and female genitalia is an important taxonomic character and is used by experts to establish the relationships among taxa of various levels and especially, in many species, for definite identification at the specific level.

The abdomen is generally covered by abundant scales and hairs, which are very highly developed, particularly at the end. In the females of some families terminal tufts of hairs are used to cover and protect the eggs.

General Anatomical Features

In this section the principal organ systems of the insects are briefly reviewed and various features unique to the Lepidoptera are highlighted (the anatomy of an adult lepidopteran is illustrated in Plate 3).

Nervous System

Typically the nervous system consists of a *brain* and a suboesophageal ganglion connected by a short *circumoesophageal connective*. The optic lobes are highly developed, particularly in butterflies. The ventral chain of ganglia consists of three (in the more primitive forms) or of two thoracic ganglia and of four abdominal ones. The sympathetic nervous system is well developed. In many families there is a sensory structure, the *chaetosema*, which is peculiar to the Lepidoptera and of unknown

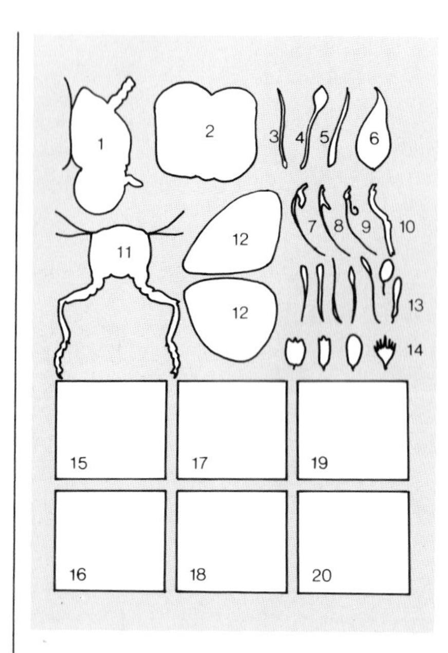

Plate 4 **Structural details of Lepidoptera**

(1) The head of an adult seen from the side with the proboscis coiled. (2) Transverse section through the proboscis. (3-6) Various types of antenna (filiform, clavate, unipectinate, bipectinate). Different types of maxilla in *Sabatinca aurella*, Micropterigidae (7), *Micropterix calthella*, Micropterigidae (8), *Mnemonica auricyana*, Eriocraniidae (9), *Sphinx*, Sphingidae (10). Note the differing development of the galeae in the three suborders: Zeugloptera (7, 8), Dachnonypha (9) and Ditrysia (10) — only part of the galea is shown in 10.

(11) Transverse section of the thorax of a generalized insect, showing the insertion of the wing, which articulates by means of axillary sclerites on the edge of the pleuron (fulcrum), and the muscular system involved in flight and locomotion.

(12) Theoretical model of wing pattern, from which it is possible to derive the majority of known patterns through the suppression or emphasis of eyespots or bands in particular parts of the wing.

(13) Various forms of androconia. (14) Four different kinds of scale (about 100 times larger than life size). (16) Different concentrations of pigment on the scales of a butterfly result in the different chemical colorations shown in Figure 15. Structure of the scales on the upper (17, 18) and lower (19, 20) wing surfaces of a male of *Morpho rhetenor*. The iridescent coloration of the upper wing surface is structural and is caused by grooves on the longitudinal ridges of the scale (shown under high magnification in Figure 18), producing interference phenomena. These grooves are absent from the scales of the under surface, whose coloration is chemical (20).

function. It consists of two papillae with numerous hairs, situated behind the antennae, close to the eyes, and linked to the brain by nerves.

Digestive System

The digestive system is characterized by extensive development of the *pharynx*, which functions as a suctorial pump because of powerful dilator muscles. The pharynx is followed by a thin oesophagus with a diverticulum, the *crop*, which varies considerably in form and development. Besides storing food, the crop can be filled with air and act as an aerostatic organ; it also seems to assist emergence from the pupa. The crop is absent in the Micropterigidae but highly developed in the Sphingidae. The middle portion of the digestive tract, or *mesenteron*, is simple and lacks diverticula.

One or two intestinal caeca are, however, present in the *proctodeum* and in particular in the *colon*. The entire digestive system may be reduced in species whose adults have an atrophied proboscis and do not feed.

Circulatory System

The dorsal blood vessel, or *heart*, runs the full length of the abdomen and extends forward as an aorta. It has seven or eight pairs of ostia, and there are generally two accessory pulsatile organs, one of which, the *mesotergal*, is more highly developed, and the other, the *metatergal*, reduced. The circulatory fluid of insects is called *haemolymph* and lacks hemoglobin.

Respiratory System

The tracheal system opens to the outside through two pairs of thoracic spiracles and six to eight pairs of abdominal ones. Some families have more or less well developed *air sacs* close to the abdominal spiracles.

Excretory System

The excretory system consists of six *Malpighian tubules* (occasionally two or four); three on each side join a duct that opens into the *proctodeum*.

Secretory System

The secretory system comprises the well-developed *salivary glands*, the *accessory glands* of the genital system, and the *tegumental glands*, which are largely odoriferous and situated on the wings, legs, and abdomen in males and on the abdominal segments in females. The principal endocrine glands are the retrocerebral *corpora cardiaca* and the *corpora allata* and, in the larvae, the *prothoracic glands*.

Reproductive System

Each of the female's ovaries are composed of four *meroistic polytrophic* (that is, the nurse cells, or *trophocytes*, lie close to the individual oocytes) *ovarioles* per side, which open into paired oviducts. These in their turn join to form a single large oviduct. The *spermathecae*, which store the sperm until fertilization, and the *accessory*, or *colleterial, glands* open into this oviduct as well. In the Monotrysia the *bursa copulatrix* also opens into this oviduct, whereas in the Ditrysia it opens to the outside through the *ostium bursae*. In the latter case the oviduct is linked to the bursa copulatrix by a *sperm duct*.

The male reproductive system comprises a pair of separate or fused *testes*, both of which generally consist of four follicles, two *vasa deferentia*, each with an *accessory gland*, and a *ductus ejaculatorius*, which terminates in a penis. During copulation the male transfers a *spermatophore*, a sperm-containing packet of variable form, to the bursa copulatrix of the female. Subsequently the spermatozoa, which have been released from the spermatophore through its opening or after it has been broken down by the bursa copulatrix, move to the spermathecae, where they remain until fertilization. The eggs are fertilized at the moment of oviposition.

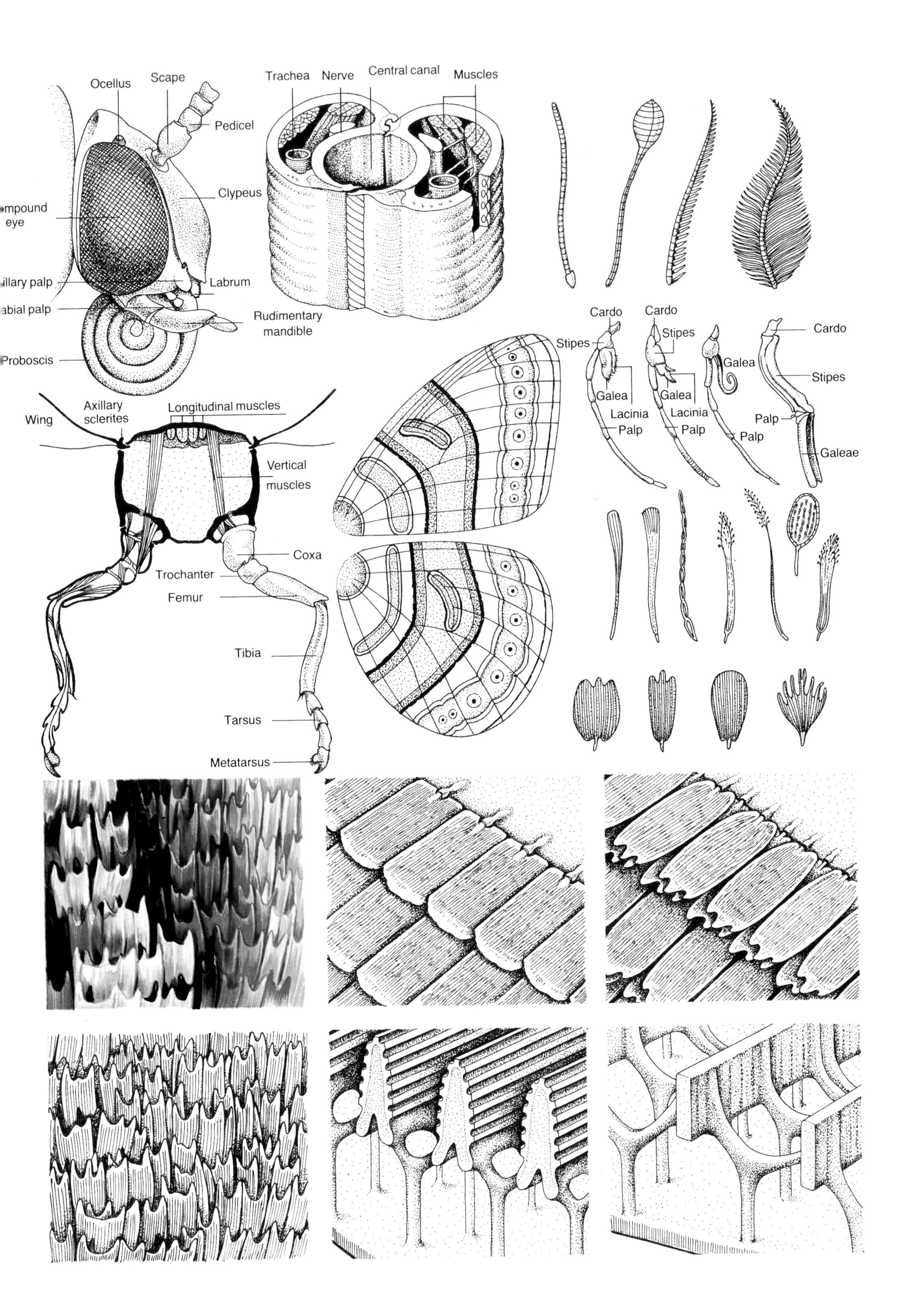
Ocellus
Scape
Pedicel
Clypeus
mpound eye
illary palp
Labrum
abial palp
Rudimentary mandible
Proboscis
Trachea
Nerve
Central canal
Muscles
Cardo
Stipes
Galea
Lacinia
Palp
Cardo
Stipes
Galea
Lacinia
Palp
Galea
Palp
Cardo
Stipes
Palp
Galeae
Wing
Axillary sclerites
Longitudinal muscles
Vertical muscles
Coxa
Trochanter
Femur
Tibia
Tarsus
Metatarsus

Life Cycle and Metamorphosis

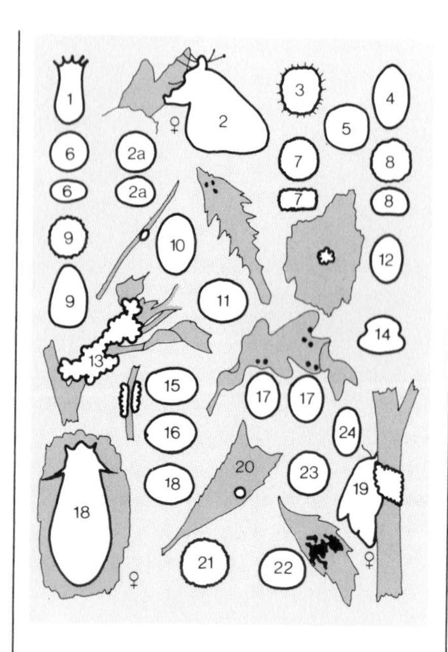

Plate 5 **Eggs**

Butterfly and moth eggs, enlarged between 20 and 50×. (1) Black-veined white (*Aporia crataegi*, Pieridae). (2) Apollo (*Parnassius apollo*, Papilionidae), ovipositing female. (2a) Apollo (*Parnassius apollo*), dorsal and side view of egg. (3) White admiral (*Limenitis camilla*, Nymphalidae, Nymphalinae). (4) Grayling (*Hipparchia semele*, Nymphalidae, Satyrinae). (5) *Melanargia psyche* (Nymphalidae, Satyrinae). (6) *Euchrysops barkeri* (Lycaenidae). (7) *Anthene butleri livida* (Lycaenidae). (8) *Epamera alienus alienus* (Lycaenidae). (9) Window acraea (*Acraea oncaea*, Nymphalidae, Acraeinae). (10) Pine hawkmoth (*Hyloicus pinastri*, Sphingidae). (11) *Amata phegea* (Ctenuchidae). (12) *Cossus cossus* (Cossidae). (13) Emperor moth (*Saturnia pavonia*, Saturniidae). (14) *Lemonia dumi* (Lemoniidae). (15) *Endromis versicolora* (Endromidae). (16) *Lasiocampa quercus* (Lasiocampidae). (17) *Gastropacha quercifolia* (Lasiocampidae). (18) Gypsy moth (*Lymantria dispar*, Lymantriidae); left, a female with an egg mass, right, a single egg. (19) *Malacosoma neustria* (Lasiocampidae), a female laying a ring of eggs. (20) *Cerura vinula* (Notodontidae). (21) *Phlogophora meticulosa* (Noctuidae). (22) The dot (*Melanchra persicariae*, Noctuidae). (23) *Plusia gamma* (Noctuidae). (24) *Biston hirtarius* (Geometridae).

As mentioned previously, butterflies and moths, like other holometabolous insects, undergo complete metamorphosis in the course of their development from the egg, passing through three quite distinct stages: larva, pupa, and adult. It is highly likely that metamorphosis became established in the course of evolution as a direct means of exploiting resources more effectively and of dividing tasks between highly specialized larval and adult stages, as a result, limiting competition among different individuals of the same species. Furthermore, metamorphosis offers particular advantages in temperate regions or environments in which there are marked seasonal changes, where a particular food resource — such as buds, leaves, flowers, or fruit — may be present at certain seasons and not at others. Such resources can be used in different ways by caterpillars, which with their chewing mouthparts mainly eat leaf parenchyma, and by the adult butterflies and moths, whose suctorial mouthparts enable them to feed off the sugary liquids produced by fruits and flowers. Many butterflies and moths fly from one flower to another and are major factors in the pollination and hence the reproduction of angiosperms, to some extent making up for the damage done by their caterpillars.

Egg

The lepidopteran life cycle begins with the egg, which is normally deposited as soon as it is fertilized. The embryo generally develops inside it within the space of a few days. However, in many species there is a period of *diapause*, during which development is halted and the egg remains in a latent condition. This adaptation enables it to survive periods of harsh weather, such as the winter of temperate regions or the dry season of tropical ones. Lepidopteran eggs are enclosed in a thick, resistant, and horny *chorion*, which is sometimes smooth but more often minutely ornamented in various ways. A polygonal grid that reflects the arrangement of the covering of follicle cells is more or less apparent on the surface, and there is also a depression with a minute hole, the *micropyle*, which marks the entrance of the spermatozoon and, once the egg has been deposited, allows the embryo to respire. The eggs may be spherical, hemispherical, or ellipsoidal or spindle, lens, or rod shaped; they also may vary on colour (Plate 5). Their coloration also varies with time; and once the embryo has completed development and consumed all the available yolk, it may be seen through the transparent chorion.

The eggs are always laid on, or in the immediate vicinity of, plants or any other food resource suitable for larval feeding. Guided by visual, olfactory, tactile, and gustatory stimuli, the female normally chooses the spot with great care. Females with ovipositors lay their eggs inside plant tissues, while others use viscous secretions to attach them firmly to a support, often the underside of the host leaf. Others still, such as the hepialids and some nymphalids, drop their eggs during flight. The number of eggs and their size vary widely and depend upon the size of the species, its adaptive strategy, and the degree of dietary specialization. Within a group of related species those with unspecialized, *polyphagous* diets (which drop their eggs in flight) produce smaller eggs in larger numbers than do those with specialized, *monophagous* diets (which carefully choose the site where they will lay their eggs).

Larva

Caterpillars are *polypodous* larvae; that is, their abdomen bears a series of

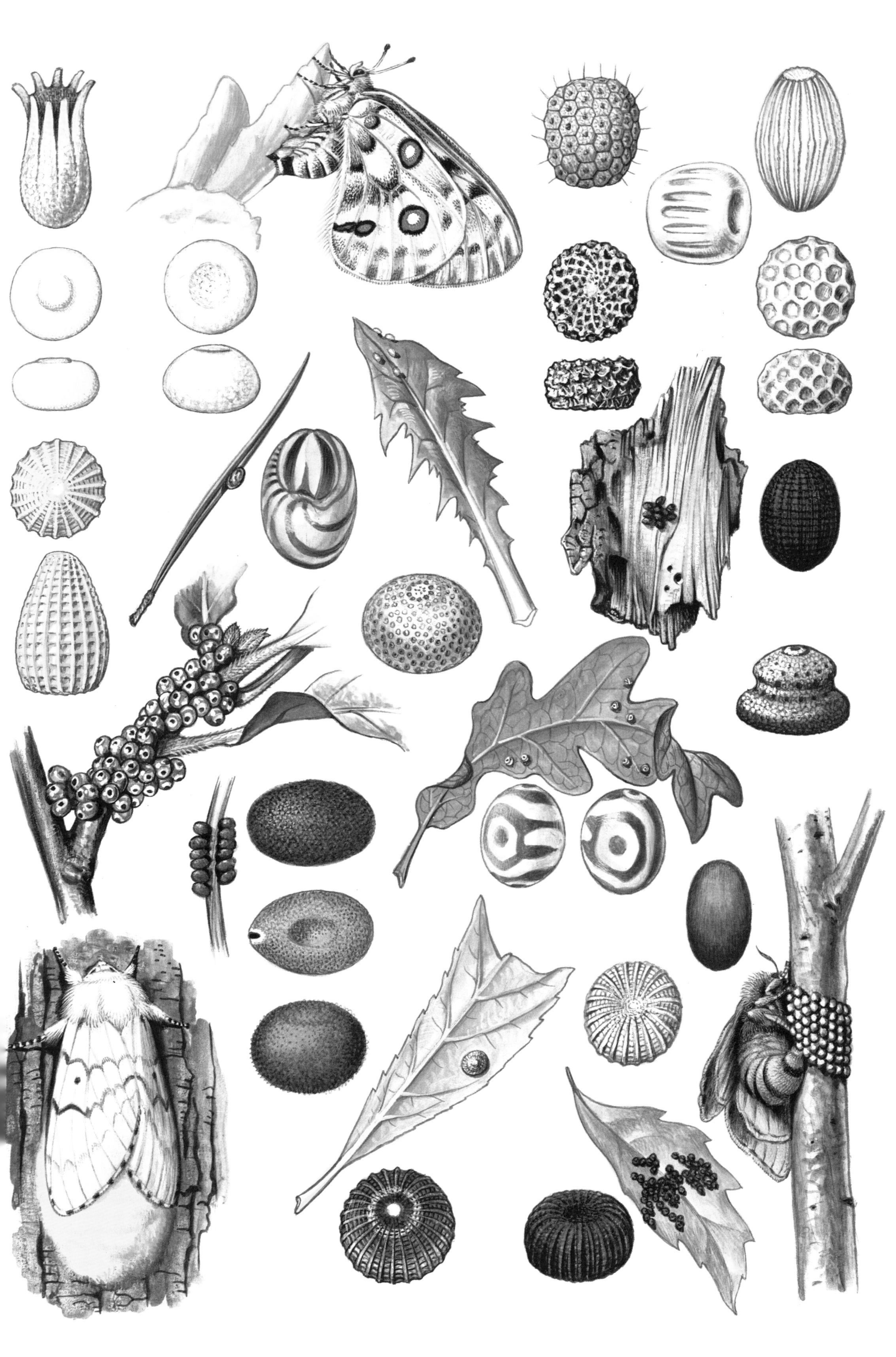

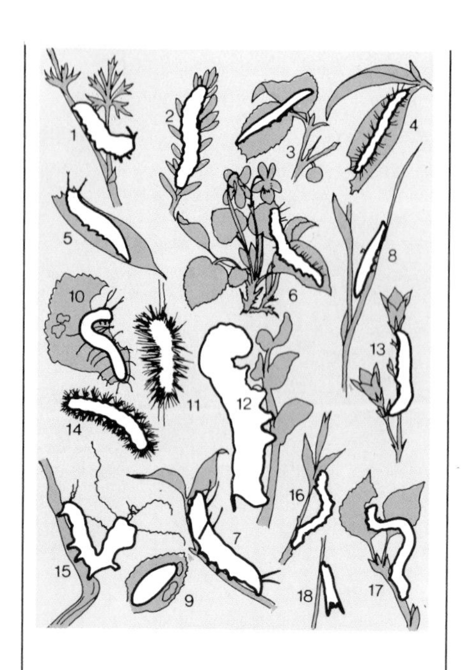

Plate 6 **Caterpillars**

The plate shows a number of representative larval stages, either life size or slightly enlarged, of butterfly and moth families. (1) Swallowtail (*Papilio machaon,* Papilionidae). (2) Apollo (*Parnassius apollo,* Papilionidae). (3) Brimstone (*Gonepterix rhamni,* Pieridae). (4) Mourning cloak (*Nymphalis antiopa,* Nymphalidae, Nymphalinae). (5) Lesser purple emperor (*Apatura ilia,* Nymphalidae, Nymphalinae). (6) Silver-washed fritillary (*Argynnis paphia,* Nymphalidae, Nymphalinae). (7) Plain tiger (*Danaus chrysippus,* Nymphalidae, Danainae). (8) Grayling (*Hipparchia semele,* Nymphalidae, Satyrinae). (9) Checkered blue (*Scolitantides orion,* Lycaenidae). (10) *Apatele alni* (Noctuidae). (11) *Apatele aceris* (Noctuidae). (12) *Acherontia atropos* (Sphingidae). (13) *Cucullia campanulae* (Noctuidae). (14) *Leucoma salicis* (Lymantriidae). (15) *Brahmaea wallichi* (Brahmaeidae). (16) *Phigalia pedaria* (Geometridae). (17) *Biston betularius* (Geometridae). (18) *Fumea casta* (Psychidae).

locomotory tubercules, or *prolegs*, in addition to the three pairs of jointed thoracic limbs, the true legs, which are homologous with those of adult butterflies and all other insects. The basic structure of the caterpillar differs little from one family to another (Plate 3): Externally a head and a trunk may be distinguished, the latter consisting of thirteen segments, three of which are thoracic and ten abdominal.

The head is generally spherical, rarely flattened, and bears a pair of antennae made up of three very short segments (except in the Micropterigidae). There are also three pairs of ocelli, situated in a semicircle to the sides and low down, as well as chewing mouthparts. The caterpillar lacks the compound eyes that are such important sense organs in the adult. The mouthparts have a more or less flattened, transverse labrum, with six pairs of setae on its dorsal surface. The mandibles are very stout, robust, and generally toothed. The maxillae are partially fused with the labium and form its lateral lobes. These consist of one cardo, one stipes, and a maxillary palp (two segments) on each side as well as an inner sclerite, the *lobarium*, which corresponds to the galea. The central part of the lower lip consists of the labium proper, which is composed of a proximal part fused to the maxillae and a free distal part, the *prementum*. The latter bears palps (two segments) and a more or less elongated median process, the *spinneret*, at the tip of which the duct of the silk glands opens. The heads of many caterpillars bear tubercules, spines, or long hairs, arranged in a variety of ways, and are consequently rather strange in appearance.

The thoracic segments each bear a pair of jointed legs, which typically consist of a coxa, trochanter, femur, tibia, a single tarsal segment, and a pretarsus with one claw, while there are prolegs — soft, cylindrical, membranous appendages, ending with a *planta* that bears a circular or semicircular crown of hooks, or *crochets* — on certain abdominal segments. The number of these varies from family to family but there are generally five pairs, those on the third, fourth, fifth, and sixth segments being known as abdominal feet and the pair on the tenth being the claspers. The Micropterigidae are an exception, having eight pairs of abdominal feet on their first eight abdominal segments; various other families have different numbers, usually fewer, of such feet. In the Geometridae, for example, there is only one pair of abdominal feet, well back on the sixth segment, and the caterpillar moves in a characteristic fashion, drawing its posterior portion up to its thorax and then moving its head forward. This pattern of looping and extending the body makes it seem that the caterpillar — which is known as a "looper" or "geometer" — is measuring its path. In *Cerura* and *Stauropus fagi* the claspers are transformed into long, thin appendages coloured at their tips. These have lost their original locomotory function and are probably used instead to confuse predators.

The integument of caterpillars may be more or less sclerotized, smooth, or wrinkled, and it may bear varying numbers of hairs and bristles. These consist of the primary setae, which are sensory and already arranged in a fixed and well-defined pattern as early as the first instar larva, and secondary setae, which act as a covering and a protection and are present in all instars. The distribution of the primary setae (*chaetotaxy*) is of great help in the systematics of the larval instars. Some setae sting.

In addition to setae the body of the caterpillar is frequently covered by warts or by tubercles carrying tufts of setae. It may bear bizarre branched processes, horns, and other structures.

The overall shape of some larvae is so radically modified that at first sight it is difficult to recognize these larvae as caterpillars. The flattened and highly sclerotized *oniscoid* larvae of many lycaenids are an example of this, as are the sluglike limacodid larvae, whose heads are completely

covered by their prothoraces. Various larvae build protective outer cases in which they spend all or part of their time. The movable outer cases of many psychids are quite characteristic. They consist of a shell of silk covered with various materials, such as twigs, fragments of leaves, and sand; and not only do the larvae never leave them, but neither does the normally wingless adult female.

The internal organization of the caterpillar (Plate 3) resembles the basic adult arrangement although it differs in detail. The nervous system is less concentrated than is that of the adult, and the sympathetic system is more highly developed. In comparison with the adult the larva has a relatively wide, short, and straight digestive system. It also has a *peritrophic membrane*, which protects its inner surface. The respiratory system is tracheal, but it usually lacks the air sacs. There are generally nine pairs of spiracles, the first being situated on the promesothorax and the others on the first eight abdominal segments. Aquatic larvae may breathe through the epidermis or by *tracheal gills*, evaginations of the tracheae, which are branched to varying degrees and absorb the oxygen dissolved in the water. Caterpillars are also distinguished by the increased development of their glandular system. The head contains the *mandibular glands*, which secrete a liquid that is possibly used for repelling enemies, and the previously mentioned silk glands, which may be enormously developed and extend almost the whole length of the caterpillar. Their secretion is initially liquid but hardens on contact with air. A thread of silk consists of two proteins, an inner core of *fibroin* and an outer sheath of *sericin*. The thread, which is produced and suitably lubricated by the spinneret, is of great importance in the life of the larva. It is used to make cocoons and protective shelters, but it is also used continually as an aid to movement and an indicator of the animal's path. The thorax bears the *osmeterium*, an unusual structure peculiar to the Papilionidae, which consists of a Y-shaped integumental invagination in the middorsal region of the prothorax, with a glandular base and tubular extremities. The osmeterium may be everted to emit a highly aromatic liquid, and it is possible that it acts as an excretory organ capable of eliminating volatile toxic substances contained in food plants although it it highly likely that it is also used as a defense against predators. In many families the thorax also houses a ventral *jugular gland*. This may sometimes be evaginated; and it is capable of secreting, and in some cases shooting out, an acidic liquid as a means of defense.

The endocrine system of the larva includes the *corpora allata*, which secrete *juvenile hormone*, and the *corpora cardiaca*, which secrete brain hormone. As in the adult the latter are retrocerebral but are less centrally situated with respect to the aorta. There are also the *prothoracic glands*, which lie close to the prothoracic spiracles and the large tracheae. These produce a steriod hormone, *ecdysone*, which, together with the other hormones, controls moulting and metamorphosis. A number of other glandular structures are to be found in the abdomen. These are often specific to a particular group, examples being the abdominal defensive gland found only in the Lymantriidae and the dorsal gland of the Lycaenidae, which produces a secretion attractive to ants (see the chapter on ecological relationships).

The caterpillar is unable to grow continuously because the chitin (a nitrogen-containing polysaccharide) and sclerotin (a hardened and darkened protein) in the integument make it rigid and inextensible. Instead it grows through a series of critical stages, known as *moults*, in which the cuticle and the integumental invaginations such as the tracheae and the initial and terminal portions of the gut are replaced.

Moulting takes place periodically and occurs once the caterpillar has become too large for its integument. The phenomenon is controlled by complex endocrine mechanisms that are triggered by the secretions of particular cells in the brain, the neurosecretory cells. These act on the prothoracic glands, which in turn secrete ecdysone, or moulting hormone. This hormone causes a rapid growth and multiplication of the epithelial cells of the epidermis; they separate from the old cuticle and begin to secrete a new one, which, since it is larger than the preceding one, is folded and wrinkled. The inner part of the old cuticle, the endocuticle, is digested by enzymes in the exuvial liquid, and it then splits under the pressure exerted by the body of the caterpillar. The caterpillar then frees itself completely from its old case, which is then known as the *exuvium*. The new cuticle is initially soft and thin, but it subsequently hardens with the addition of chitin and sclerotin, moulding itself to the caterpillar, which in the meantime has increased its own size as much as possible by swallowing air or water. Another hormone, the juvenile hormone secreted by the corpora allata, plays an important part in this process. So long as this substance is present in the hemolymph, the epidermal cells continue to produce a larval cuticle during moulting. However, if it is absent, the moult leads directly to the formation of a pupa. The action of juvenile hormone has been shown experimentally by removing the corpora allata from young silkworms, which as a result turn into minute pupae and subsequently emerge as tiny moths. On the other hand, if the juvenile-hormone concentration in a larva's hemolymph is increased, it will continue to moult and grow indefinitely. Under natural conditions it appears that the cessation of juvenile-hormone production is directed by the brain. This research suggests that the complicated process of metamorphosis is controlled in the final analysis by the dosage of juvenile hormone and by its interaction with moulting hormone. When the latter acts in the presence of high concentrations of juvenile hormone, the moult leads to a larval form; but when ecdysone acts with low concentrations of juvenile hormone, a pupa is formed. If juvenile hormone is completely absent, an adult insect develops. However, many questions remain to be answered about the genetic control of the phenomenon.

The intervals between moults are known as instars, and in butterflies and moths the number of larval instars normally varies between three and five. Frequently there are marked differences in coloration and overall appearance between one instar and another; indeed it can be difficult to identify two caterpillars of different ages as belonging to the same species. The number of instars is generally constant for a given species, but temperature and feeding may have an effect. For example, if the amount of food is experimentally restricted in a laboratory, it is possible to obtain as many as forty larval moults in *Tineola bisselliella*, the common clothes moth. The growth rate is extremely variable, and it too depends greatly on temperature and on the quantity and quality of food. In many species larval development is completed in fifteen to thirty days under normal conditions, while in others, such as the Cossidae, which eat wood, the larval cycle takes two years. *Diapause*, in which growth periods, is a common phenomenon, having become established as a means of overcoming periods of adverse weather or the seasonal absence of a food plant.

Pupa

When full grown, the mature larva ceases feeding and searches for a suitable spot in which to transform itself into a pupa. Generally it will hide under a stone or bark, deep in the soil, or by rolling up a leaf on its food plant, which it then secures with silk. The larva may also use silk to build suitable supports or pads to which it attaches itself head downwards. It may spin silk bands around its thorax to attach itself to a vertical support,

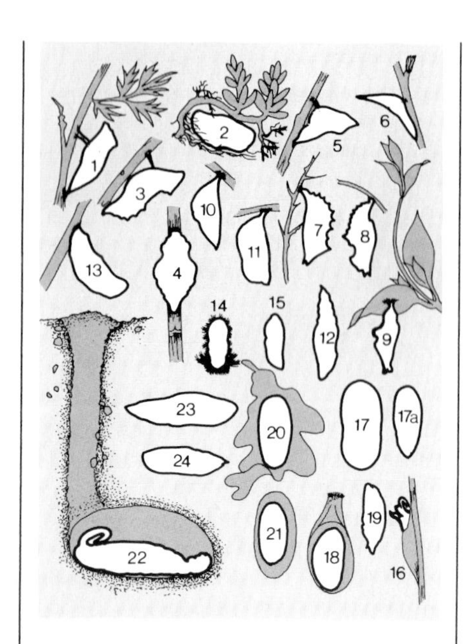

Plate 7 **Pupae**

The plate shows a number of pupae, enlarged to varying degrees (Figures 12, 14 and 15 are enlarged about 2.5×), representing some Lepidopteran families. (1) *Papilio machaon* (Papilionidae). (2) Apollo (*Parnassius apollo,* Papilionidae). (3) *Ornithoptera priamus* (Papilionidae). (4) *Atrophaneura alcinous* (Papilionidae). (5) Brimstone (*Gonepteryx rhamni,* Pieridae). (6) Orange tip (*Anthocharis cardamines,* Pieridae). (7) Mourning cloak (*Nymphalis antiopa,* Nymphalidae, Nymphalinae). (8) Silver-washed fritillary (*Argynnis paphia,* Nymphalidae, Nymphalinae). (9) White admiral (*Limenitis camilla,* Nymphalidae, Nymphalinae). (10) Lesser purple emperor (*Apatura ilia,* Nymphalidae). (11) Two-tailed pasha (*Charaxes jasius,* Nymphalidae, Charaxinae). (12) *Acraea cabira* (Nymphalidae, Acraeinae). (13) *Dynastor darius* (Nymphalidae, Brassolinae). (14) *Alaena amazoula* (Lycaenidae). (15) *Poecilmintis nigricans* (Lycaenidae). (16) *Zygaena lonicerae* (Zygaenidae). (17) Silk moth (*Bombyx mori,* Bombycidae), cocoon. (17a) Silk moth (*Bombyx mori*), pupa removed from the cocoon. (18) Emperor moth (*Saturnia pavonia,* Saturniidae), section through a cocoon to show the pupa. (19) *Aglia tau* (Saturniidae). (20) *Lasiocampa quercus* (Lasiocampidae), section through cocoon. (21) *Cosmotriche potatoria* (Lasiocampidae), section through cocoon. (22) *Herse convolvuli* (Sphingidae) in its underground chamber. (23) *Catocala fraxini* (Noctuidae). (24) *Mormo maura* (Noctuidae).

or finally it may construct various forms of protective cocoon. Once it is suitably positioned, the caterpillar moults for the last time. The old cuticle splits and curls backwards. In the Nymphalidae and other butterflies that pupate with their head downwards the newly formed *chrysalis* (butterfly pupa) must perform a delicate maneuver to free itself of the exuvium and resuspend itself from the pad of silk without falling; whereas the caterpillar was suspended from the support by means of its claspers, the chrysalis will have to do so by means of the *cremaster*, a projecting lobe at the tip of the abdomen bearing spines or bristles curved like hooks. The maneuver is performed rapidly and involves an intermediate attachment point, the old exuvium, which the chrysalis grips by briefly pinching it between the margins of the terminal segments, hooking the pad of silk at the end of the operation and so reaching its final position. Naturally it is easier for free pupae, which are not attached to supports, or for ones that produce cocoons to shed the larval exuvium.

During nymphosis, the stage of relative immobility during which the insect does not feed, the transformation and replacement of the larval organs and tissues by adult ones are completed although the processes of histolysis and histogenesis that cause these changes had already begun at different stages during the larval development. This transitional condition is also apparent in the external morphology. Pupae display adult features—such as the division of the body into the head, thorax, and abdomen, the structure of the appendages, and the presence, although hidden, of the wings—alongside such larval characters as verrucae, tubercles or horns, and traces or scars of the prolegs. In addition there are unique pupal characters, such as spines, hooks, and other structures for attachment or protuberances used for breaking the cuticle during emergence.

There are two basic types of pupa: the *exarate*, which has free adult appendages, the antennae, mandibles, legs, and wings being individually encased and protected by the cuticular integument, and the *obtect*, in which the adult limb buds are stuck to the head and covered by a single cuticle. The first type occurs in the more primitive groups and may be either *decticous*, having articulated and mobile mandibles and found only in the Micropterigidae and Eriocraniidae, or *adecticous*, with immobile mandibles and occurring in the Incurvariidae and the Cossidae. In most lepidopterans the pupa is obtect although it may sometimes have, as in the case of the Sphingidae, a projecting proboscis with its own cuticular covering. (As implied above, the term chrysalis is limited to the obtect pupae of butterflies, that is, Papilionoidea and Hesperioidea.) The pupa is occasionally dark brown, the colour of sclerotized cuticle from which certain pigments are absent. However, those normally exposed to light are pale grey, yellowish, off-white, or more rarely green in colour. Nymphalid pupae are often silvery or golden as a result of having pockets of air in the cuticle.

The pupa displays reduced activity although sudden movements of the abdomen often occur as it defends itself or adjusts its position. Respiration takes place by means of spiracles, which are situated in the same positions as they are in the caterpillar, with the exception of the terminal pair on the eighth abdominal segment, which are rudimentary. The anus and the genital apertures are closed. Excretion is interrupted, and water loss is minimized by the impermeability of the cuticle. The pupal stage varies in length and may undergo diapause. This phenomenon is genetically determined and controlled by complex hormonal mechanisms: The retrocerebral endocrine glands secrete hormones that circulate in the blood, or haemolymph, and once these reach the various organs and tissues, they inhibit or stimulate particular activities. The interruption of diapause depends essentially on ecdysone (produced by the prothoracic glands), whose activity, as has already been explained, is governed by

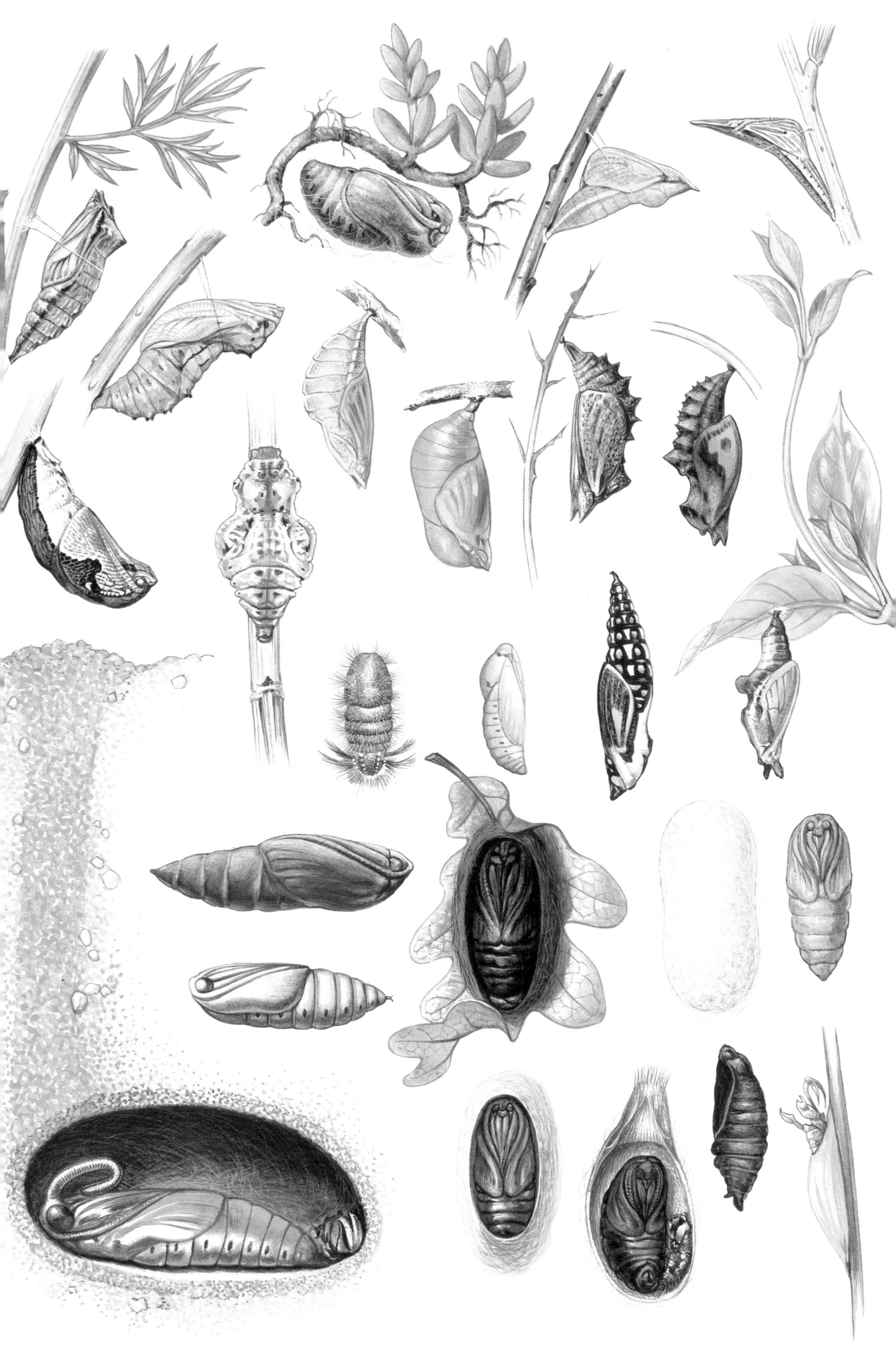

Plate 8 **Stages in the development of a butterfly (*Papilio xuthus*)**

(1) Female laying an egg on the food plant. (2,3) Development of embryo. (4,5) Emergence of first-instar larva; (6) second-instar caterpillar; (7) moulting caterpillar; (8) third-instar caterpillar with evaginated osmaterium. (9) Suspended caterpillar preparing for pupation; (10-12) last larval moult and formation of the chrysalis. (13) Rupture of the pupal case. (14,15) Progressive emergence of the adult; (16) fully formed adult.

neurosecretory cells in the brain. These in turn are activated by external stimuli, such as day length (*photoperiod*) or humidity.

The transition from pupa to adult is known as emergence. It is a somewhat delicate phase, during which the imago must succeed in splitting the pupa's cuticular case and possibly the cocoon as well. Under the influence of the appropriate environmental stimuli the insect forces its body fluids into its head and thorax, swelling them and splitting the cuticle. The adult then gradually emerges from this, freeing its legs first and then the abdomen. The cocoon may be shed in various ways. Some cocoons already have a natural opening, whereas in other cases the adult itself makes an opening. It may do so mechanically, using cutting spines or points situated on its head or the base of its wings, or chemically, dissolving the silk with a strong alkaline substance. With its wings still soft and only partly unfolded, the adult then hangs with its back downwards. The wings gradually extend as the hemolymph is pumped through the veins. Subsequently, when the blood pressure has fallen, the veins harden and sclerotize, and the adult is readly to fly. At this point the uric acid accumulated during nymphosis is eliminated in the form of a liquid, known as *meconium*, that is frequently pink or red. In some parts of the world numerous drops of meconium sometimes produced following the mass emergence of certain species have given rise to such popular beliefs as the "rain of blood."

Diversity and Evolution of Butterflies and Moths

Kinds of Diversity

The incredible variety of living creatures is a never-failing source of amazement. Why is the world not populated by just a single "kind" of organism? Why are there so many differences among organisms? How are they produced? Although elementary, these questions lie at the heart of the concept of *diversity*, and the answers to them are the substance of an important part of biology.

The diversity of butterflies and moths is familiar to everybody, and although they are but a small sample of organisms, they may be used as an example. If one simply walks through a meadow, a wood, or even a garden, it is possible to collect several butterflies and moths. Careful observation reveals that they differ to a greater or lesser extent in size, shape, colour, pattern, and the development of particular structures, such as antennae or palps.

Next one quickly finds that differences may exist between individuals of the same species, whether of the same or opposite sex, between different but related species (belonging to the same genus, for example), or between more distantly related species from different genera or families. If the sample being examined has been collected at two or more separate sites, it will also be possible to observe racial differences between individuals of the same species (Plate 19). These aspects of diversity are summarized in Plate 9. The first section (A) shows two particularly marked examples of differences between the sexes (*sexual dimorphism*). In the South American nymphalid *Catonephele numilia* the male and female differ in the colour, pattern, and wing shape, while in the European lymantriid *Orgyia antiqua* the two sexes differ in the presence or absence of wings as well as in the shape of the antennae, abdomen, and so on. Section B shows differences between individuals of the same sex coming from the same population: a sample of females of the lycaenid *Plebicula escheri*, collected at a site in Liguria, Italy, which vary in the extent of their blue coloration. Section C illustrates differences between *geographical races* of the nymphalid *Hypolimnas pandarus*, which live on separate islands in the Moluccas. This species displays sexual dimorphism, and racial differences are apparent between both the males and females. The fourth section (D) is of six markedly different *Milionia* species (Geometridae). The final section (E) shows various butterflies and moths which differ substantially in many respects and which represent families that are only very distantly related.

In summary there are essentially two kinds of differences: The first is between individuals from the same population, whereas the second is between individuals belonging to separate populations or species.

These two terms need to be defined before proceeding further. A *population* is a community of individuals that live in a given area and are linked by sexual breeding relationships and kinship. A *species* is an interfertile group of populations whose members may interbreed to give fertile offspring but which are reproductively separate from other groups of populations. The population and the species are the chief actors in the drama of *evolution*; and to understand the causes of the diversity between the members of a population on the one hand and of the diversity between species on the other, one also has to understand the mechanisms and processes of evolution.

This chapter will be devoted to the first aspect, namely, at the level of populations the causes of individual variation, which are essentially due to the genetic makeup of the various individuals and their interaction with the environment, while the following chapter will analyze how the differences between species arise.

Plate 9 **Variability of Lepidoptera**

No two butterflies are identical. Whenever two individuals are compared, it is always possible to find differences to a greater or lesser degree in at least some characters. Such variation is basically determined by the individual's genetic makeup but is also influenced by the environment. Various classes of variation may be recognized even if only the morphology, pattern, and colour of various body parts are considered. The plates of this chapter and the next will illustrate in detail all these kinds of diversity, and their causes and significance for individual survival and for the adaptation of populations will be analyzed.

This introductory plate provides examples of variation between the sexes (A), between individuals of the same sex belonging to the same population (B), between two different populations or races of the same species (C), between different species from the same genus (D), and finally between species belonging to different genera and families (E).

A: *Catonephele numilia* (Nymphalidae), male (1) and female (2); *Orgyia antiqua* (Lymantriidae), male (3) and female (4).

B: *Plebicula escheri* (Lycaenidae), individual variations between females belonging to a Ligurian population.

C: *Hypolimnas pandarus* (Nymphalidae) from the Molucca Islands, male (5) and female (6) of the *pandarus* race (islands of Ambon, Serang, and Saparua); male (7) and female (8) of the *pandora* race (islands of Buru).

D: Various *Milionia* (Geometridae) from New Guinea and neighbouring islands, *M. grandis* (9), *M. basalis* (10), *M. exultans* (11), *M. elegans* (12), *M. plesiobapta* (13), *M. ovata* (14).

E: Species belonging to different genera and families. (15) *Himantopterus fuscinervis* (Zygaenidae). (16) *Pterophorus pentadactylus* (Pterophoridae). (17) *Rothschildia orizaba* (Saturniidae). (18) *Melittia gloriosa* (Sesiidae). (19) *Micropterix allionella* (Micropterigidae). (20) *Graphium weiskei* (Papilionidae). (21) *Composia credula* (Hypsidae). (22) *Greta quinta* (Nymphalidae).

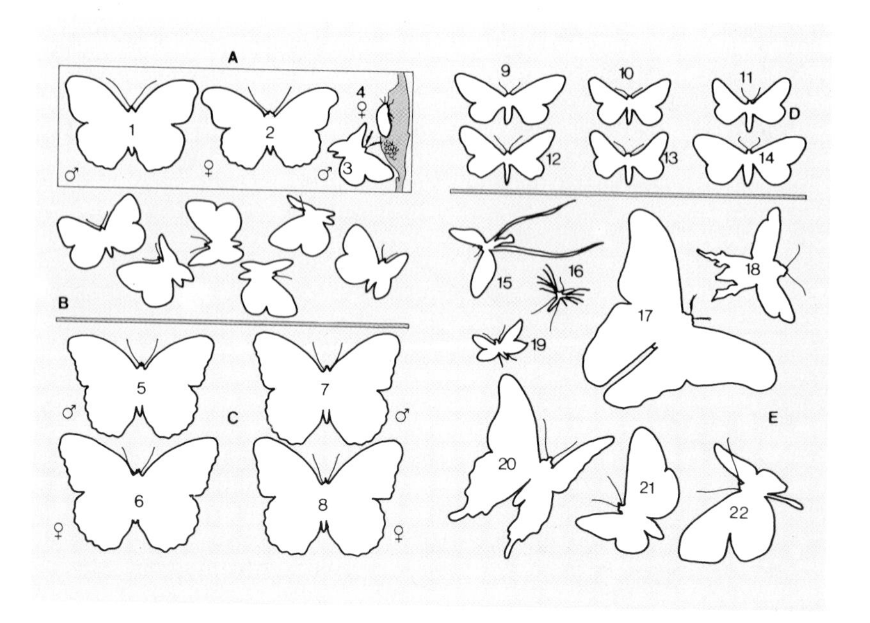

Variability in Populations

The existence of variations among individuals of the same species or, rather, the same population is a fundamental feature of all sexually reproducing organisms. This variability may be more or less marked, depending upon the characters and the organisms concerned, but it is always present. The first important point to remember is that two distinct components act on the characters that distinguish individuals: One, the genetic makeup, is internal, and the other, the environment, is external. Each individual is the result of the combination of a unique assortment of genes (the genotype), which is expressed in its structural and functional characteristics (the phenotype), with the more or less significant influence of the environment during the course of development. Therefore the *genotype* represents the genetic constitution, or in other words the program of the individual, while the *phenotype* is the sum of characters actually expressed. It is essential to distinguish hereditary genetic variability from nongenetic variability, the latter being the sum of the differences between individuals which are not under genetic control and which may occur between two individuals with the same genotype (as, for example, between identical twins) because of the effect of the environment. Butterflies and moths offer many opportunities for a more detailed discussion of variability. The shape and size of the wings and above all the complexity of the patterns and coloration are sufficiently varied to reveal even small differences between individuals. For any given character the variation is *continuous* or *discontinuous* according to whether or not a series of intermediate forms exists between one variant and another.

Polymorphism

Discontinuous genetic variability, or *polymorphism*, exists where there are individuals belonging to two or more distinct "types," known as *forms* or *morphs*. These differ in their genotype and reproduce in succeeding generations in accordance with Mendel's laws. There are numerous examples of polymorphism in the coloration and patterning of the wings of butterflies and moths, but the phenomenon is not confined to these characters. In this group, as in other organisms, physiological and biochemical differences, such as the molecular composition of a particular protein or enzyme, are still more common. Examples of biochemical polymorphisms are well known in man and include the blood groups and the altered structures of the hemoglobin molecule that are the basis of certain inherited illnesses, such as sickle-cell anemia.

Origin of Variability: Mutations and Genes

In any event variability is a result of phenotypic differences that initially arose from chance variations (*mutations*) occurring in the inherited factors, or genes. *Genes* are segments of the molecule of deoxyribonucleic acid (DNA), and it is this substance that makes up the *chromosomes* that lie inside the nucleus of the cells. The DNA molecule is made up of building blocks, known as *nucleotides*, each of which consists of a phosphate group, a sugar (deoxyribose), and one of four possible different nitrogenous bases. These fall into two groups, the purines (comprising *adenine* and *guanine*) and the pyrimidines (comprising *thymine* and *cytosine*). The nucleotide units link together in a chain to form a long, helically coiled filament, the polynucleotide. The DNA molecule is made up of two of these filaments (the double helix), which are held together by hydrogen bonds between complementary bases (one purine and one pyrimidine) located in the two polynucleotides, the precise pairing being adenine and thymine and guanine and cytosine. Genetic instructions can be transmitted through a special language derived from the sequence of the different nucleotide bases, which correspond to the letters of an alphabet. DNA is therefore the storehouse of all inherited information, and its structure meets two fundamental requirements: First of all, it is capable of duplicating itself exactly in the course of cell division and can therefore transmit the genetic message unaltered to subsequent generations; secondly it is able to pass on genetic instructions to other cellular components, which can then interpret and use the instructions in metabolism, growth, and development. The structure and function of living cells are largely determined by proteins, and the primary structure of these — their sequence of amino acids — faithfully translates the sequence of triplets of bases in the DNA molecule. Therefore a gene, or genetic locus, may be defined as a segment of DNA corresponding to a sequence of bases that encode a particular protein or at least a certain sequence of amino acids, a polypeptide, that form part of a protein.

If the mechanism of transmitting genetic information through the replication of DNA were perfect, life on earth would be much more monotonous since there would probably be just one type of organism, which would endlessly reproduce identical copies of itself through succeeding generations. In reality occasional "error" occurs in the transmission of the genetic message, and the errors, known as mutations, have a more or less marked effect on the phenotype of the new individual, which consequently differs from its parent in certain characters. A change in just one base of a triplet may be enough to lead to the incorporation of a different amino acid in a protein. This in turn may profoundly affect the nature and function of the protein and therefore of the cell. At this point it is not difficult to imagine, at least in general terms, how mutations may give rise to variability in such characters as the coloration and patterning of the wings of a butterfly or moth. Consider, for example, the different effect that the normal and mutant forms of an enzyme might have on the

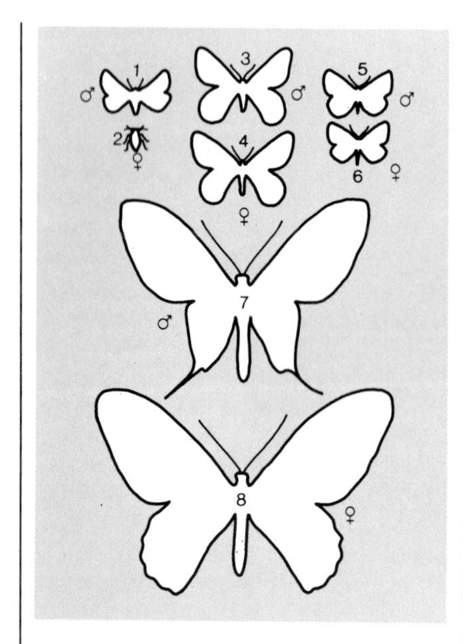

Plate 10 **Sexual dimorphism I**

It is not uncommon among Lepidoptera for the two sexes to differ strikingly in their colour pattern, in the shape and development of their wings, and in the structure of their antennae. These secondary sexual characters are genetically controlled and often have adaptive significance. (1, 2) Male and wingless (apterous) female of *Nyssia florentia* (Geometridae); note that the antennae also differ in form. (3, 4) Male and female of *Parnassius autocrator* (Papilionidae). (5, 6) Male and female of *Plebicula amanda* (Lycaenidae). (7, 8) Male and female of *Ornithoptera paradisea* (Papilionidae).

oxidation of certain substances found on scales and the resultant change in coloration. However, in the Lepidoptera and other higher organisms it is very difficult to follow experimentally all the stages of the series of phenomena that make up the process known as *amplification*, by which the original genetic information is expressed in the phenotype.

Not all mutations are of the type described above (genetic mutations), in which there is a variation in the nucleotide sequence of a segment of DNA. More extensive mutations may involve whole chromosomal regions and hence a large number of genes. These may take the form of *deletions* or *duplications* (respectively the loss or repetition of a chromosomal segment), *inversions* (the rotation through 180° of a segment of chromosome), and *translocations* (the transfer of a segment from one chromosome to another). Together these are known as chromosomal mutations, and they involve the alteration of individual chromosomes while leaving unchanged the basic chromosome number typical of each species of eukaryotic (having a clearly marked nucleus) organism. A further class of mutations, genomic mutations, is caused by errors in the duplication of chromosomes or in the distribution during the processes of cell division. Genomic mutations result in the doubling, tripling, or quadrupling of the entire chromosome set (*polyploidy*) or individual elements of it (*aneuploidy*). Mutations are therefore the cause of genetic variability and the evolutionary process. Many of the possible mutations produced are deleterious and reduce the fitness of individuals carrying them or cause their deaths at some stage in their development. Others are neutral and neither advantageous nor disadvantageous to their carriers. Some, under particular environmental conditions, are advantageous to their carriers, and hence these individuals have a greater probability of surviving and breeding than other individuals do. The resultant effect will be an increase in the frequency of the mutant characters in the population. Many rare forms of butterflies and moths, which in the past were known as varieties, aberrations, or "sports" and were sought by collectors, are in fact mutants. Others are present at a low but constant level and produce stable or transient polymorphisms. The term polymorphism should be applied only when the frequency of the rarer form is sufficiently high — above 1 percent, for example — for it not to be due simply to recurrent mutations but to have spread into the population through inheritance. Mutations are in fact very rare, and the probability that one will occur at a given genetic locus averages one-millionth per generation, or one in every million gametes. It is obvious that the mutations capable of having significant evolutionary effect are those in the cells of the germ line and in the final analysis the gametes, that is, the eggs and sperm.

Genetic Basis of Polymorphism

The next step is to see how genetic variants may be inherited. In the majority of cells of a eukaryotic organism, such as a butterfly, the chromosomes are present in pairs (*diploid* set), and the two chromosomes in each pair are said to be *homologous*. However, the sex cells, or gametes, have only half the chromosome complement and are *haploid*. The reason for this is that, although somatic cells divide by means of a process known as *mitosis*, during which the chromosomes are first duplicated and then divided between the daughter cells, the gametes are formed in the gonads by a special form of cell division known as *meiosis*. During meiosis each cell divides twice, but the chromosomes are duplicated only once so that the resulting gametes have only half the chromosomes present in other cells. The diploid chromosome complement is restored at fertilization, when the sperm and egg unite to form the

Plate 11 **Sexual dimorphism II**

(1, 2) Male and female of the dryad (*Minois dryas*, Nymphalidae, Satyrinae). (3, 4) Male and female of the Greek mazarine blue (*Cyaniris antiochena*, Lycaenidae). (5, 6) Male and female of the holly blue (*Celastrina argiolus*, Lycaenidae). (7, 8) Male and female of *Morpho rhetenor* (Nymphalidae, Morphinae).

zygote. In an individual the genes, like the chromosomes, are present in duplicate, one copy being maternal in origin and the other paternal. The two genes, known as *alleles*, that form a pair are situated on homologous chromosomes and can be present in different forms. If the two alleles at a particular genetic locus are identical, the genotype is said to be *homozygous*; if they are different, the locus is *heterozygous*. Normally an individual is homozygous for certain genes and heterozygous for others. The separation of the copies of a gene into different reproductive cells is known as *segregation*. Segregation is one of the fundamental laws of genetics that were formulated in 1865 by Gregor Mendel, following a now-famous series of experiments involving the crossing of sweet peas, before the cytological and molecular basis of heredity was known. Prior to Mendel it was believed that heredity was transmitted by parents through fluids similar to blood, which mixed in the offspring. Mendel provided the first proof that heredity was based on the transmission of elementary particles (which today are known to be the same as genes) carried by the reproductive cells, which remain separate without any blending or contamination.

We may explain Mendelian inheritance and the law of segregation by using as an example *Abraxas glossulariata*, a widespread European geometrid. The most common form of this species (illustrated in Plate 48:4) is white with conspicuous black spots and two narrow orange bands on the forewings. In the rather rare *lutea* form the white pigment is replaced by yellow. This difference in coloration is controlled by two alleles, C^w and C^y, C representing the genetic locus that controls coloration and ^{w} and ^{y} standing for the alternative alleles that code respectively for white and yellow. The normal form of *Abraxas glossulariata* is homozygous, and its genotype is therefore C^wC^w, whereas the *lutea* form is homozygous for C^yC^y. In the gametes, the sperm and eggs, each of the two forms will be present as just a single allele. So, for example, a male of the normal form will produce spermatozoa that are all identical in having a haploid C^w set, and the eggs of a *lutea* female will all be C^y. If these two individuals are taken as the parental generation (P_1, or simply P) and crossed to produce the first filial generation (F_1), the diploid progeny will all be C^wC^y. All F_1 individuals will therefore be heterozygous, having one copy of each of the alternative alleles. Such individuals are pale yellow (the *semi-lutea* form) and are intermediate between their parents. It should be pointed out that this situation, in which the phenotype of the heterozygote displays intermediate characteristics (*incomplete dominance*) is not the only possible one; and indeed, as will subsequently be seen, it is not even the most common. All the F_1 individuals, both male and female, are of the genotype C^wC^y, and as a result of Mendelian segregation they will produce gametes that contain just one of the alleles C^w and C^y. Statistically 50 percent of the gametes, whether they are eggs or sperm, will be C^w and 50 percent C^y, and so a C^w spermatozoon will have an equal chance of meeting a C^w or a C^y gamete and forming C^wC^w or C^wC^y zygotes. The same argument is true for C^y sperm, which will produce equal numbers of C^wC^y and C^yC^y zygotes. Therefore, if F_1 *semi-lutea* individuals are crossed, one would expect the F_2 offspring to contain normal (C^wC^w), *semi-lutea* (C^wC^y), and *lutea* (C^yC^y) individuals in the ratios 1:2:1 (that is, a genotype frequency of 0.25, 0.50, and 0.25). Obviously the actual proportions of the phenotypes observed in the crosses will not exactly match those expected although the discrepancy will be smaller the larger the number of crosses and F_2 individuals produced. Now that the rules have been established, it is possible to see the effect of crossing a heterozygous *semi-lutea* individual with one or another homozygous form (a back cross). As shown previously, *semi-lutea* produces equal numbers of C^w and C^y gametes. If such individuals are crossed with the normal form

Plate 12 **Gynandromorphs**

Individuals that display male and female sexual characters in different parts of their bodies are known as gynandromorphs. Gynandromorphism is generally caused by genetic anomalies arising during gametogenesis or the development of the zygote. It may be bilateral (one-half being male and the other female) or it may be a mosaic of characters. (1) The large white (*Pieris brassicae*, Pieridae), bilateral gynandromorph, male left, female right. (2) The Cleopatra (*Gonepteryx cleopatra*, Pieridae), mosaic gynandromorph; the background coloration of the wing is female, while the orange of the male is irregularly distributed. (3) The mazarine blue (*Cyaniris semiargus*, Lycaenidae), bilateral gynandromorph, female left, male right. (4) The holly blue (*Celastrina argiolus*, Lycaenidae), bilateral gynandromorph, female left, male right. (5) Androgeus swallowtail (*Papilio androgeus*, Papilionidae), bilateral gynandromorph, female left, male right, with traces of female mosaic on the right hind wing. (6) *Ornithoptera priamus admiralitatis* (Papilionidae), bilateral gynandromorph, male left, female right. *Morpho aega* (Nymphalidae, Morphidae): normal female (7), blue female (8), mosaic gynandromorph (9). *Parnassius autocrator* (Papilionidae): male (10), female (11), bilateral gynandromorph (12).

(C^wC^w genotype), which produces only C^w gametes, then C^wC^w and C^wC^y zygotes will be produced in a ratio of 1:1 (that is, each will occur at a relative frequency of 0.50).

It was mentioned earlier that situations in which the phenotype of the heterozygote is intermediate between two homozygotes are not very common. More often the heterozygote is totally indistinguishable from the phenotype of one of the two alternative homozygotes. The character common to the heterozygote and one of the two homozygotes is said, in this case, to be *dominant*, while the other one, which is expressed only in the other homozygote, is said to be *recessive*. Examples of such a situation may be found in hybrids produced by crosses between different races or species (Plate 21). By convention it is usual to use a capital letter to indicate an allele coding for a dominant character and a lower-case letter for one controlling a recessive character.

Mendel's second law, the law of recombination, deals with the case in which forms are crossed that differ in two or more pairs of alternative characters rather than just one. Depending upon the number of pairs of characters being considered, crosses are referred to as monohybrid (as in the case of *Abraxas glossulariata*), dihybrid, trihybrid, and so on. The law of recombination may be checked by means of dihybrid crosses, as in the following example involving the zygaenid *Zygaena ephialtes*. This moth, which will later be discussed extensively in connection with mimicry, has some polymorphic populations with widely different phenotypes (Plate 14). Such phenotypes are produced by the variation of two characters, the red or yellow of the coloured zones of the wings and the abdominal ring and the *peucedanoid* or *ephialtoid* pattern. For a long time the peucedanoid and ephialtoid forms were regarded as separate species. The latter are distinguished by the melanin pigmentation that covers the whole of their hind wings with the exception of one or two small white spots and by the coloration of the spots on the forewings, which apart from the basal ones are white. Peucedanoid forms are not melanic, and the colour of all the spots on the forewing, whether basal or not, is the same as the hind wing and the abdominal ring. The two characters, coloration and patterning, are controlled by two separate gene loci, each of which has two alleles. The peucedanoid pattern is dominant to the ephialtoid, and the red coloration is dominant to the yellow. Therefore the two alleles that control the peucedanoid and ephialtoid patterns may respectively be designated *P* and *p*, and those that control the red or yellow coloration as *R* and *r*. If a red peucedanoid individual with a genotype that is homozygous for both characters (*PPRR*) is crossed with an ephialtoid yellow individual that is homozygous for both the alternative characters (*pprr*), the F_1 generation consists exclusively of red peucedanoid individuals, the dominant characters being expressed, with a heterozygous *PpRr* genotype. Such a result would be expected from the fact that each of the two parents produces only one type of gamete (*PR* or *pr*). If the dihybrids are crossed with each other the F_2 will be composed of four distinct phenotypes in the following proportions: red peucedanoid 9/16, yellow peucedanoid 3/16, red ephialtoid 3/16, and yellow ephialtoid 1/16. The dihybrids can produce four classes of gametes: *PR*, *Pr*, *pR*, and *pr*; and upon fertilization these give rise to the sixteen genotype combinations shown in Plate 14. If the genotypes that, because of dominance, display the same phenotype are grouped together, they occur in the ratio of 9:3:3:1. One would expect to obtain F_2 phenotypes in these proportions from such crosses each time the two pairs of characters in question segregate independently. In general this occurs when the loci controlling them are on different chromosomes. Mendel's second law does not apply when the genes are on the same chromosome because in such circumstances they are more or less closely linked, depending upon their distance apart. This condition is known as

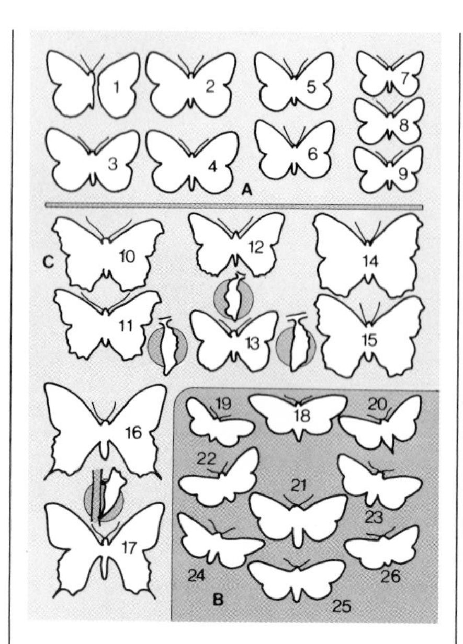

Plate 13 **Varieties: mutants and ecophenotypes**

Varieties originate for one of two main reasons: The first, purely genetic, is *mutation*, an error in the inherited transmission of characters that directly affects the genotype of the individual and is sometimes expressed in a clearly visible way. The following are examples in (A) the butterflies and (B) the moths of forms probably due to mutation: Marbled white (*Melanargia galathea*): (1) normal form, upper side and underside; (2) female, *leucomelas* form; (3) female, *albina* form; (4) female, *quasilugens* form. Spotted fritillary (*Melitaea didyma*): (5) normal form; (6) melanic form. Chalk-hill blue (*Lysandra coridon*): (7, 8) normal form, upper side and underside; (9) *striata* form, underside. *Lymantria monacha*: (18) female, normal form; (19, 20) aberrant forms. *Arctia caja*: (21) female, normal form; (22–26) aberrant forms found in England.

The second reason why varieties arise is environmental and produces reaction changes in individuals. The same genotype may be different in appearance according to the conditions that prevailed during development. In butterflies and moths such variations, known as *ecophenotypes*, may be readily obtained by subjecting the caterpillars or pupae to abnormal temperatures or humidity for a certain period of time. Examples of ecophenotypes obtained by subjecting the pupae of four butterfly species to warmth or cold are illustrated in C: Peacock butterfly (*Inachis io*): (10) normal form; (11) form produced from low-temperature pupae. Queen of Spain fritillary (*Issoria lathonia*): (12) normal form; (13) melanic form obtained from low-temperature pupae. Mourning cloak (*Nymphalis antiopa*): (14) normal form; (15) form produced from low-temperature pupae. Swallowtail (*Papilio machaon*): (16) normal form; (17) form obtained from pupae subjected to 40°C.

association, or *linkage*, and it causes significant deviations from the expected proportions of phenotypes in dihybrid crosses.

Such genetic mechanisms as those described underlie the genetic polymorphisms that are known in large numbers of butterflies and moths. Some further examples of polymorphisms in adult butterflies and moths are shown in Plate 15. Caterpillars, too, frequently display polymorphic characters, one example being the red, black, or bicoloured caterpillars of the American nymphalid *Chlosyne lacinia*, whose coloration is genetically controlled by two loci.

Polymorphism and Evolution

Until now two fundamental aspects of genetic variability have been examined: namely, how it originates through mutations and how it is transmitted from one generation to another. This section will deal with how the genetic makeup of a population changes over time. If polymorphism were exclusively controlled by the genetic process, then one would not expect changes in the frequency of the various forms with the passage of time. Imagine a polymorphic population of butterflies that has no links with other populations and in which the crosses between males and females occur by chance, that is, without any preference shown for a particular phenotype or genetic makeup during mate selection. Imagine also that this population is very numerous and that mutations do not occur in it. Now in such a hypothetical situation one would expect, on the basis of Mendel's laws of inheritance, that the frequencies of the alleles and of the homozygous and heterozygous genotypes would remain unchanged over the generations. Such a population is said to be in equilibrium; this principle (known as the Hardy-Weinberg law after the names of the two scientists who first formulated it mathematically) is very important and forms the basis of the genetic theory of evolution. As long as a population is in equilibrium, its variability is perpetuated from generation to generation. A population is no longer in equilibrium when there is an increase or decrease in the frequency of certain genotypes and alleles under the influence of such factors as *natural selection*, the immigration of new individuals from other populations (*gene flow*), and the occurrence of mutations or chance factors (*gene drift*), which may have a greater or lesser effect, depending upon the size of the population. Such variations are the essence of the evolutionary process.

Butterflies and moths are a good means of checking the existence of such variations because they have been collected for many years. If one examines collections of a polymorphic species made in the same locality in different years, one can see how the frequency of a particular allele may change over the generations by examining the polymorphic phenotypes. The studies of E.B. Ford, P.M. Sheppard, and other British geneticists on the moth *Callimorpha dominula* is a classic record of this phenomenon. This species is polymorphic for two alleles that control the number of white spots on the forewings. The normal homozygote, *dominula*, has six main white spots distributed over the whole wing, while the rare homozygote, *bimacula*, has only two basal spots. It is a semidominant character, and so the heterozygote can be distinguished from the homozygotes since it has an intermediate number of spots (form *medionigra*). Careful estimates made annually at Cothill in Berkshire, England, showed a significant variation in the allele for the melanic coloration, which in the homozygous condition is expressed as the *bimacula* form. Its frequency rose from around 1 percent in 1929 to over 11 percent in 1940 and then gradually dropped back to its original level in

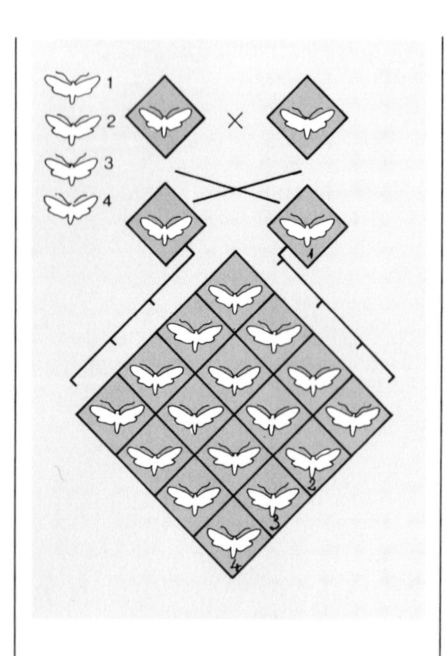

Plate 14 **Polymorphism in *Zygaena ephialtes*: independent assortment of characters**

Zygaena ephialtes displays marked polymorphism in certain of its populations. There are four principal forms: red peucedanoid (1), yellow peucedanoid (2), red ephialtoid (3), and yellow ephialtoid (4). The genetic basis of the polymorphism has been clarified with numerous crosses carried out independently over more than 20 years by the Swiss biologist P. Bovey and the Polish engineer A. Dryia. The four forms are determined by two genetic loci, each with two alleles, which control respectively the melanic patterning on the hind wing (peucedanoid or ephialtoid) and the coloration (red or yellow).

The peucedanoid character is dominant over the ephialtoid one, and red coloration is dominant over yellow. The illustration summarizes the results of a cross made by Dryia between a red peucedanoid male with a *PPRR* genotype, and an ephialtoid female of the genotype *pprr* (see the text). All the F_1 offspring display the dominant red peucedanoid characters. If two of these are crossed, the F_2 generation again contains red peucedanoid individuals; but in addition yellow ephialtoid individuals reappear, and two new combinations, red ephialtoid and yellow peucedanoid, are also present. In this type of cross, where there are two pairs of alleles on separate chromosomes, an independent assortment of the characters give rise to the four forms in the ratio 9:3:3:1, where 9 is the proportion of individuals bearing the two dominant characters (peucedanoid, red), 3 and 3 the proportion having a combination of recessive and dominant characters, and 1 the proportion in which both the characters are recessive (ephialtoid, yellow). (Adapted from Dryia, 1959.)

1955. In subsequent years it fluctuated between 0.8 and 4.6 percent (Plate 15). The causes of this variation have still not been clearly identified although it is probable that natural selection played a part.

Natural Selection

Natural selection is the process by which individuals with a certain genotype have an advantage under certain environmental conditions and a greater chance of surviving and breeding than do individuals with a different genotype. Put more simply, natural selection may be defined as a difference in the breeding success of alternative genetic variants. The concept of natural selection, which forms the basis of the processes of adaptation and evolution, was first advanced — in a slightly different version — by Charles Darwin and Alfred Russel Wallace in 1858 and then exhaustively explained in Darwin's famous *The Origin of Species*, published in 1859. Darwin was unaware of the laws of genetics and based his idea of selection on the principle of "the survival of the fittest" and on the struggle for existence that organisms of any species must carry on because on the average many more individuals are born than can survive and breed.

Industrial Melanism

Industrial melanism is one of the phenomena that most clearly illustrates the role of natural selection in the evolution of polymorphism and in the adaptation of a species. It has been studied in detail by a number of English geneticists and principally by H. B. D. Kettlewell. His subject was *Biston betularius*, a geometrid displaying what is essentially discontinuous variability although an occasional individual with intermediate coloration occurs. There are three principal forms: *typicus*, which is white with fine black spots, *carbonarius*, which is completely black, and the grey, partially melanic, *insularius*. Presumably a number of gene loci underly this variability, one of these forms having several alleles and determining the three principal forms. The *carbonarius* character is dominant to the *insularius* one, which in turn is dominant to *typicus*. The melanic forms were extremely rare until the middle of the last century, as is shown by old English collections. In particular the *carbonarius* form was first found in Manchester in 1848. However, only fifty years later it has become extremely abundant in the same area, making up 95 percent of the population. In the period 1952 to 1956 its frequency rose still further to 98 percent. Careful estimates made by Kettlewell throughout Great Britain between 1952 and 1964, following the examination of a further 20,000 moths, showed that throughout eastern Great Britain the frequency of the *carbonarius* form has increased to over 90 percent. The phenomenon was closely associated with the level of industrialization and atmospheric pollution in the various localities; the populations in the non-industrial regions of the southwest and the north of Scotland were exclusively of the *typicus* form. Subsequent research in Scandinavia and central Europe has confirmed that melanic forms are more abundant in industrialized areas. Many authors have tried to provide an explanation for the phenomenon of industrial melanism. Some, not believing in the inherited nature of melanism, have thought that this character might be induced by the environment and in particular by pollution's having an effect through the larva's food or the breathing of toxic gases by the pupa. Others, recognizing the inherited nature of melanism, thought that the *carbonarius* form was at an advantage in polluted zones because it was physiologically

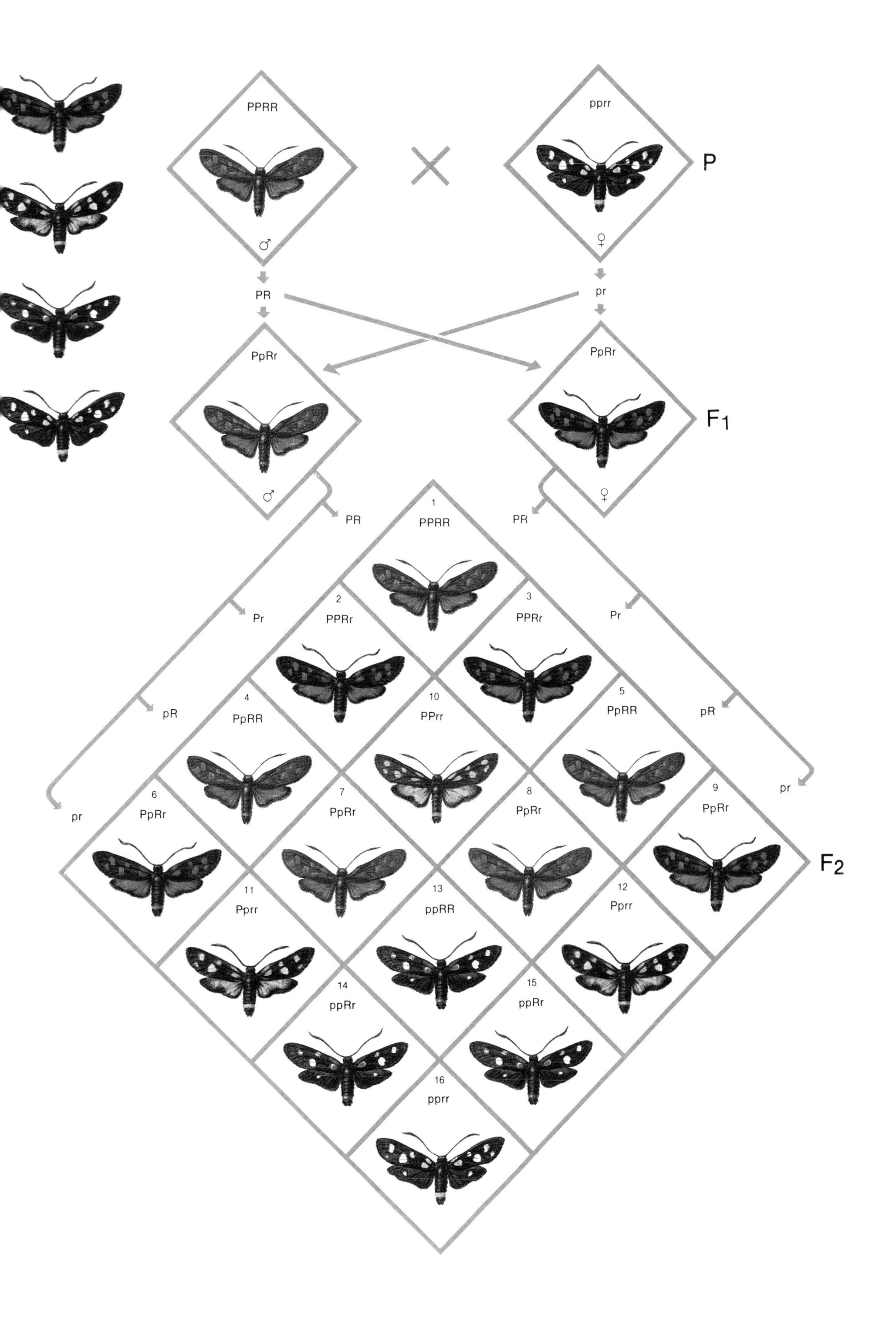

PPRR
♂
pprr
♀
P
PR
pr
PpRr
♂
PpRr
♀
F1
PR
PR
Pr
Pr
pR
pR
pr
pr
1
PPRR
2
PPRr
3
PPRr
4
PpRR
10
PPrr
5
PpRR
6
PpRr
7
PpRr
8
PpRr
9
PpRr
F2
11
Pprr
13
ppRR
12
Pprr
14
ppRr
15
ppRr
16
pprr

Plate 15 **Other examples of polymorphism: temporal and spatial variation**

Genetic polymorphism is relatively common among butterflies and moths in terms of such characters as coloration and pattern. This plate illustrates a number of cases of polymorphism in butterflies (A, B) and moths (C, D); it shows how the frequency of a polymorphic character can vary in time in a population (E) and how individuals belonging to different forms may prefer distinct environmental situations (F).

A: Lesser purple emperor (*Apatura ilia*, Nymphalidae): Some populations from France, central Europe, and the Italian peninsula are characterized by the presence, at differing frequencies, of the two forms illustrated, *ilia* (1) and *clytie* (2).

B: Four forms displaying polymorphism in the patterning of the upper wing in an Italian population of the lunar yellow underwing (*Noctua orbona*, Noctuidae).

C: Four of the numerous polymorphic forms of an East African population of the white-barred acraea (*Acraea encedon*, Nymphalidae, Acraeinae), studied by D. F. Owen.

D: Some polymorphic forms of a population of *Zygaena transalpina* (Zygaenidae) from central Italy, characterized by red or yellow coloration and by melanism on the hind wing.

E: Polymorphism in an English population of the scarlet tiger (*Callimorpha dominula*) and the annual variation in the frequency for the allele for melanic coloration, which in the homozygote is expressed as the *bimacula* form (3). The nonmelanic homozygote (*dominula* form) and the heterozygote (*medionigra*) are illustrated in Figures 4 and 5 (see text).

F: Polymorphism in the female of *Argynnis paphia* (Nymphalidae, Nymphalinae): (6) male; (7) female, normal form; (8) female, *valezina* form. In certain localities the two forms show a different choice of habitat.

stronger. However, it was Kettlewell once again who in a brilliant series of experiments succeeded in showing that the advantage of the melanic forms— and *carbonarius* in particular — in the industrialized zones is tied to greater protection from predatory birds. Under natural conditions the *typicus* form is camouflaged when resting during the day on tree trunks covered by lichens; but where there is heavy pollution, the melanic forms are better able to avoid being seen by birds since there are no lichens and tree trunks and other supports have been blackened by coal smoke. Kettlewell's experiments consisted in releasing a certain number of suitably marked males of the three forms into polluted and unpolluted regions and recapturing them after a few days with the help of light traps and virgin females, which are a powerful attraction for the males over long distances. The results showed that in the polluted area (Birmingham) the proportion of recapture of the melanic forms amounted to 53 percent, whereas that of the pale forms did not exceed 25 percent of the individuals released. In an unpolluted area the opposite was the case, a higher percentage of the *typicus* form (13.7 percent) being recaptured than of the melanic forms (4.7 percent).

Consequently it is clear that the increase of the melanic forms, which has occurred in a surprisingly short time for an evolutionary process, is the result of the selective pressure applied by predators, which in the polluted areas operates against the *typicus* form. It is interesting to note that over the last twenty years, following the more rigorous control of pollution in many British cities, a marked reduction has occurred in the black coating on tree trunks, and in recent years the *typicus* form has again increased in relative abundance.

The phenomenon of industrial melanism is widespread in Europe, and besides *Biston betularius* it is to be found in a further 100 species of Lepidoptera, including *Lasiocampa quercus* and *Phigalia pedaria*, illustrated in Plate 16. Only widely scattered reports of industrial melanism are known in North America.

Balanced Polymorphisms

Natural selection does not always operate in favour of a particular genetic variant, increasing its occurrence at the expense of others present in a population (*directional selection*). A certain level of polymorphism may be maintained under particular environmental conditions. If, for example, the heterozygous genotype has a selective advantage compared with individuals homozygous for the alternative characters, it will have a greater chance of breeding, and so both genetic variants will be passed on to the succeeding generations. This, together with other ways of maintaining a polymorphism, is defined as *balanced selection*, and the resultant polymorphism is said to be balanced. In other situations it is possible to observe balanced polymorphisms determined by different and sometimes complex mechanisms. For example, the males of the nymphalid *Argynnis paphia* are monomorphic while females from certain localities are dimorphic, occurring as the normal form and a dominant melanic-green variant known as the *valezina* form. The green form varies in frequency, and in certain parts of the Alps it is very common although it is less efficient than the normal form in attracting males. It appears that it prefers to fly in woods, and under these conditions it may be at a reproductive disadvantage compared with the normal female. However, the green form is less visible in woods and less subject to attack by predators. This polymorphism appears to be maintained by the interplay of these two factors, which vary with habitat heterogeneity.

%
11
10
9
8
7
6
5
4
3
2
1
Frequency of *medionigra* allele
1939
1941
1943
1945
1947
1949
1951
1953
1955
1957
1959
1961
Years

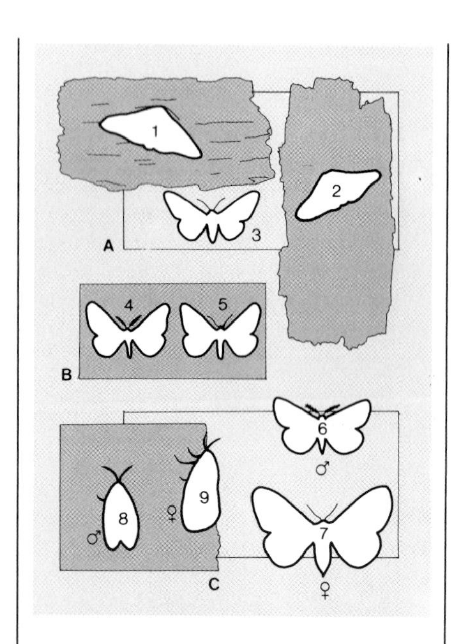

Plate 16 **Adaptive significance of polymorphism: industrial melanism**

Many moths possess polymorphic populations with melanic forms. When the frequency of the melanic forms is correlated with the level of environmental pollution, the phenomenon is referred to as *industrial melanism.* The plate shows three species studied by British authors in which this phenomenon is now well known.

A: *Biston betularius* (Geometridae): (1) typical form, camouflaged from predatory birds on surfaces covered with lichens; (2) *carbonarius* form, camouflaged on trunks blackened by soot; (3) *insularius* form (see text).

B: *Phigalia pedaria* (Geometridae): (4) typical form; (5) *monacharia* form. The frequency of the latter rises as high as 75 percent in the urban areas of Lancashire and southern Wales, whereas it is uncommon (about 20 percent) in the rural areas of southern England. Studies by D. Lees have shown that the geographical distribution and the relative abundance of this form are correlated with the reflective power of the surface, in particular oak trunks, which varies with the level of pollution.

C: *Lasiocampa quercus* (Lasiocampidae): (6) male, typical form; (7) female, typical form; (8) male, *olivacea* form; (9) female, *olivacea* form. The melanic *olivacea* form is uncommon in certain industrialized areas of England but it is present at high frequency (about 50 percent) even in the unpolluted areas of northeastern Scotland. H. B. D. Kettlewell has been able to show that in Scotland the *olivacea* form is at an advantage with respect to the typical form because it is less visible to the black-headed gull, its main predator, against the dark background of heather.

Seasonal Variation

Various *Colias* (subspecies Pieridae), especially the common European *C. croceus* (Plate 17), are well-known examples of female-linked polymorphism — which is itself a common phenomenon. In *C. croceus* normal males and females are yellow; females of the *helice* form are white. The locus coding for this character is not on the sex chromosomes but on one of the autosomes (as in the *valezina* character in *Argynnis paphia*); however, the phenotypic expression is confined to the female. Thus homozygous *HH* and heterozygous *Hh* females have a white *helice* phenotype, whereas *hh* homozygotes and all males, whether *HH*, *Hh*, or *hh*, are yellow. This butterfly produces several generations in the course of a year, and the frequency of the *helice* form tends to increase in the colder months. This agrees with the experimental observation that the normal form is favoured by higher temperatures. Consequently the polymorphism is controlled by natural selection, favouring or acting against one or other form at different seasons.

Variability Controlled by the Environment

It has already been stressed that not all the variability in the coloration of butterflies and moths is genetic. The same genotype can produce quite different phenotypes if it is subject to different environmental conditions. This phenomenon is known as the *law of reaction*, and the forms expressed in different situations are *ecophenotypes*. When an ecophenotype closely resembles a known genetic variant, it is termed a *phenocopy*. It is relatively easy to obtain unusual ecophenotypes or phenocopies by subjecting pupae to heat or cold. If pupae of the small tortoiseshell, *Aglais urticae*, are kept at low temperatures for several days, they will produce butterflies that resemble the *polaris* form that occurs naturally in Lapland. Similar phenomena have been observed in other Nymphalinae. Likewise, pupae of the European *Papilio machaon*, when subjected to high temperatures, produce butterflies with reduced black patterns similar to the southern races from Turkey and Syria (Plate 13).

One of the best known cases of ecophenotypic variation concerns the small European map butterfly, *Araschinia levana*, which produces a spring generation (form *levana*) and a quite different summer one (form *prorsa*). The two forms, illustrated in Plate 17, are so dissimilar that in the past they were described as separate species. The principal factor that determines these differences is the duration of diapause in the pupal stage, and this in turn is determined by the photoperiod (the relative lengths of day and night) that the larva previously experienced. In the wild the spring form, *levana*, develops from the previous autumn's larvae (short day length), which produce pupae with a winter diapause, while the summer form, *prorsa*, is derived from summer larvae (long day length), which give rise to pupae with no diapause.

Instances of seasonal variation that are not genetically determined are also known in various tropical butterflies, such as *Precis octavia* (Plate 17), the widely differing forms of which are seasonally controlled by ambient humidity. In North America examples include *Polygonia comma*, *Eurema daira*, and *Rothschildia lebeau*.

Sexual Dimorphism

The most common variation within a population is that between individuals of different sexes. It is known as *sexual dimorphism* when it

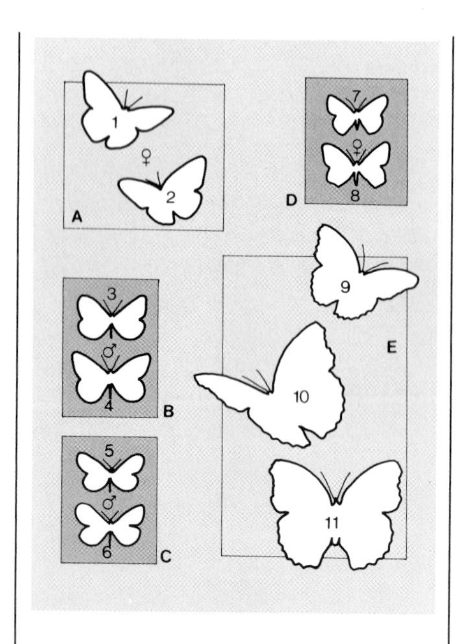

Plate 17 **Seasonal variation**

Over a year a butterfly or moth species may vary considerably in average size and wing shape, pattern, and colour. In a given population the individuals of the spring generation may differ to a greater or lesser extent from those of the summer or any later generation. Two kinds of mechanism underlie seasonal variation: One is genetic (A), in which the frequency of the polymorphic forms vary seasonally; the other is environmental (B–E), in which individuals belonging to different generations display ecophenotypic differences to varying degrees.

A: Clouded yellow (*Colias crocea*, Pieridae): (1) female, normal form; (2) female, *helice* form, determined genetically and more common in generations flying during cold months.

B: Southern small white (*Pieris mannii*, Pieridae), underside: (3) male, spring generation; (4) male, summer generation. Note the difference in the intensity of the green coloration of the two generations.

C: Wood white (*Leptidea sinapis*, Pieridae), underside: (5) male, spring generation; (6) male, summer generation. The seasonal differences in coloration are produced ecophenotypically and are similar to those of the preceding species and many other white pierids.

D: Map butterfly (*Araschnia levana*, Nymphalidae): (7) female, *levana* form, spring generation; (8) female, *prorsa* form, summer generation.

E: Gaudy commodore (*Precis octavia*, Nymphalidae): (9) wet-season form; (10) dry-season form; (11) intermediate form. The coloration of the individuals illustrated is duller than in nature, where the blue is more brilliant.

involves more or less marked differences between females and males in characters that are not connected with the reproductive apparatus (*secondary sexual characters*). Butterflies and moths often display sexual differences in the colour, shape, and size of the wings and antennae and even in the presence or absence of wings (Plates 10 and 11). Sometimes the differences are so marked that in the past the two sexes were attributed to separate species or genera. The females are often larger than the males and cryptically coloured, in order to hide from predators, or are mimetic, so that predators are unable to distinguish them from other, inedible species. The males, on the other hand, are frequently brilliantly and strikingly coloured, which helps in intraspecific recognition and courtship.

Well-known sexually dimorphic North American Lepidoptera include *Speyeria diana*, *Neophasia terlootii*, *Callosamia promethea*, and *Orgyia gulosa*.

Sex is an inherited character with two alternative forms that normally occur in a 1:1 ratio. It is under the direct control of genes situated on particular chromosomes, the so called *sex chromosomes*. As explained previously, chromosomes are usually present in cells in the form of matching pairs (homologous chromosomes), with each parent's contributing one chromosome of each pair. In various animal species the sex of an individual may be determined by the presence of identical or different sex chromosomes or the absence of one of the two chromosomes of a pair. Two kinds of sex chromosome are recognized, the X and Y chromosomes, but in reality the genes responsible for the determination of sex are situated on the X chromosomes. In the Lepidoptera the presence of two X chromosomes gives rise to a male (*homogametic sex*), while an X and a Y chromosome or simply the presence of just one X chromosome results in a female (*heterogametic sex*). Therefore each male germ cell will necessarily contain an X chromosome; and if it fertilizes an egg carrying another X, it will give rise to a male (XX), whereas if there is a Y chromosome in the egg or no sex chromosome at all, it will produce a female (XY or XO). This method of sex determination is peculiar to butterflies and moths and a few other organisms, such as birds. In most animals, including man, the male sex is heterogametic, and the XX combination determines an individual of the female sex.

Gynandromorphism

Aberrant individuals that simultaneously display both male and female characters in different proportions in their bodies have occasionally been observed in the Lepidoptera and other insect orders. Such individuals are known as *gynandromorphs*. Although it is not a common phenomenon, gynandromorphism has been observed in many species of butterflies and moths, including ones displaying marked sexual dimorphism. *Ornithoptera priamus* and *Parnassius autocrator* provide some of the most spectacular known examples, with instances in which there is an almost perfect longitudinal division of the body into two halves, one side male and the other female (Plate 12). However, the alternative sexual characters are not always equally divided, and sometimes a small part of the animal may display characteristics peculiar to the other sex. In other cases the gynandromorph may appear as an intersexual mosaic. The males of *Morpho aega* are normally an iridescent blue in colour, but the females occur in two colour forms, orange-brown and blue. Plate 12 shows the two female forms and a gynandromorph in which the female component, in the brown form, is confined to a portion of the right hind wing. Anomalies of this kind may occur in insects because the genetic control of sexual characters acts directly at the level of each individual cell in the body.

To be a gynandromorph, a butterfly or moth must have some cells with XX sex chromosomes (male characters) and others with XY or XO chromosomes (female characters). There are essentially two mechanisms that can create such a situation: The first is an error in gametogenesis, resulting in an egg with two nuclei, one containing an X and the other a Y, both of which are fertilized and give rise to two, sexually opposite halves and an offspring that is a perfect bilateral gynandromorph; the second possibility involves an error in cell division during the development of a male (XX) zygote and the consequent loss of an X chromosome. All the cells that are derived from the normal one will be male (XX), and the others will be female (XO). In this case gynandromorphs will be bilateral only if the error occurs in the first cell division.

Intersexuality

The processes described above are not the only ways in which individuals with sexually intermediate characteristics may occur. *Intersexuality* is a widely known phenomenon. The affected individual is often the product of a cross between members of widely separated populations or geographical races and differs from a gynandromorph in that male and female characters are not present simultaneously but expressed one after the other. Consequently male and female characters do not necessarily occupy separate portions, and some parts of the body may be intermediate between the sexes. This phenomenon has been particularly studied in *Lymantria dispar* by means of numerous crosses between different European and Asian races that have produced individuals displaying varying degrees of intersexuality. The explanation is rather complex, but it has been suggested that a male factor linked to the X chromosome varies between the races.

Origin of Species

To understand the diversity of species, one has to be aware of the natural processes that lead to their formation, and this is one of the key problems of evolution. Before analyzing in detail the mechanisms by which two or more species originate from a single ancestor, we must stress some characteristics of animals:

Each species may be regarded as an isolated collection of genes, known as a *gene pool*.

Species are generally divided into populations that are or could be linked by gene flow, that is, the exchange of genetic factors by individuals migrating from one population to the other.

Each species is continuously adapting to its environment.

Each species occupies a particular *ecological niche* — the array of environmental conditions with which it interacts — which is not used by any other species.

Each species has a set of mechanisms that isolates it reproductively and either directly or indirectly prevents the exchange of genes with closely related species.

An important contribution to the problem of the origin of new species has been made by studies on the geographical variation that exists between the various populations of a species. These populations generally differ to a greater or lesser degree, and often the differences are gradual so that they are more marked the farther apart the populations are. Geographical variation is present in the vast majority of animal species; it may be either continuous or discontinuous.

Geographical Variation: Clines and Subspecies

Continuous variation is also known as a *cline*, and it occurs when the populations in the area of distribution are not fragmented into groups that can be distinguished and are extensively interconnected by gene flow while still displaying a continuous and gradual change in their characters along gradients of one kind or another, such as latitude, altitude, environment, and so on. Plate 18 illustrates two of the many known examples. The common satyrine *Pararge aegeria* (Plate 18A) displays a gradual increase in the darkness of the tawny spots on its wings as one observes the species from the northern Scandinavian populations to the southern Mediterranean and North African ones. *Acraea natalica* (Plate 18B), a widespread African species, has a continuous distribution with no isolated populations and also displays clinal variation. The wings have a reddish-orange background coloration with black spots and shading. This varies along the northwest-southeast cline so that the northwestern forms are gradually replaced by lighter forms. In addition the dark margin of the hind wing is weakly developed in the northern forms but considerably extended in the southeastern ones.

Discontinuous geographical variation occurs when the exchange of genes among populations is limited or nonexistent. Groups of fairly isolated populations may exist within a species and display unique characteristics because the flow of genes to and from other populations is blocked by geographical barriers. When the differences between populations or groups of populations are so marked that the majority of individuals from one population can be distinguished from those of another with certainty, such isolated units may be defined as *subspecies*. A species may consist of several subspecies, and it is then termed polytypic. Obviously the subspecies occupy different parts of the range of the species, and the differences between them are primarily due to the impetus of natural selection's acting differentially under the different environmental conditions that characterize the various areas. One of the most marked

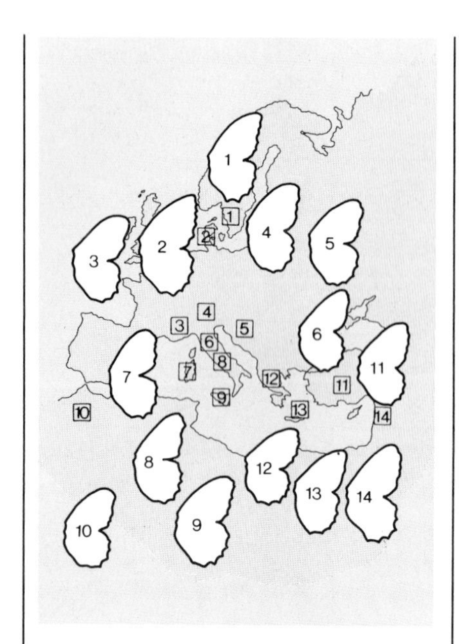

Plate 18A **Continuous geographical variation: a cline in the speckled wood (*Pararge aegeria*)**

The continuous geographical variation of one or more characters through several populations of a species is known as a *cline*. Examples of clinal variation are common among those lepidopteran species that have an extensive range but no geographically isolated populations. The tawny coloration of the wing of the Palaearctic species *Pararge aegeria* (Nymphalidae, Satyrinae) is a good example of clinal variation. The pale colour gradually increases in intensity from the Scandinavian populations to those of the Mediterranean region: (1) southern Sweden, (2) Denmark, (3) Piedmont, (4) Trentino, (5) Yugoslavia, (6) The Marches, (7) Sardinia, (8) Lazio, (9) Sicily, (10) Morocco, (11) Turkey, (12) Greece, (13) Crete, (14) Lebanon.

and commonly referred-to examples of clearly separate subspecies concerns the island populations of the *Ornithoptera priamus* species group (Plate 19). These occur in the Australian region, from Queensland to the Solomon Islands and the Moluccas, and they differ principally in the iridescent background colour of the male's wings. This varies discontinuously from the green of *priamus* and other races to blue (*urvillianus*) and golden (*croesus*). The first is found on the islands of Ambon and Seram (similar races occur in Australia, New Guinea, and other islands to the west of New Guinea). The second occurs in the Solomon Islands, New Ireland, and New Hanover, while the third is typical of the island of Bacan and, with similar forms, the islands of Halmahera, Ternate, and Tidore (all in the Moluccas). However, the last of these forms (*croesus*) is so different that experts have now recognized it as a separate species.

In some cases there exist circular clines, the end either overlapping or not. Usually the butterflies at the two ends are not genetically compatible. This sibution is called a *Rassenkreis*, or "circle of races." In California the sylvan hairstreak (*Satyrium sylvinus*) presents just such a picture: It forms a circle of races around the great Central Valley, occurring along drainages leading from the coastal ranges and Sierra Nevada. Just north of San Francisco Bay the subspecies *dryope* occurs; it has no taillike projections from the trailing edge of its hind wings and ranges in this form south through the coastal mountains to about Los Angeles. Going around the southern end of the Central Valley, the subspecies *desertorum* occurs; its populations consist of a mixture of tailless individuals like *dryope* and of individuals with normal hairstreak tails. North along the western foothills of the Sierra Nevada one finds the populations to be composed of individuals all of which are tailed. This condition persists around the northern end of the valley and south through the coastal ranges to just north of San Francisco Bay. There the completely tailed and tailless populations exist within about 50 to 60 km (30 to 40 mi) of each other; the Petaluma River is the dividing line. There is no exchange of genetic material at this disconformity.

Geographical Speciation

In the wild, therefore, variations in the adaptive characters of populations exist within a single species, and geographical isolation may be decisive in keeping populations genetically separate as they adapt continuously to particular environmental conditions. This makes it much easier to understand how a single ancestral species can give rise to several new ones. The term *speciation* is applied to the whole process of divergent evolution, from the first appearance of mechanisms of reproductive isolation (which, as will be seen later, enable two species to maintain separate gene pools) to the final break in the genetic continuity of the two populations.

The most usual way that speciation occurs in the wild is known as *geographical* or *allopatric speciation*, in which a new species develops when one population is geographically isolated over time and acquires characters that lead to or ensure its reproductive isolation from its closest relatives.

To make it easier to understand how allopatric speciation works, consider the hypothetical case of a single species occupying a given range and initially present as several populations. Plate 20A shows an imaginary species, *Utopiella ancestralis*, in which different selective pressures operating in the various parts of its range produce small genetic differences among some populations. These are countered to a greater or lesser degree by gene flow (see stage 1 in the plate). Imagine now that, as a result

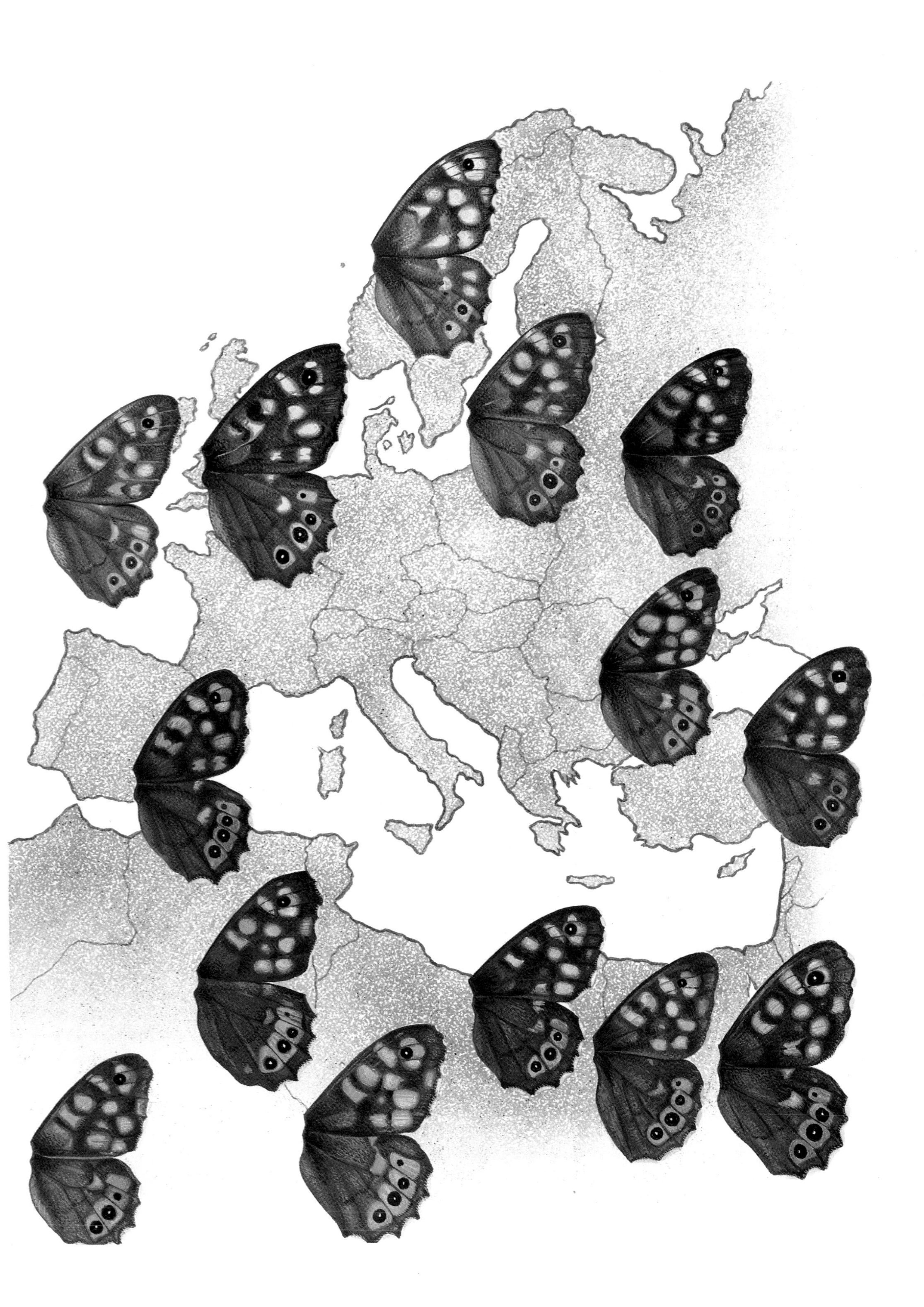

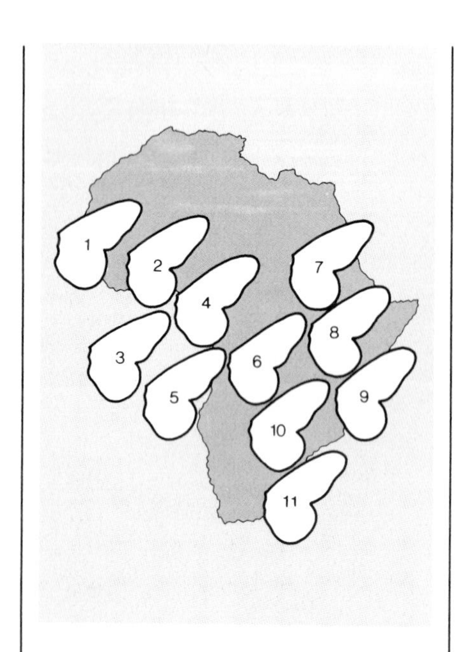

Plate 18B **Continuous geographical variation: a cline in the Natal acraea (*Acraea natalica*)**

Described by D. F. Owen, an example of clinal variation occurs in the African *Acraea natalica* (Nymphalidae, Acraeinae), which is continuously distributed over Africa south of the Sahara. The West Africa populations show a high degree of melanism, whereas melanism is greatly reduced in the eastern and southern ones. The intermediate populations are transitional between these two extremes. (1) Sierra Leone, (2) Ghana, (3) Fernando Po, (4) southern Nigeria, (5) Angola, (6) Zaire, (7) Ethiopia, (8) Uganda, (9) Tanzania, (10) Zambia, (11) South Africa.

of geological or climatic events, a geographical barrier — an area that the species cannot inhabit, such as an estuary, a major river, or a desert — is produced between one of these populations and the others. In such a case the two groups of populations will continue to evolve independently and will diverge to the level of subspecies (stage 2). If after a relatively short period (for example, a few centuries or even a few thousand years) the barrier disappears, some individuals of the isolated subspecies may return to the original area and coexist with individuals from the mother population (a condition known as *sympatry*). In such a situation crosses will probably occur between members of the two populations capable of producing fertile hybrids (stage 2*a*), and these will lead to a greater or lesser extent to their being mixed together genetically again (*introgression*). If the barrier persists for much longer — on the order of hundreds of thousands of years — then the isolated population will continue to evolve independently, increasing its degree of genetic divergence from the mother population (stage 3). At this stage the isolated population group would probably already have accumulated enough genetic differences to ensure its reproductive isolation from the original population. This could only be verified following contact between the two populations. If they don't cross or fail to produce offspring or if the hybrids are sterile or have reduced fertility, then the process of species formation is complete (stage 3*a*).

The various stages that have been described above in a schematic way and in a hypothetical case can be seen in a series of natural examples (Plate 20B).

Stage 1 (genetically linked populations) occurs in the satyrine *Coenonympha tullia*, which is widespread in Britain as a series of populations differentiated to varying degrees, with marked differences existing in the development of the eyespots on the underside of the hind wings; these were linked in the recent past by a series of intermediate populations (now extirpated), resulting in clinal geographical variation.

The pierid *Euchloe ausonia* provides an example of stage 2 (allopatric subspecies). It is present in southern continental Europe with more or less differentiated populations and races, while a markedly different subspecies, *E. ausonia insularis*, occurs in Sardinia and Corsica.

Stage 2*a* (secondary hybridization between subspecies in zones of contact or overlap) is to be found in *Pieris napi napi* and *P. napi bryoniae*. These have become symparic again over a vast belt of territory in the Swiss and Italian Alps after having been previously separated by climatic barriers. The position of this group is, however, further complicated by the fact that there are other contact zones in the Alps where the two forms remain reproductively isolated.

Battus devilliers is one of the many examples of stage 3. It is a papilionid endemic to Cuba and is derived, although markedly different, from *B. philenor*, a widely distributed continental American species.

Finally stage 3*a* has been reached by *Papilio demodocus*. This is widely distributed in Africa, and three species derived from it have arisen in Madagascar. *P. demodocus* now also occurs again in Madagascar, but it is reproductively isolated from its three relatives.

Factors in Speciation

It should now be clear that geographical isolation functions to enable populations to adapt to their respective environments by barring gene flow from neighbouring populations and the consequent mixing of genes. Viewed from this perspective, gene flow is a conservative influence, blocking the tendency of populations to diverge by adapting promptly to

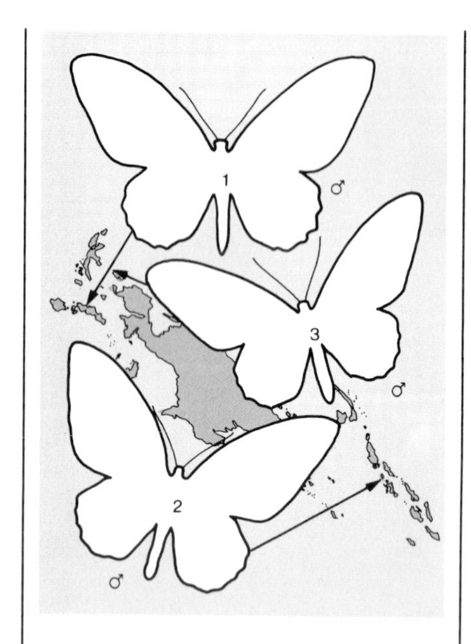

Plate 19 **Discontinuous geographical variation: subspecies**

When populations of a species are more or less isolated geographically, with gene flow reduced or absent, they often display marked differences in various characters. Such populations or groups of populations are termed *subspecies*. Subspecies that have been geographically isolated for some time (tens or hundreds of thousands of years) may evolve into separate species. In the *priamus* group of the *Ornithoptera* (Papilionidae), found in the Australian region from Queensland to the Solomon islands and the Moluccas, a number of subspecies have been recognized. These include (1) *Ornithoptera priamus priamus* on the islands of Ambon and Serang, whose males are characterized by a magnificent green, and (2) *O. priamus urvillianus* from the islands of New Ireland and New Hanover in the Bismarck Archipelago, in which the males are largely iridescent blue. Other groups of populations occur on islands in the Moluccas (Bachan, Ternate, Tidore, and Halmahera) and are sufficiently different to be classed as (3) a separate species, *O. croesus*, in which the males are golden.

their respective environments. Indeed it opposes, although weakly, natural selection and counteracts its effects. Therefore gene flow considerably slows down the speed with which populations may diverge.

Another factor shown to be fundamental in determining the rapidity of speciation is population size. Small, isolated, peripheral populations are often derived from only a few individuals that have colonized a new marginal area or an island, and they carry only a small fraction of the parent population's original genetic variability. In such small populations genetic drift — chance variation in the frequencies of alleles — may play a major role, enabling new genotypes or genetic combinations to become more common than natural selection would allow. The result can be a rapid and sometimes drastic reorganization of the genetic composition of the population, with resultant morphological and physiological changes and the acquisition of isolation mechanisms.

For example, in the wild it is easy to find instances of island populations (of species with extensive continental ranges) that have diverged as separate species. In Italy there is a group of closely related satyrines belonging to the genus *Hipparchia*, which are probably derived from *H. semele*, a widespread continental species. The group contains species endemic to small islands — such as *H. sbordonii* on Ponza or *H. leighebi* on the Aeolian Islands — or to promontories that were once separate from the mainland — such as *H. ballettoi* on Monte Raito on the Sorrento peninsula.

Obviously the model of speciation described above is highly schematic, and in the wild there are a variety of intermediate paths.

It has been shown how geographical isolation can lead, in particular situations, to the development of mechanisms that immediately ensure reproductive isolation. However, as mentioned previously, two populations which have been isolated but which have come into contact again in a particular area may produce hybrids. It is possible to distinguish at least two classes of such hybrids: If hybridization is complete and the hybrids continue vigorously for many generations, then it is probable that the two populations were no more than subspecies of the same species; on the other hand, if hybridization is incomplete with individuals' tending to mate with others belonging to the same population or with hybrids that are less successful in breeding or survival, then the first stage has been reached and the isolation mechanisms can develop until two separate species are formed. In fact reproductive isolation is not necessarily completely achieved in the course of geographical isolation since no selective force acts directly on it while the populations remain isolated. Moreover, one would expect that it would more often be precisely when two populations again became sympatric that reproductive isolation would be fully developed. So selection can act in the following way: Individuals that tended to mate with the members of a foreign population would produce less vigorous and fertile offspring than would the offspring of crosses within a population. Consequently mutations or genetic combinations that favoured mating between similar individuals within a population would be at a considerable selective advantage and would spread throughout the population.

Incomplete Speciation: The Semispecies

There are many natural instances where speciation has been only partially completed. There are, for example, cases of populations that are reproductively isolated from others that are morphologically similar, whereas they are completely or partially interfertile with other populations in different parts of the range. In such instances the population has

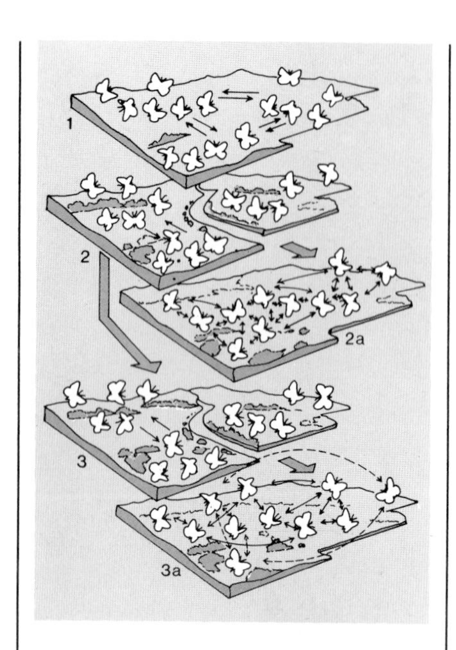

Plate 20A **Stages in geographical speciation: a hypothetical case**

Geographical (or allopatric) speciation is the most common way in which two or more new species may arise from an ancestral one. Several stages can be distinguished in this process. They have been described at length in the text and may be summarized here by reference to the imaginary *Utopiella ancestralis*.

In stage 1 three populations of *U. ancestralis* occupy separate areas of the range and differ slightly although linked by gene flow, as shown by the arrows.

In stage 2 the creation of a natural barrier, such as a large river, separates two groups of populations, which then evolve independently and develop into two distinct subspecies, which we shall call *U. ancestralis ancestralis* and *U. ancestralis nigroflava*. Now imagine that the geographical barrier disappears after a relatively short (hundreds or thousands of years) or long period (tens or hundreds of thousands of years). In the first case, stage 2*a*, the two population groups that come back into contact (sympatry) will probably still be interfertile and will interbreed to produce fertile hybrids.

In the second case, stage 3, the two subspecies will probably have accumulated enough genetic differences to prevent successful crosses. The existence of reproductive isolation between the two groups will be tested only when the geographical barrier disappears and they are again sympatric (3*a*). If there are no crosses

(continues on page 66)

reached an intermediate stage in the process of speciation, in which the genetic divergence that has occurred during the period of geographical isolation has not been sufficient to prevent hybridization completely. Such populations cannot be regarded as good species and are referred to as incipient species or *semispecies*.

It is not difficult to find areas in the wild where two morphologically different semispecies are sympatric and still produce a certain number of hybrids. For example, the North American nymphalids *Limenitis arthemis* and *L. astyanax* respectively occur in Canada and in the northeastern and southern United States. The semispecies overlap in several areas, and in one zone in particular, about 200 km (125 mi) across, intergradation and hybridization occur; such zones are known as *hybrid zones*. These forms are both treated as if they were subspecies of *L. arthemis*: *Arthemis* has a broad white band running across its fore- and hind wings, while *astyanax* does not; it, however, has a blue-green iridescent area on the upper surface of its hind wing, whereas *arthemis* generally displays a series of dull red spots between the white band and the margin of the hind wing. The hybrid closely resembles *astyanax*, but it does not have the iridescent area on the hind wing (Plate 21).

A similar situation occurs between *Papilio zelicaon* and *P. polyxenes* in a different part of the United States (Plate 21). The hybrids are introgressive with *P. zelicaon* and are referred to as form *nitra*. It is interesting to note that several hybrid or secondary contact zones exist in North America, in which occur many semispecies or closely related species, representing a wide variety of organisms. Remington has termed these *suture zones*, and they show that now-vanished geographical barriers once existed in these areas.

Mechanisms of Reproductive Isolation

It has been shown how the biological species concept is not sufficient to deal with populations at intermediate stages in the speciation process. To be certain that a group of populations constitutes a clearly differentiated species, it must maintain its genetic individuality through mechanisms that either directly or indirectly prevent the exchange of genes with related species. Therefore mechanisms that allow the isolation of gene pools of independent evolutionary units play a prime role in speciation.

Mechanisms of reproductive isolation are a property of organisms, and all species, even closely related ones, are kept separate by various, often complementary, routes.

Classically the routes to reproductive isolation were divided into the following two principal categories:

Mechanisms that *prevent* interspecific crosses, that is, stop hybridization between the members of different populations and hence the formation of hybrid zygotes (*precopulatory* mechanisms).

Mechanisms that *reduce the success* of interspecific crosses, that is, do not stop mating but reduce the vitality or the fertility of the hybrid products (*postcopulatory* mechanisms).

Pre- and postcopulatory mechanisms perform the same function but with an important difference in the amount of energy wasted by an organism. Energy loss is much greater if there are only postcopulatory mechanisms since these do not prevent the useless release of gametes, whereas precopulatory mechanisms do so. When two organisms that are already isolated by postcopulatory mechanisms coexist in the same area and there is the possibility of hybrid-zygote formation, natural selection will most probably favour the development of precopulatory mechanisms.

between members of the two groups, if they do not produce progeny, and if the hybrids are sterile or have reduced fertility, then speciation will be completed and two separated species will result, which we might call *U. ancestralis* and *U. nigerrima*. The arrows in the figure indicate that gene flow is confined to individuals of one of the species (continuous arrows) or the other one (broken arrows) but does not occur between members of the two species.

There are several precopulatory barriers that promote reproductive isolation, including the following:

1. Ecological Isolation. This occurs when the species occupy different habitats within the same region; it is common among closely related, or *sibling*, species (morphologically indistinguishable). For example, in the northeastern United States and adjacent southern Canada two satyrine sibling species, *Satyrodes appalachia and S. eurydice*, are sympatric over large areas but display markedly different ecological preferences. The first flies in swampy woods, while the second flies in open marshes. Butterflies and moths frequently display precise habitat preferences, but ecological isolation by itself rarely provides a sure guarantee of reproductive isolation.

2. Seasonal or Temporal Isolation. This is to be found in populations the members of which attain sexual maturity at different times or (sometimes) seasons. A good example is provided by two ctenuchids, *Amata phegea* and *A. ragazzii*, sympatric in several parts of central Italy. However, *A. phegea* flies between the beginning of June and mid-July, while the first *A. ragazzii* adults appear in July and fly until the middle of August. As a result the adults of the species can encounter each other only in early July. The chances of interspecific crosses are further reduced by the fact that in both species the males emerge before the females. Temporal isolation also occurs between species that prefer only certain hours for mating, as is the case with three North American *Callosamia* (Saturniidae). The ranges of these three species overlap extensively, and in a given locality and habitat it is possible to collect individuals of *C. promethea*, *C. angulifera*, and *C. securifera*. *C. angulifera* always mates at night between dusk and midnight, with a peak of activity around 10 o'clock. *C. promethea* and *C. securifera*, on the other hand, are diurnal. In the former the males begin to fly about 1:00 or 2:00 P.M. and are most active between 4:00 and dusk, when the females increase their release of attractants. The females of *C. securifera* begin their sexual activity around 10:00 A.M., four or five hours after sunrise, and the males stop flying between 3:00 and 4:00 P.M., before the activity peak of *C. promethea*. Hybrids have never been observed in the wild between these species although hybridization has occurred under experimental conditions. This shows quite clearly that the specific differences in mating times form an effective barrier. Similar situations have recently been studied in other American saturniids, such as between several *Hemileuca*.

3. Behavioural or Sexual Isolation. This is the result of the weakening or the absence of sexual attraction between the males and females of different species. Coloration, behaviour, or the production of special chemicals (pheromones) are very often crucial as stimuli in the attraction of a mate. In butterflies and moths the presence of a certain kind of wing coloration is sometimes closely linked to the choice or rejection of a partner for mating. Ultraviolet reflections, normally invisible to the human eye, are very important in the interspecific isolation of many pierids. In *Colias*, *Eurema*, *Phoebis*, and other genera there are marked differences in the type of ultraviolet reflection between the males of sympatric congeneric species. For example, the females of *C. eurytheme*, a common species in North America, accept conspecific males that display a strong ultraviolet reflection. If the male is experimentally deprived of his ability to reflect ultraviolet, the visible colours remaining unaltered, there is a reduction in mating frequency compared to normal individuals. On the other hand, in *C. philodice*, which is now sympatric with *C. eurytheme* over a large part of its range, the males absorb the ultraviolet component and the females are not capable of discriminating between normal conspecific males and ones whose wings have been masked with pieces of the nonreflective *C. eurytheme* wings. Courtship and the emission of

Plate 20B **Stages in geographical speciation: some examples**

The various stages in the process of geographical speciation were illustrated in Plate 20A by a hypothetical species since obviously no existing species has populations representing all the stages described. However, with several examples of different species it is possible to show the various stages in natural situations. The stages are numbered in the same way as in the previous plate.

A: Stage 1: Scottish populations (1–3) of the large heath (*Coenonympha tullia*, Nymphalidae, Satyrinae) differ from those of southern England (7–9) in the intensity of their coloration and the development of the eyespots. Geographically and morphologically intermediate populations (4–6) link, or have linked in the recent past, the extreme populations, resulting in an almost clinal type of variation. Figures 1, 4, and 7 are males and 2, 5, and 8 are females. Figures 3, 6, and 9 show the underside of the wing.

B: Stage 2: In the dappled white (*Euchloe ausonia*, Pieridae) the continental populations (10, 11) differ to varying degrees, but *insularis* from Sardinia and Corsica (12, 13) is markedly different. It is an allopatric subspecies.

C: Stage 2*a*: The two subspecies *bryoniae* (16, male, 17, female) and *napi* (14, male, 15, female) of the green-veined white (*Pieris napi*, Pieridae) have diverged allopatrically but have come back into contact in certain parts of the Alps. In some areas (blend zones) they hybridize and produce intermediate forms (18, males, 19, females); in others the two forms remain reproductively isolated. Some authors regard the two forms as semispecies (see the text).

D: Stage 3: There are many examples of distinct species separated by geographical barriers. The one illustrated is the Cuban Devilliers swallowtail (*Battus devilliers*, 20, 21). This evolved in isolation from the ancestral species the pipevine swallowtail (*B. philenor*, 22, 23), which is widely distributed in continental North America.

E: Stage 3*a*: An example of the sympatric coexistence of species that are reproductively isolated from the ancestral species is provided by the presence in Madagascar of *Papilio erythonioides* (24), *P. morondavana* (25), *P. grosesmithi* (26), and *P. demodocus* (27). The citrus swallowtail (*P. demodocus*) is widely distributed in Africa and is thought to be the ancestral species, from which the other three species endemic in Madagascar were isolated, one after the other. Its appearance on the island is probably due to its recent introduction by man.

specific peromones are also very important among butterflies and moths in determining the success of a mating attempt. The behavioural sequences that precede mating are sometimes very complex and require a specific sequence of stimuli of various kinds (see Plates 65 to 67).

4. Mechanical Isolation. This occurs when there are structural or mechanical barriers to mating between individuals belonging to different species. It is principally the result of differently shaped or sized genital structures, making union during copulation or the transfer of sperm difficult or impossible. In butterflies and moths, as in many other insects, there are generally marked differences in the structure of the external genitalia even between congeneric species. Plate 21 shows the varied shapes and sizes of the male valves in seven *Eurema* species (Pieridae) and the differences in the morphology of an unusual male structure (the organ of Julian) in two closely related *Hipparchia* species (Satyrinae). The extent to which such differences may hinder copulation is not known, but the ease with which it is possible to obtain crosses between different species in the laboratory suggests that these differences cannot always guarantee isolation.

Postcopulatory reproductive-isolating mechanisms occur in all those cases where mating takes place but fails to produce offspring capable of giving rise to successive fertile generations. They may be grouped together under the following headings:

1. Gamete or Zygote Mortality. In this category the gametes or the zygote die before the process of forming a new individual (embryogenesis) has begun.

2. Lack of Hybrid Vitality. Here all or almost all the hybrids are eliminated before they reach sexual maturity. Death may occur at any stage of the life cycle, from embryonic development to just before breeding.

3. Hybrid Sterility. The process of gamete formation is interrupted in the hybrids so that they are incapable of producing effective gametes.

4. Lack of Vitality or Sterility in Hybrid Progeny (*Hybrid Breakdown*). The initial product of an interspecific cross is fully or partially fertile, but subsequent crosses give rise to sterile individuals or ones lacking vitality.

Two lycaenids, *Lysandra bellargus* and *L. coridon*, are widespread in continental Europe, and hybridization has been observed in the wild in

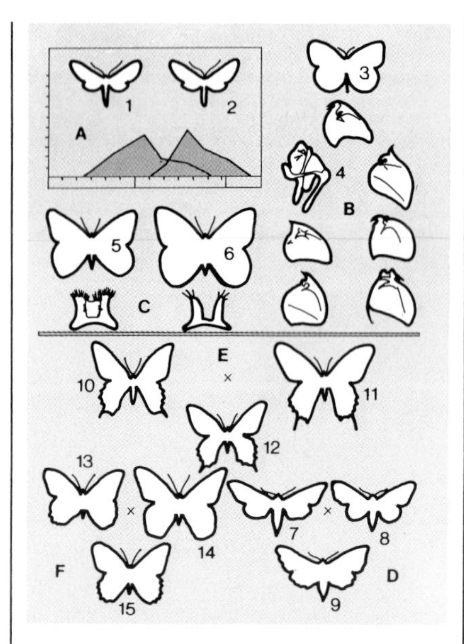

Plate 21 **Mechanisms of reproductive isolation and hybridization**

The types of isolating mechanism illustrated are seasonal isolation (A) and mechanical isolation (B, C).

A: Two ctenuchids, (1) *Amata phegea* and (2) *A. ragazzii*, are sympatric in parts of central Italy but fly at different seasons, reducing the chances of the two sexes' meeting.

B: Differences in the structure of the genitalia may play a role in preventing mating. This can be seen by comparing the valvae of the male genitalia in a group of related *Eurema* (Pieridae), of which the common grass yellow (*E. hecabe*) is illustrated (3). The position of the valvae with respect to the other male copulatory apparatus is shown in Figure 4.

C: The reproductive isolation of the related rock grayling (*Hipparchia alcyone*, 5) and the woodland grayling (*H. fagi*, 6), which are sympatric in Europe, appears to be connected with structural differences in the organ of Julian. This organ is located with the male copulatory apparatus and is probably sensory. Reproductive isolation often acts after mating has taken place and is the result of poor gamete, zygote, or hybrid viability. Three pairs of species are illustrated with their hybrids, which differ in their degree of sterility.

D: The eyed hawkmoth (*Smerinthus ocellata*, 7) and poplar hawkmoth (*Laothoe populi*, 8) were crossed in the laboratory and produced completely sterile hybrids (9).

E: The American swallowtails *Papilio zelicaou* (10) and *Papilio polyxenes* (11) have been studied by C. L. Remington. They cross naturally in a narrow band along the eastern Rocky Mountain front and give rise to hybrids (12) that in only one sex have poor fertility.

F: The white-banded admiral (*Limenitis arthemis arthemis*, 13) and red-spotted purple arthemis (*L. astyanax*, 14) come into contact in parts of the eastern United States and southern Canada and produce extensively interfertile hybrids (15).

some southern areas. Natural hybrids between the two species collected in the Abruzzi (central Italy) have been named *L. italoglauca*; hybrids from the Pyrenees have been classified as *L. polonus*. However, the two parental species are very different cytogenetically. *L. bellargus* has a haploid chromosome number, *n*, of 45, while *L. coridon* has an *n* of 88 (87 in the Italian peninsula). The hybrids are therefore probably all sterile in both sexes since it is known that, if the germ cells contain very different chromosome sets, homologous chromosomes are unable to pair at meiosis, blocking the formation of functional gametes.

The sterility of hybrids is not necessarily associated with their vitality. It is experimentally possible to obtain between distant species perfectly vigorous hybrids that are completely sterile. If *Smerinthus ocellata* and *Laothoe populi*, two morphologically very different sphingids, are crossed (Plate 21), hybrids are readily obtained that display characters intermediate between the two species, with the eyespots of *S. ocellata* and the background coloration of the hind wings of *L. populi*. Such hybrids are extremely vigorous but sterile. Although the chromosome number of the two species is not very different (in *S. ocellata* the haploid number is 27 and in *L. populi* 28), meiosis is again not regular, and functional gametes are not formed.

Glacial Refuges and Speciation in Alpine Butterflies

The climatic changes that have occurred in the course of various geological ages and during the Pleistocene in particular have proved to be important factors in the geographical isolation and formation of animal and plant species in such large massifs or mountain chains as the Alps, the Himalayas, the Rockies, or the Andes. In many ways mountains are like islands or archipeligoes separated by valleys or plains that today, as in previous interglacial periods, form ecological barriers to species adapted to cold climates. Many butterflies and moths that live at high altitudes and have discontinuous ranges in the Alps, the Carpathians, the Pyrenees, and the Apennines would, during periods of glacial expansion, have had a much more extensive and continuous distribution than at present. Indeed these species still occur over vast areas of the plains of Eurasia and, sometimes, North America but only at northern latitudes (see the chapter on geographical distribution and Plates 99 and 100).

In the case of species adapted to Alpine tundra biomes, conditions favourable to speciation by geographical isolation persisted even during the periods of maximum glacial expansion. These species would have taken refuge on the summits (*nunataks*) of uplands left uncovered by the glacial layer. In such instances the ice formed a barrier that separated populations of a given species taking refuge on nunataks from other populations that had been forced down to the base of the massif, where, owing to the harshness of the climate and a suitable flora, they found conditions that favoured their expansion over the vast plains.

Such isolating phenomena can explain the differentiation of populations of the nymphalid *Euphydrys aurinia* at high altitudes in the Alps, Pyrenees, and Caucasus. These differ from the typical form in their small size and in the extensiveness of their black patterning (form *debilis*). The subsequent improvement in the climate and the end of glacial activity enabled *aurinia* populations to recolonize the valleys, again coming into contact with the *debilis* form (Plate 22) but remaining ecologically separate. It is therefore probable that the two forms have completed the process of speciation, at least in some areas.

A similar situation exists in *Coenonympha* between the Alpine *C. gardetta*, which flies between 1,500 and 2,300 m (4,900 and 7,550 ft), and *C. arcania*, which is widespread in Europe and the Near East, flying from

5
10
15
20
25
30
5
10
15
20
25
30
June
July
August

Plate 22 **Species formation in the Alps and the Pyrenees**

The alternation of glacial and interglacial periods, with the consequent appearance and reappearance of ecological-cum-geographical barriers, has influenced — particularly in the massifs of the Alps and the Pyrenees — the evolution of numerous species, especially butterflies and moths.

A: The high-altitude [2,000 to 2,500 m (6,560 to 8,200 ft)] populations of the marsh fritillary (*Euphydryas aurinia, debilis* form; 1, male, 2, female) differ markedly from the low-altitude populations in the Alps and Pyrenees (*provincialis* form; 3, male). The two forms appear to have completed the process of speciation in at least some areas where they are again in contact. The dark shading on the map shows the range of the *debilis* form; the paler shading represents the rest of the range.

B: A similar situation arises in *Coenonympha* (Nymphalidae, Satyrinae). The *gardetta* form (4, male, 5, female) is confined to the Alps, where it flies between 1,500 and 2,300 m (4,920 and 7,545 ft); the *arcania* form (6, 7, male) occurs from sea level to about 1,300 m (4,265 ft). The two forms are regarded as separate species even though there is some local hybridization at intermediate levels and some transitional forms are produced (*darwiniana*, 8, male, 9, female).

C: The false dewy ringlet (*Erebia sthennyo*) from the Pyrenees (Nymphalidae, Satyrinae; 12, male, 13, female) presumably evolved from relict western populations of the dewy ringlet (*E. pandrose*, 10, male, 11, female), which was extensively distributed during the ice ages but which now has a discontinuous range in Finland and Scandinavia to the north and, south of 50°N, in the high mountains of Europe.

sea level to 1,300 m (4,250 ft). The existing ecological separation and the marked difference in coloration and patterning suggest that the two forms are reproductively isolated and that speciation has already been completed. This is in spite of the fact that at moderate altitudes in certain regions of the Alps a number of individuals displaying intermediate characteristics (form *darwiniana*) occur. These are probably hybrids resulting from crosses between the two species, but some also consider them to be a separate species.

Geographical isolation resulting from interglacial climatic barriers has caused the separation of species or races between different mountain ranges or even between parts of the same range. So, for example, *Erebia sthennyo*, from the central Pyrenees, is derived from relict populations of *E. pandrose*, a discontinuously distributed species from Finland, Scandinavia, and the high mountains of Europe and Asia (the Altai Mountains). The two species have the same number of chromosomes but differ clearly in the male genitalia. It is interesting to note that the marginal populations of the western Pyrenees have evolved into *E. sthennyo*, while those of the eastern Pyrenees have retained the characters of typical *E. pandrose* (Plate 22).

Evolution of the *Erebia tyndarus* Species Group

The best known situation, again with *Erebia*, concerns the *E. tyndarus* group. Entomologists long regarded it as a single species, *E. tyndarus*, but following detailed research by two specialists, a Frenchman, H. de Lesse, and a Yugoslav, Z. Lorković, it was realized that the group consisted of at least eight species or semispecies, differing slightly in coloration and wing pattern and more markedly in their chromosome set and ecological habits. There are also more or less marked differences in the morphology of their genitalia. The group as a whole is very widely but also discontinuously distributed in the mountains of Europe and Asia, from the south of the Iberian peninsula to the Alps and the Apennines, to Iran and the Altai Mountains, and as far as the Rocky Mountains in North America.

The haploid chromosome number ranges from an n of 8 in *E. calcarius* from the Julian Alps and Monte Cavallo in the Venetian pre-Alps to the 51 chromosomes of *E. dromulus* from the Pontus and Armenia and of *E. iranica* from the Elburz Mountains in Iran. The alpine species *tyndarus* (central Alps), *nivalis* (eastern Alps), and *cassioides* (widely distributed in the western and eastern Alps as well as the Pyrenees, the Apennines, the Carpathians, and the Balkans) have a chromosome number of 10 or 11. *Erebia hispania* has 25 chromosomes in the race typical of the Sierra Nevada and 24 in the Pyrenean one (subspecies *rondoui*). Finally the related American populations of *E. callias* have a set of 15 chromosomes. The situation is summarized in Plate 23, and it is clear that the differences have been acquired in the course of geographical isolation.

The most interesting aspect of the *E. tyndarus* complex is the ecological behaviour and the degree of sexual isolation in alpine populations that at least potentially may come into contact with clearly allopatric species. Lorković has conducted a series of crossing experiments, summarized in Plate 23, which shows how sexual isolation is marked more in geographically contiguous populations than in widely allopatric ones. Further, when two species are present on the same mountain, they are found at markedly different altitudes. Thus on the Tauern, at the eastern end of the Alps, *E. nivalis* and *E. cassioides* are differently distributed, the first occupying the alpine zone between 2,300 and 2,700 m (7,550 and 8,850 ft) and the second flying in the subalpine zone between 1,400 and 2,400 m (4,600 and 7,875 ft). In a few cases there is a narrow strip where they overlap, no more than 100 m (328 ft) wide; but both species are

Altitude (m)
2,500
2,000
1,500
1,000
500
arcania
darwinia
gardetta

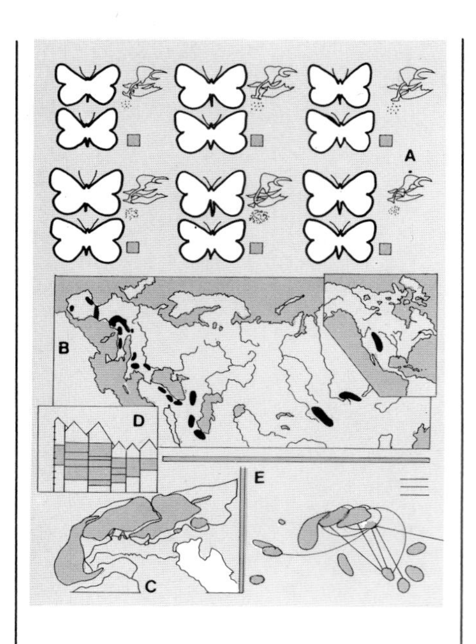

Plate 23 Evolution in the *Erebia tyndarus* group

The *E. tyndarus* species group (Nymphalidae, Satyrinae) is discussed in the text and various aspects of the situation are illustrated here.

A: Six species from the group. Each figure shows a male, a female, a side view of the male genitalia, and the haploid chromosome number. The coloured square beside each species provides the key to B, C, D, and E.

B: Ranges of the various species.

C: Geographical ranges of the alpine forms in greater detail.

D: Vertical distribution of De Lesse's brassy ringlet (*E. nivalis*) and common brassy ringlet (*E. cassioides*) on certain mountains in the Tauern massif (eastern Alps, Austria). The deeper the shade of blue, the greater the abundance of *E. cassioides*. The two species occur at different altitudes as a result of competition. On mountains where only one species occurs, it is found at intermediate levels.

E: Z. Lorkovič was able to establish the degree of reproductive isolation between certain species on the basis of the proportion of nonviable or sterile hybrids among the progeny of crosses. The coloured areas correspond roughly to the distribution of the species and the various subspecies of *E. cassioides (neleus, macedonica, illyrica, illyromacedonica, majellana)*. Note how sexual affinity is greatest between geographically distant populations of different species (for example, *E. calcarius* × *E. iranica*) as opposed to nearby ones (for example, *E. nivalis* × *E. cassioides*). (C, D, and E adapted from Lorkovič, 1958.)

uncommon there, and they do not appear to hybridize. Differences in food plants are not the cause of this separation since, as larvae, both species feed on two grasses, *Festuca* and *Nardus*, which are widespread in both zones. The separation is therefore clearly the result of some form of interactive competition since in the parts of the Tauern where only one of the two species, *nivalis* or *cassioides*, flies it tends to occupy the zone between 1,800 and 2,600 m (5,900 and 8,525 ft). A similar situation has been observed in the Pyrenees between *E. cassioides* and *E. hispania*. These examples clearly show that two or more related mountain species, which originated allopatrically but which have come into contact again, under the influence of natural selection, tend to evolve mechanisms to avoid competition and hybridization by living at different altitudes.

Formation of Pronophilinae Species in the Andes

The phenomenon of species formation has recently been observed in the Andes among an interesting group of mountain satyrines, the Pronophilinae (Plate 24). In the last few years the British entomologist M. J. Adams and his colleague G. I. Bernard have visited various high-altitude zones in the Andes chain, such as the Sierra Nevada de Santa Marta and the Sierra Valledupar in Colombia and the Cordillera de Mérida in Venezuela. There they have discovered many new species of Pronophilinae. Most of these butterflies fly at altitudes between 1,800 and 4,200 m (5,900 and 13,775 ft) in habitats that range from high-altitude tropical forest (*cloud forest*) to alpine meadows (*páramos*) above treeline at about 3,000 m (9,850 ft).

The two entomologists observed that within each group of species there was a vertical stratification similar to that in the alpine *Erebia* and that the low-altitude species from the different mountain ranges resembled each other to a much greater extent than did the ones living at high altitudes. The explanation for the phenomenon, according to Adams, lies yet again in the climatic effects of the Pleistocene. The succession of cold glacial and warm interglacial cycles would have offered opportunites for the dispersal of species among the various massifs and, alternatively, for the isolation of the populations at the higher altitudes. When two or more species evolved during an interglacial and then came back into contact during the subsequent glaciation, they would tend to occupy different altitudes, reducing competition for the same resources. Therefore, gradually, the older species began to be trapped at high altitudes on individual massifs, unable, even during the subsequent glacial periods, to descend to lower altitudes and thence to colonize other mountain massifs.

Sympatric Speciation

In conclusion it must be emphasized that allopatric speciation is not the only way in which species can be formed and that others exist although they are possibly less common. For example, situations in the wild suggest that reproductive isolation may evolve between two populations while both occupy the same area. This is known as *sympatric speciation*.

The existence of sympatric speciation has yet to be fully proved, and it appears to be confined to animals with unusual habits, such as zoophagous or phytophagous parasites. It occurs as the result of two subpopulations' adapting to different ecological niches. In the yponomeutid microlepidoteran *Yponomeuta padella* the initial stages of such a process can be observed. The moth's caterpillar lives and feeds on the apple and hawthorn. The adults display a certain degree of polymorphism in wing coloration; the darker forms are more common among moths developing

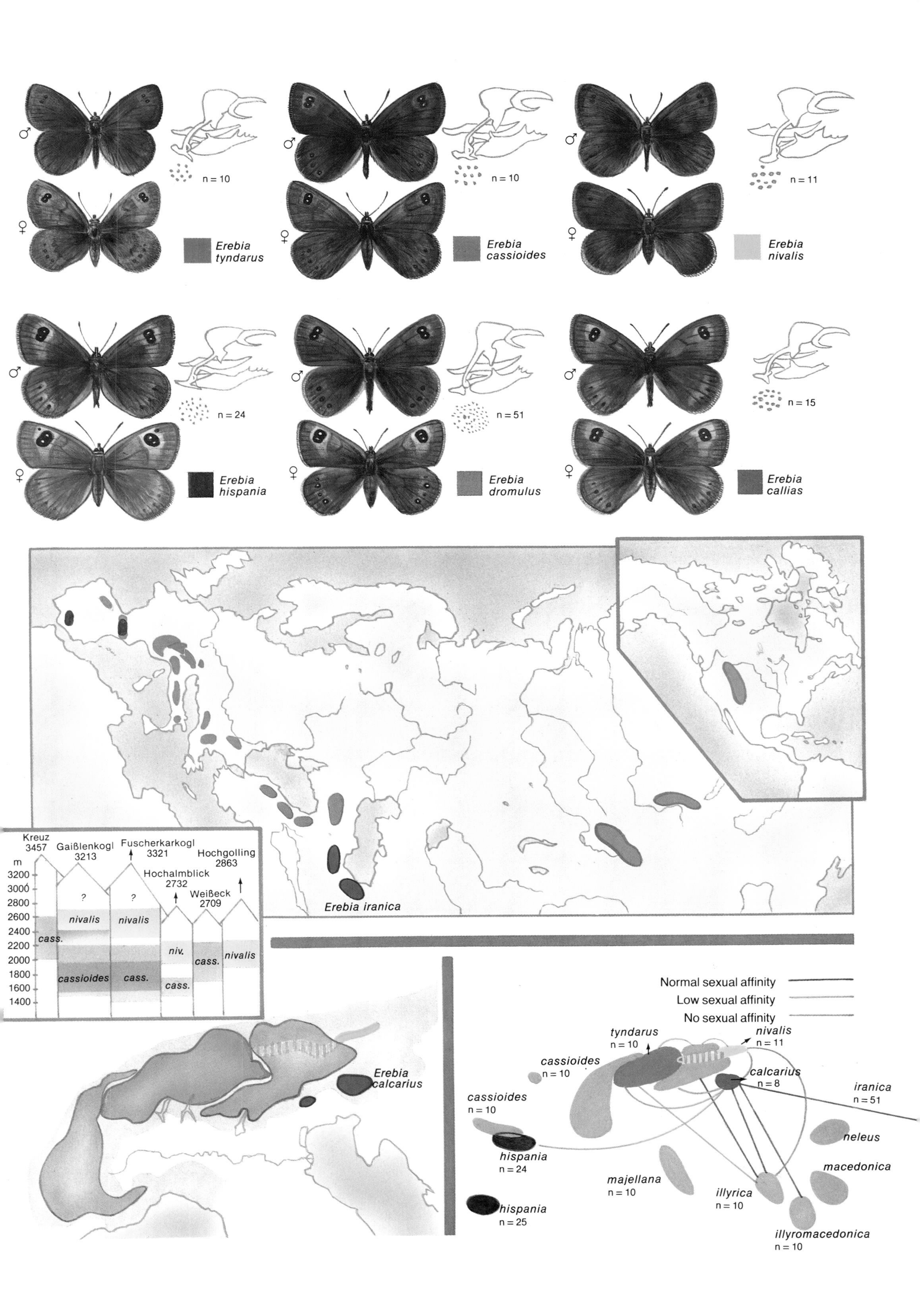

n = 10
Erebia tyndarus
n = 10
Erebia cassioides
n = 11
Erebia nivalis
n = 24
Erebia hispania
n = 51
Erebia dromulus
n = 15
Erebia callias
Erebia iranica
Kreuz 3457
Gaißlenkogl 3213
Fuscherkarkogl 3321
Hochgolling 2863
Hochalmblick 2732
Weißeck 2709
m
3200
3000
2800
2600
2400
2200
2000
1800
1600
1400
?
?
nivalis
nivalis
cass.
niv.
cass.
nivalis
cassioides
cass.
cass.
Erebia calcarius
Normal sexual affinity
Low sexual affinity
No sexual affinity
tyndarus n = 10
nivalis n = 11
cassioides n = 10
calcarius n = 8
iranica n = 51
cassioides n = 10
neleus
hispania n = 24
macedonica
majellana n = 10
illyrica n = 10
hispania n = 25
illyromacedonica n = 10

from caterpillars that lived on hawthorn, whereas caterpillars that feed on apple tend to produce white forms. It has been observed, furthermore, that individuals prefer to mate with others that develop on the same plant and also to deposit their eggs on these plants. Finally the caterpillars display a marked tendency to feed on the plant on which their mother fed. So although some gene flow occurs, it is clear that the prerequisites exist for the establishment of more effective mechanisms of reproductive isolation. Recent studies seem to indicate that in some areas at least the two forms of *Y. padella*, known as *padella* and *malinella*, are genetically isolated and have already completed the speciation process.

Evolution of Host Races on California Oaks

In their classic treatise on island biogeographic principles R. H. MacArthur and E. O. Wilson postulated that habitat patches on continents could act as ecological islands; each patch would harbour a number of species in accord with its size. P. A. Opler demonstrated that eighteen California oaks and their relatives each had a number of host-specific leaf-mining moths in proportion to the geographic area occupied by its range. Thus an evolutionary equilibrium of species number was kept in balance by the host plant's individual range. When the host range increases, new species evolve; and when the host range decreases, extant species become extinct. When new oak leaf-miner species evolve, their antecedents are species found on cooccurring oaks — not necessarily their closest relatives. Speciation takes place when the two oaks are isolated in subsequent periods.

Examples of this sort of speciation occur in the primarily western North American *Cameraria* (Phyllocnistidae) and are discussed by D. R. Davis and Opler.

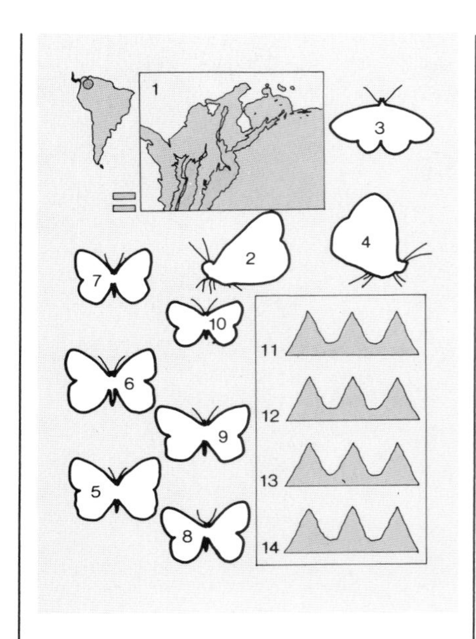

Plate 24 **Glaciations and species formation in the Andes**

Climatic changes caused by the Pleistocene glaciations also played a crucial role in species formation in the Andes. Recent entomological studies carried out by M. J. Adams and G. I. Bernard in the northern Andes in Colombia and Venezuela (1) have revealed the existence of relict species confined to high altitudes. These include the pierid *Reliquia santamarta* (2) and numerous pronophilines (Nymphalidae, Satyrinae), which occur at different altitudes in the same way as does *Erebia* (3, 4). Species 5, 6, and 7 (*Lymanopoda*) fly at progressively higher altitudes in the Sierra Nevada de Santa Marta (Colombia), whereas 8, 9, and 10 are the corresponding species in the nearby Valledupar chain.

(11) During a glacial period the first *Lymanopoda* species colonized the base of the various massifs. (12) In the subsequent interglacial period the various populations spread to higher altitudes, becoming isolated and differentiating into separate species (1A, 1B, and 1C). (13) With the onset of a new glaciation the species, by now distinct, descended to lower altitudes and moved freely from one massif to another, occupying different levels on each massif because of competition with other species. (14) The improvement in the climate during the subsequent interglacial period allows the colonization of higher altitudes. The various species maintain their stratification and diverge further from the corresponding forms on the other massifs. (Adapted from Adams, 1977.)

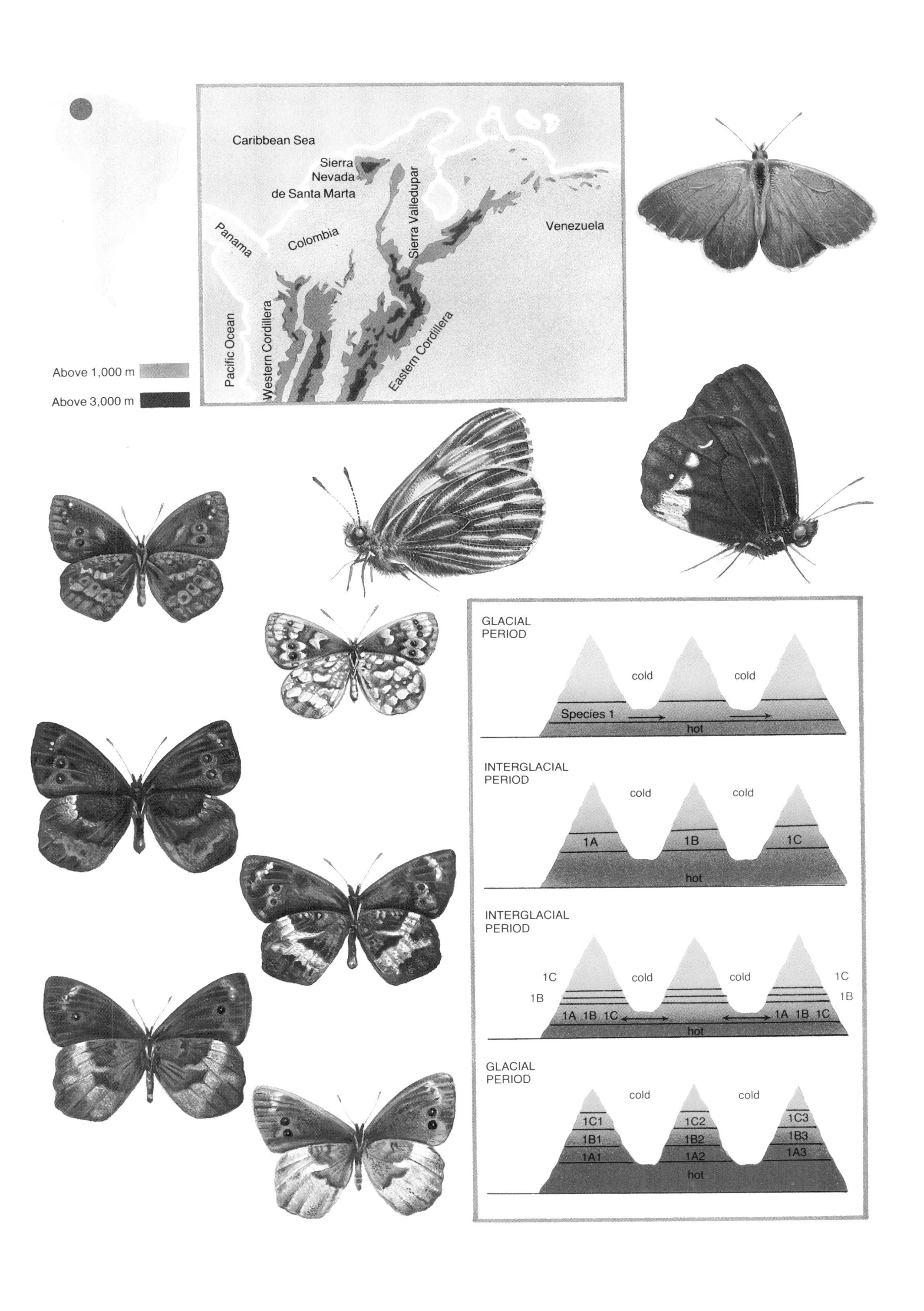
Caribbean Sea
Sierra
Nevada
de Santa Marta
Sierra Valledupar
Venezuela
Panama
Colombia
Pacific Ocean
Western Cordillera
Eastern Cordillera
Above 1,000 m
Above 3,000 m
GLACIAL
PERIOD
cold
cold
Species 1
hot
INTERGLACIAL
PERIOD
cold
cold
1A
1B
1C
hot
INTERGLACIAL
PERIOD
1C
cold
cold
1C
1B
1B
1A 1B 1C
1A 1B 1C
hot
GLACIAL
PERIOD
cold
cold
1C1
1C2
1C3
1B1
1B2
1B3
1A1
1A2
1A3
hot

Systematics: Classification and Phylogenetics

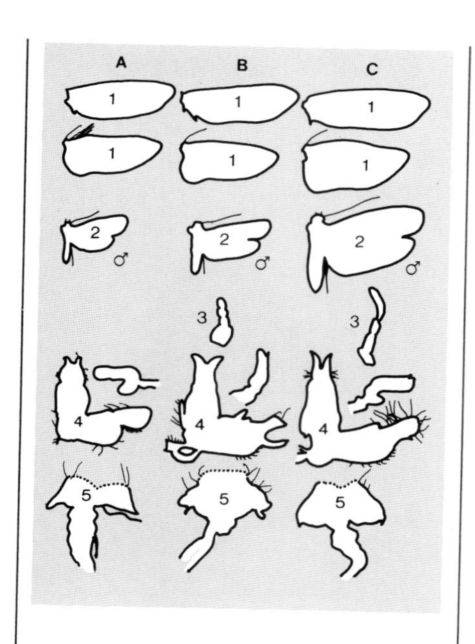

Plate 25 **Taxonomic characters in the genus *Ethmia***

Taxonomic characters are used by systematists for two purposes: One is the diagnosis and classification of specimens, and the other is the evaluation of the degree of phylogenetic relationship between different taxa. A butterfly can often be correctly identified by means of just a few "key" characters, but to place the members of a group in a system of classification that reflects their evolution and relationships as closely as possible, the careful analysis of numerous characters is necessary.

This plate illustrates certain structures that are used extensively in the classification of Lepidoptera, as exemplified by three Palaearctic *Ethmia* (Ethmiidae): *E. dodecea* (A), *E. aurifluella* (B), and *E. dentata* (C). Five characters are shown for each species: wing venation (1), male colour pattern (2), maxillary palp (3), posterior view of male genitalia, showing inner face of right valva (4), and the female genitalia with the proximate portion of the ductus bursae (5). Note, for example, on the costal (front) margin of the forewing the spacing between the radial veins (blue) and, on the hind wing, the development of the posterior cubital vein (CuP, orange), as well as the presence (A:1) or the absence (B:1, C:1) of a tuft of bristles linked to the frenulum. There are also conspicuous differences in the coloration of the wings and body and in the shape of the valvae and the uncus in the male genitalia (these are enlarged about 20×). The wingspan of the species illustrated is about 18, 20, and 26 mm (0.7, 0.79, and 1.02 in), respectively, for *E. dodecea*, *E. aurifluella*, and *E. dentata*. (Adapted from K. Sattler, 1967.)

In biology systematics is the science concerned with the diversity of living beings, their *classification* and *phylogenetics*, the latter being the genealogy of organisms in the course of their evolutionary history. Systematics began in a rather sketchy way, being confined to the recognition and classification of species, but it has gradually developed into a complex science closely linked to many other areas of modern biology. The term is often used as a synonym for *taxonomy*, but the latter should be strictly applied to the study of the methods and procedures of classification.

The roots of systematics lie in prehistory, and its development can be said to have begun with the birth of language in primitive human beings. Language exists when there are objects to talk about, and such objects include many organisms in the natural surroundings. Language could not have developed very far without the ability to group individual phenomena into more general concepts. This type of mental operation is identical with a process of classification. Early systematic concepts may have been ecologically based, grouping animals as edible or dangerous or berries as bitter or sweet. Tribal life encouraged the evolution of language, and *Homo sapiens* gradually began to develop the concept of the species and correlate phenomena concerning hundreds of plants and animals.

The system of classification used by biologists today is based on a method developed over several centuries. Its framework was laid down by the Swedish naturalist Carolus Linnaeus (1707–1778) in the tenth edition of his book *Systema Naturae*, published in 1758. In this work Linnaeus gave a *scientific name* to each animal or plant species that he described. Such scientific names consist of two Latin or latinized words: The first is the name of the genus to which the organism has been assigned and is written with an initial capital letter; the second is the name of the species and is never capitalized. So, for example, *Papilio machaon* unambiguously designates a particular species of butterfly, the common swallowtail. Many other butterflies belong to *Papilio*, and each obviously has its own specific name. These have been described by Linnaeus or numerous later authors.

The species and the genus are just two of the numerous taxonomic categories used in classification. The most important of the other categories are the *family*, the *order*, the *class*, the *phylum*, and the *kingdom*. They are arranged in an ascending hierarchy from species to kingdom so that several species are grouped together in a genus, several genera in a family, several families in an order, and so on. Consequently the categories define the rank or level in the hierarchy of classification. A *taxon* (plural *taxa*) is any group of organisms sufficiently distinct to form a category at whatever level. Thus, all true butterflies (Papilionoidea) or the pierids (Pieridae) or the whites (*Pieris*) or the large white (*Pieris brassicae*) are all taxa occupying various levels in the hierarchy. It is important to remember that all systematic categories are artificial and that the limits between one genus and another, between families, or between phyla are established by convention.

It is only in the case of the species that nonsubjective criteria can be used. The definition of a species (see the chapter Origin of Species) is based on the biological principle of interfertility, which can be verified experimentally in individual cases. In theory, at least, an expert should be able to define all the species of a given group in a particular area or, in other words, those which are sympatric — precisely because the various species are not interfertile. However, in practice this may be difficult and complicated in cases of secondary contact or hybridization among groups of populations that were previously separated geographically.

One of the subsequent changes made to Linnaeus's original classification has been the addition after the specific name of a third Latin name designating the subspecies. This trinomial nomenclature is necessary for polytypic species. The binomial or trinomial is followed by the full or

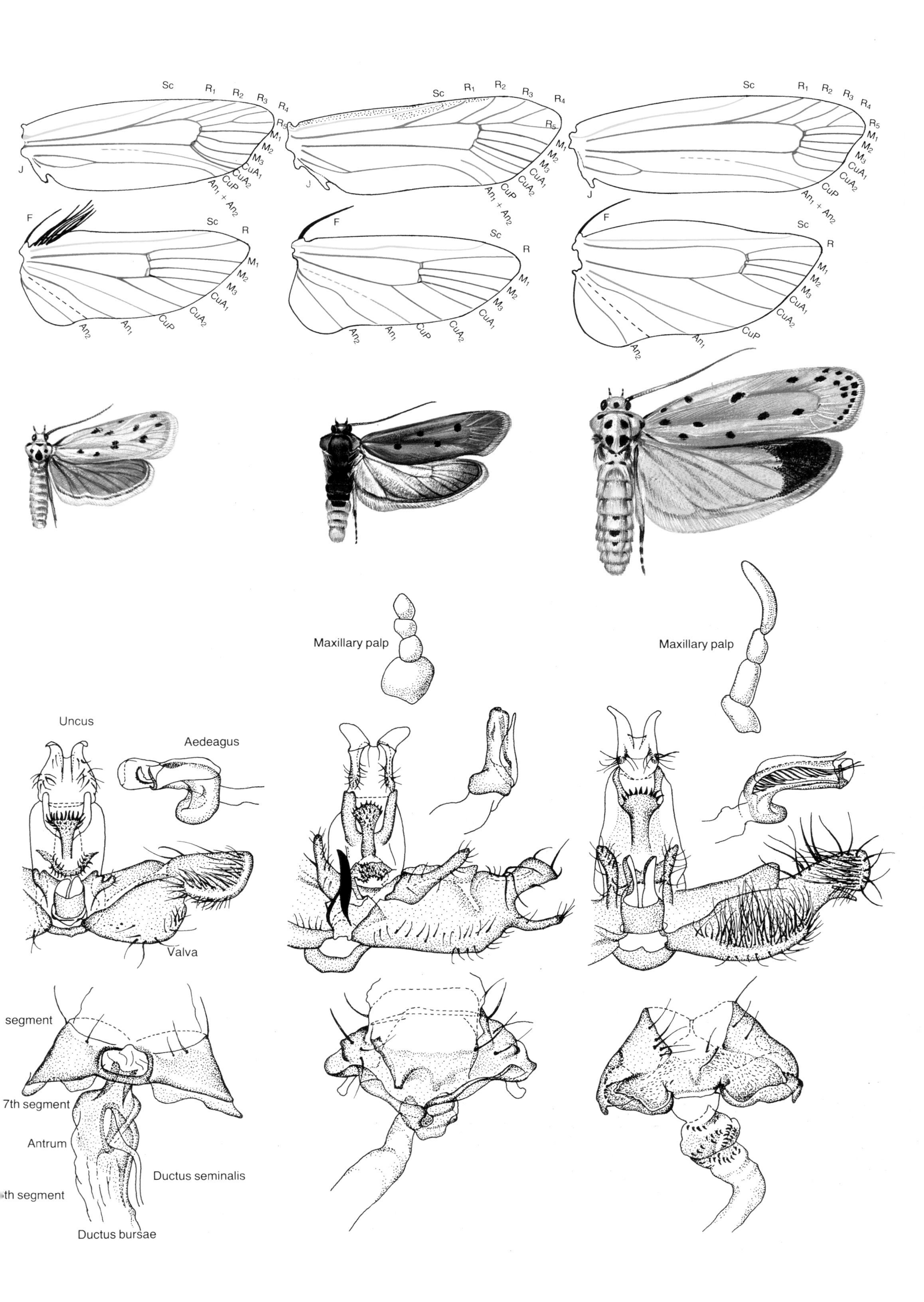

Sc
R1
R2
R3
R4
R5
M1
M2
M3
CuA1
CuA2
CuP
An1 + An2
J
F
R
An1
An2
Maxillary palp
Maxillary palp
Uncus
Aedeagus
Valva
segment
7th segment
Antrum
Ductus seminalis
th segment
Ductus bursae

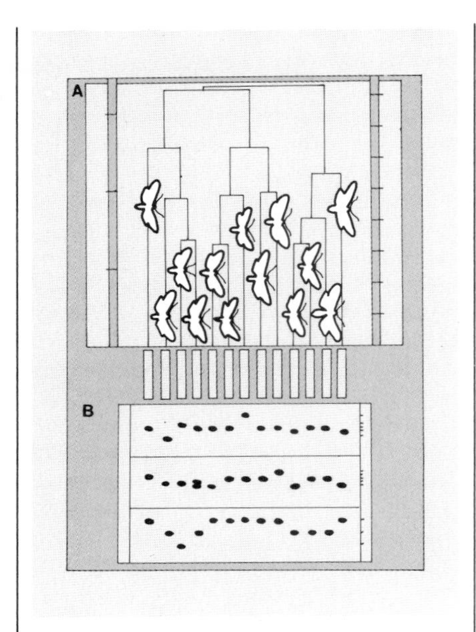

Plate 26 **Biochemical systematics in the genus *Zygaena***

Affinities and differences between organisms may be revealed by biochemical characters as well as by morphological ones. Electrophoresis (see the text and its accompanying figure) enables molecular differences to be detected among certain enzymes or other proteins, which serve as indices of the genetic differences between individuals and species. By the use of numerous enzymes, which are like characters, the degree of genetic divergence between the various members of a genus can be measured. The plate shows the results of research on thirteen European *Zygaena* species (Zygaenidae) carried out by a team under one of the authors and by means of the electrophoretic study of eleven enzymes. The similarities or differences among the species are shown in the dendrogram in Figure A.

For example, *Z. exulans* appears more closely related to *Z. oxytropis* ($D \simeq 0.7$) than to *Z. filipenduale* ($D \simeq 1.3$), which means that the third species began to evolve away from the hypothetical common ancestor earlier than the other two. In fact the degree of genetic correspondence approximately correlates with the length of evolutionary divergence shown on the scale.

Figure B shows differences among the species in terms of enzymes, malase (ME), phosphoglucomutase (PGM), phosphohexoseisomerase (PHI). The position of each band shows the electrophoretic mobility of the enzyme, indicated on a conventional scale. So, for example, *Z. exulans* differs from the neighbouring *Z. oxytropis* in the different mobility of the enzyme PGM, while the two species show identical mobility in the other two enzymes.

abbreviated name of the author who first described and named a particular species or subspecies in a scientific publication. There are various rules that govern the use and priority of names, and these were gathered together in the *International Code of Zoological Nomenclature*, published in London in 1961.

The scientific names given to butterflies and moths reflect various themes. Many are mythological names: *diana*, *polyxena*, *priamus*, *agamemnon*, *hercules*, and the like. Others refer to the extraordinary beauty of these insects: *paradisea*, *pulcherrima*, *miranda*, *elegans*, *excelsissima*, and the like. Another group highlights distinctive characteristics of a species, unusual morphological, behavioural, ecological (such as a food plant, for example), or geographical features. In addition unusual names have gained currency among experts specializing in particular groups. For example, in many microlepidopteran families the specific name is always a diminutive, and thousands of species with wing spans of a few millimeters have names such as: *ochraceella*, *obscurella*, *conturbatella*, *subdivisella*, *lacteella*, *propinquella*, and so on. In the family Sesiidae there are a large number of Batesian mimics, and many species mimic bees, wasps, hornets, and other Hymenoptera. As a result taxonomists have created such names as *apiformis*, *vespiformis*, *crabroniformis*, and *scoliaeformis*. Some playful entomologist has taken advantage of this to make a joke at a colleague's expense and named a species *Chamaesphecia schmidtiiformis*, suggesting an unlikely resemblance between Mr. Schmidt and the moth.

When any scientific name is established, it is referred to as a "type." This is a taxon (if the name is for a genus or a higher category) or an example (in the case of a species or subspecies), which becomes the official holder of the name. This practice is very important since it allows the identity of a given taxon to be guaranteed absolutely. Hence each named species or subspecies — all those dealt with in this book, for example — has a particular reference example (known as the type, or *holotype*), which may be kept in any collection, public — generally a natural history museum — or private. The holotype is the specimen representing the name, just as certain objects in platinum or other substances preserved in the International Bureau of Weights and Measures at Sèvres are specimens of units of weight and linear measure.

It may be necessary to examine a type when describing a new species, to verify the classification of a taxon, or to resolve a particular problem of nomenclature. Scientific names are not always fixed. As more gradually becomes known about a group, it may become necessary to make changes. For example, it may be discovered that two species that had long been thought to be different are in fact the same. In such a case the *International Code of Zoological Nomenclature* lays down that the name that was first assigned will be used and the more recent one discarded. Likewise, during the study of a group of butterflies or moths it might be found that the diversity of species attributed to a genus is actually greater than had been thought and that these species should be actually placed in two or three genera. Research like this often results in changes in nomenclature and classification, sometimes making it difficult to compare old and modern works.

A correct analysis of *taxonomic characters* is fundamental to both the theory and the practice of classification. It is impossible to take in and memorize simultaneously all the details of a butterfly or moth by observing it. For this reason in the discussion of a particular example or species individual details or specific characters that require careful observation are emphasized rather than a general view. A *character* is defined as any attribute of an individual of a taxon by which it differs or may differ from an individual of another taxon. Examples of characters are the shape of

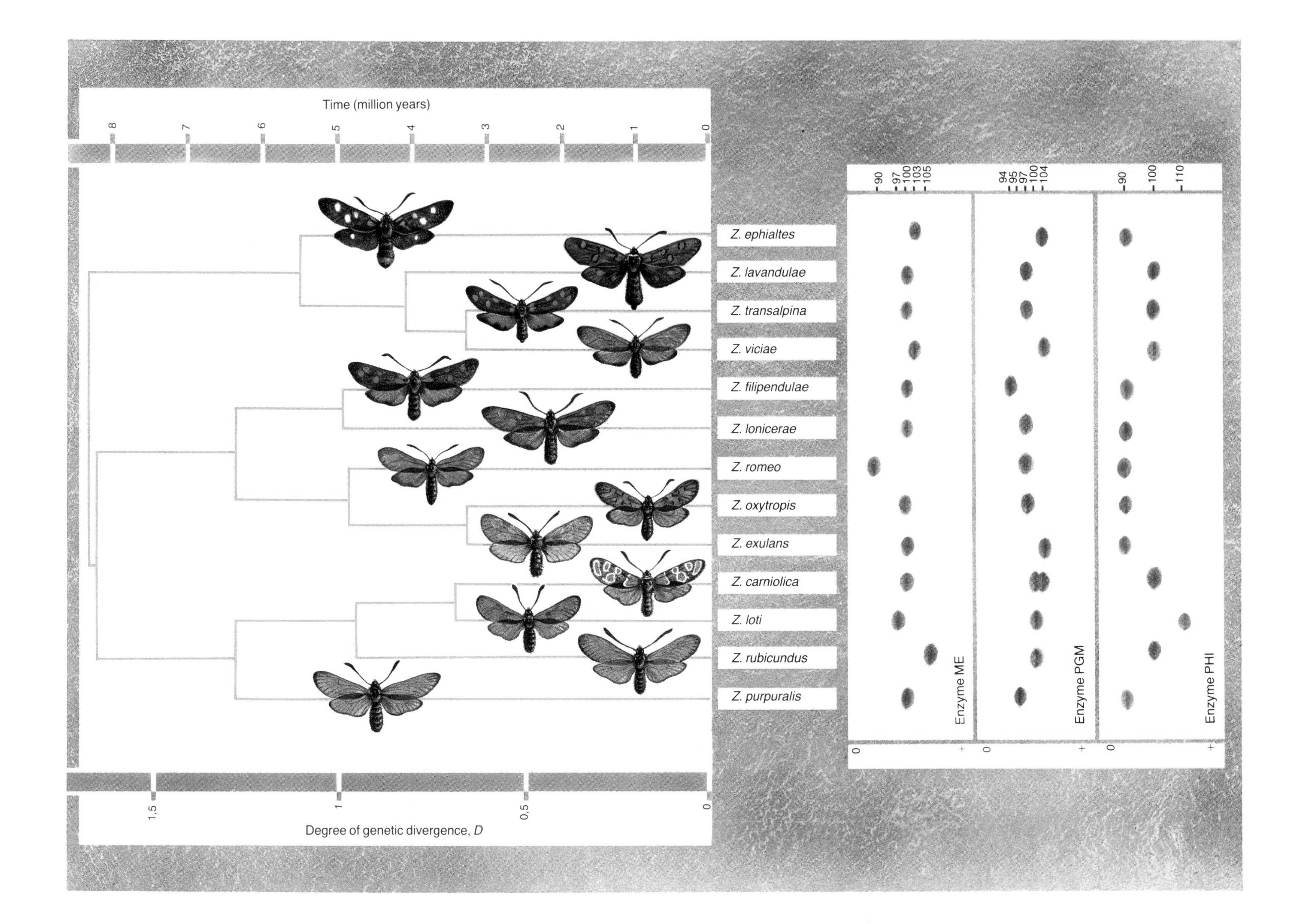

Time (million years)
Z. ephialtes
Z. lavandulae
Z. transalpina
Z. viciae
Z. filipendulae
Z. lonicerae
Z. romeo
Z. oxytropis
Z. exulans
Z. carniolica
Z. loti
Z. rubicundus
Z. purpuralis
Degree of genetic divergence, D
Enzyme ME
Enzyme PGM
Enzyme PHI

1
2
3
4
5
10
6
7
8
11
9
12
17
13
15
14
16
18

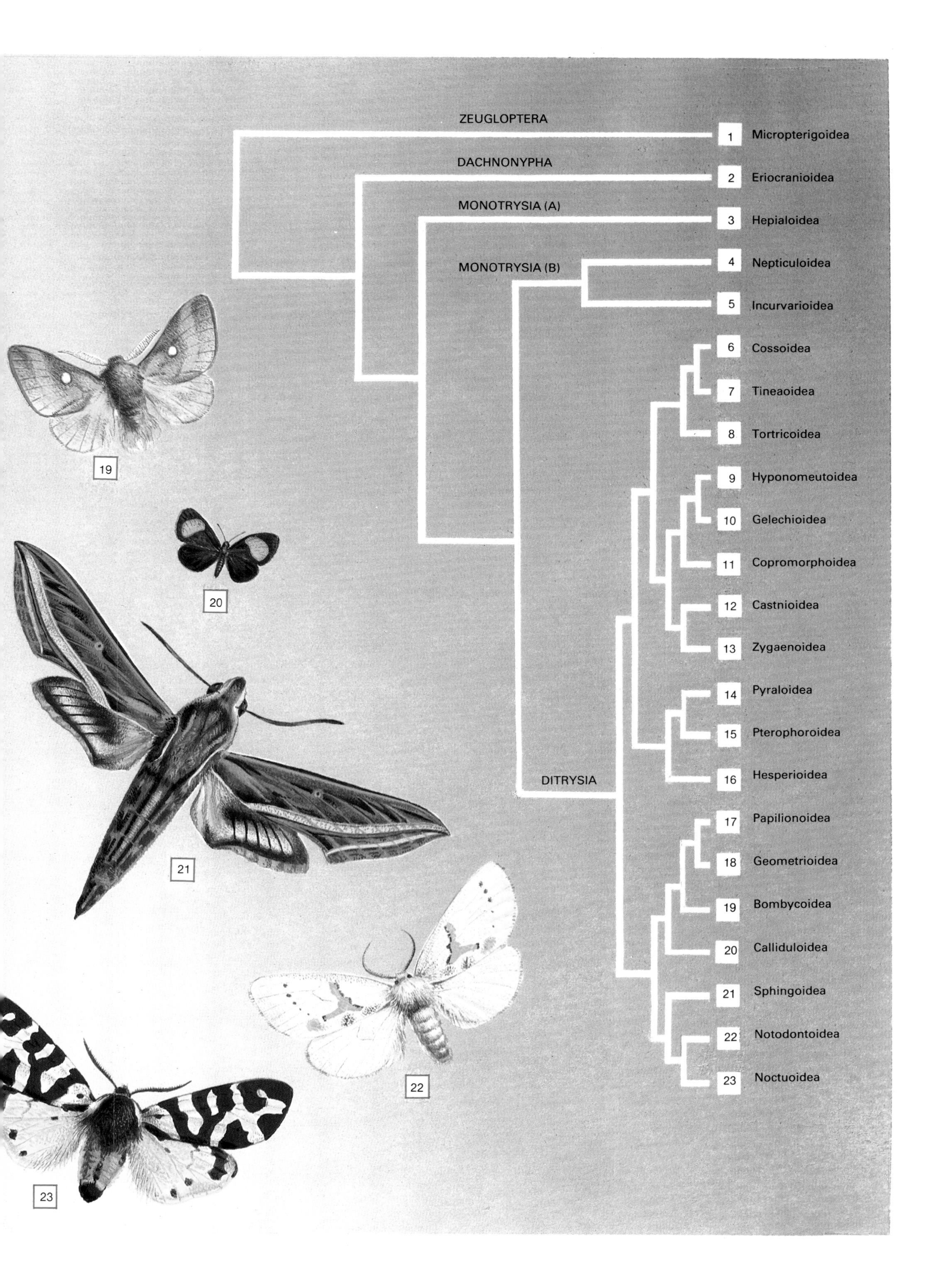
ZEUGLOPTERA
DACHNONYPHA
MONOTRYSIA (A)
MONOTRYSIA (B)
DITRYSIA
1 Micropterigoidea
2 Eriocranioidea
3 Hepialoidea
4 Nepticuloidea
5 Incurvarioidea
6 Cossoidea
7 Tineaoidea
8 Tortricoidea
9 Hyponomeutoidea
10 Gelechioidea
11 Copromorphoidea
12 Castnioidea
13 Zygaenoidea
14 Pyraloidea
15 Pterophoroidea
16 Hesperioidea
17 Papilionoidea
18 Geometrioidea
19 Bombycoidea
20 Calliduloidea
21 Sphingoidea
22 Notodontoidea
23 Noctuoidea
19
20
21
22
23

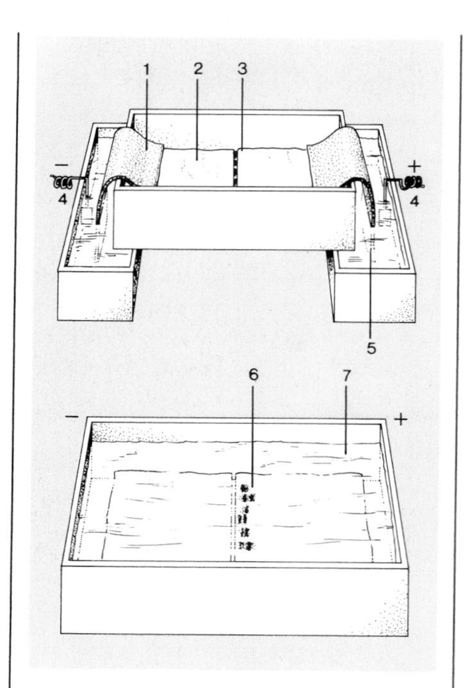

Diagram showing the apparatus used in the electrophoresis of enzymes and other proteins for the purposes of measuring the degree of genetic divergence among populations and species of Lepidoptera and other organisms.

A number of strips — one per individual — are inserted into a gel plate. Each of these strips has been impregnated with a homogenate of a particular tissue, for example, muscle. The gel plate is then attached to two electrodes, and a current is passed through appropriate buffer solutions for several hours. This causes the proteins to migrate across the plate; each one will move towards the anode or cathode (positive or negative pole) at a speed dependent on its net electric charge and the size of its molecule. The gel plate is then immersed in a solution containing a specific substrate of the enzyme being studied and a salt. The enzyme catalyzes a specific reaction that transforms the substrate into a particular product. This combines with the salt and forms a coloured substance. The result is a coloured band on the plate corresponding to the position of the enzyme. From the number and position of the bands it is then possible to determine the genotype of the individual for the gene locus coding for the enzyme in question. The analysis of numerous enzymes — and hence of gene loci — enables the degree of genetic divergence among individuals of different species to be measured. (1) Contact sponge (filter), (2) gel, (3) row of samples, (4) electrode, (5) container with buffer solution, (6) coloured band indicating the position of the enzyme, (7) solution with substrate, coenzymes, and dye salts. (Adapted from Ayala and Valentine.)

androconia, the morphology of the antennae, and the colour of the discal cell.

The term character should not be confused with characteristic, the latter referring to the condition or the particular alternative *character state*, such as *pyriform* androconia, *plumose* antennae, or *red* discal cells. The importance of various characters depends upon both the taxonomic group and the level in the classification heirarchy. In the major divisions of the Lepidoptera (suborders and superfamilies) characters connected with the wing-coupling apparatus, the number of female genital apertures, the resemblance between the venation of the fore- and hind wings, and so on, are often used. At the species level the most widely used characters are, classically, the coloration and patterning of the wings and the morphology of the genitalia. Plate 25 shows the morphological differences for various characters between *Ethmia* species (Gelechiidae). The illustrations are taken from *Microlepidoptera Palaearctica*, one of the classic works used by specialists in the Microlepidoptera.

Until now only morphological characters — and traditional ones in particular — have been discussed, but today the range of characters available for differentiating between species and for establishing their phylogenetic relationships has been considerably increased by developments in the theories and techniques of modern biology. So, for example, the scanning electron microscope has opened up an immense field of investigation for the study of such characters as the ultrastructure of various types of scales or the fine ornamentation of eggs, which display extraordinary and diverse microarchitecture. In the case of Pieridae the observation of wing coloraton and pattern under ultraviolet light has proved particularly useful. It has been found that the absorption and reflection of the ultraviolet component of the spectrum vary greatly not only between species but also between individuals of the same species but different sexes, even in cases where under normal light the colour and pattern are virtually identical. The ultraviolet image is obtained by photography through special filters and/or particular light sources. This technique provides an image of the butterfly or moth that is similar to that seen by insects themselves, while without such devices what we see is close to what many vertebrate predators see. In this way a butterfly or moth may make itself attractive to its potential mate without the risk of being too striking or may even protect itself from predators by mimicry.

Another important character, in use for several decades, is the karyotype, that is, the number and form of the chromosomes, which, particularly in certain lycaenid and satyrine genera, may vary substantially between one species and another (Plate 23). Although research into the chromosomes of butterflies and moths has not been so detailed or sophisticated as it has been in other insects — *Drosophila* and mosquitoes, for example — it has still made a useful contribution to the study of such genera as *Lysandra* and *Agrodiaetus*, where morphological characters are of little help.

In recent years the use of biochemical characters has become widespread, particularly the use of series of enzymes that act as markers for genetic differences between individuals of the same or different species. The technique used is *electrophoresis*, a method employing the different speeds at which enzymes and other proteins migrate in an electric field to reveal differences between them. The differences in electrostatic charge that affect the enzymes' mobility are the result of substitutions, deletions, or additions in the amino acid sequences of the polypeptides making up their structure. These in turn are caused by *mutations* in the nucleotide sequence of the DNA forming the structural gene.

Electrophoresis has led to highly important advances in several areas of biology and especially in the study of evolutionary processes. Its

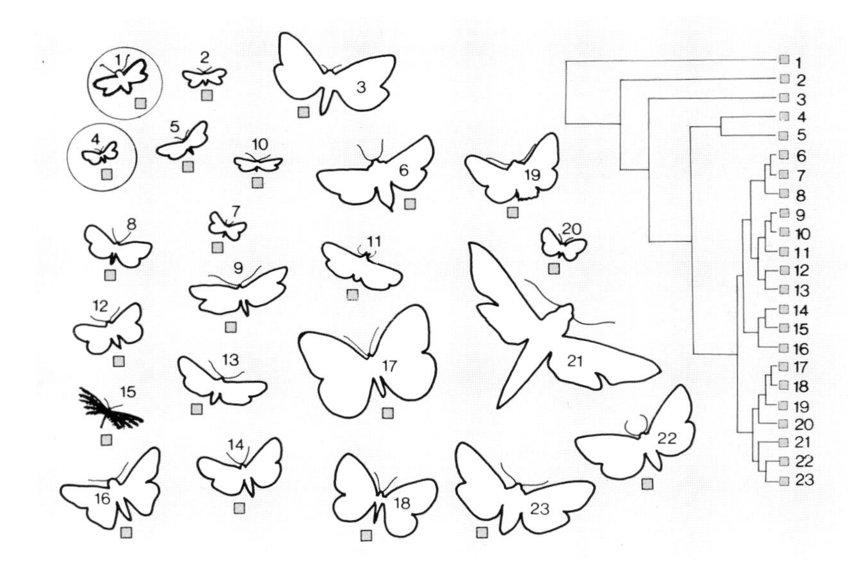

Plate 27 **Systematics and relationships of Lepidoptera**

The enormous number of known Lepidoptera and the extreme variability that they display as the result of adaptation to the most diverse conditions make the development of an overall classification system that reflects the group's interrelationships a difficult task. The problems are increased by the lack of fossils. Experts are essentially in agreement over classification below the subfamily level, but there is still much uncertainty over the higher classification, especially between the superfamilies. A tentative classification of the superfamilies is put forward here on the basis of a series of morphological characters of different levels and the comparison of various hypotheses advanced by different authors. The hypothetical phylogenetic line is expressed in a cladogram, which groups together taxa of progressive degrees of similarity. In addition to the twenty-three superfamilies the five suborders most often used in classification systems are also included. A representative butterfly or moth is illustrated from each superfamily: (1) *Micropterix ammonella* (Micropterigidae). (2) *Eriocrania sparmanella* (Eriocraniidae). (3) *Hepialus fusconebulosus* (Hepialidae). (4) *Stigmella acrimoniae* (Nepticulidae). (5) *Incurvaria capitella* (Incurvariidae). (6) *Hypopta caestrum* (Cossidae). (7) *Apterona pusilla* (Psychidae). (8) *Philedone gerningane* (Tortricidae). (9) *Ethmia pusiella* (Ethmiidae). (10) *Uliaria rasilella* (Gelechiidae). (11) *Copromorpha tetracha* (Copromorphidae). (12) *Synemon sophia* (Castniidae). (13) *Procris statices* (Zygaenidae). (14) *Loxostege aeruginalis* (Pyralidae). (15) *Alucita tetradactyla* (Alucitidae). (16) *Pelopidas mathias* (Hesperiidae). (17) *Teracolus eupompe* (Pieridae). (18) *Cidaria hastata* (Geometridae). (19) *Eriogaster catax* (Lasiocampidae). (20) *Callidula lunigera* (Callidulidae). (21) *Hippotion celerio* (Sphingidae). (22) *Leucodonta bicoloria* (Notodontidae). (23) *Arctia fasciata* (Arctiidae).

application to systematics is based on the production of suitable indices of genetic similarity between populations and species that take into account the average degree of biochemical similarity for each individual enzyme. In the study of various species the similarities are shown by means of cladograms or dendrograms, hierarchically arranged diagrams that group together progressively more similar taxonomic units (Plate 26). Plate 26 shows the genetic relationships among various *Zygaena* on the basis of differences in enzymes revealed by electrophoresis (gene-enzyme systems).

Earlier it was stated that in addition to the classification of organisms systematics may also include the study of their phylogenetics. The two activities are closely linked; indeed, in principle, good classification should reflect phylogenetics both in general and in individual groups. (*Numerical taxonomy*, which relies strictly on quantitative relationships among selected characters, does not consider phylogeny.) One such attempt is shown in Plate 27, which illustrates the phylogenetics of Lepidoptera on the basis of numerous morphological characters. These are assigned a level of significance in a hierarchy according to Common's criteria for classification, which is one of the best thought out and up to date. The diagram used here is crude, but it gives an idea of the relationships and relative ages of the various superfamilies. Better results can be obtained by using such rigorous taxonomic criteria as *cladistic analysis*, which by careful evaluation of the primitive or evolved state of each character enables systems of classification to be developed that very nearly approach phylogenetics.

Systematic Survey of Butterfly and Moth Families

This chapter is devoted to a survey of the suborders, superfamilies, and families of Lepidoptera. The system followed here has been outlined in the chapter on the classification of butterflies and moths, to which the reader is referred for an overall view of the whole group.

This book does not aim to treat each family exhaustively but rather to illustrate in a general way the taxonomic diversity among butterflies and moths and to account for it in terms of evolution. A comparative analysis of the superfamilies should give the reader a clear idea of the differences between a nepticulid and, for example, a gelechiid or a noctuid — at least for certain stages of their development — or on a more general level of the differences between the principal "models" of butterfly and moth. If only morphological characters are considered, these models in fact correspond to the structural features characteristic of each superfamily. It is inevitable that the descriptions of families will also rely heavily on morphological characters, which are fairly constant within groups and in some cases easy to observe.

The amount of space devoted to the individual familes reflects their relative importance, and the discussion will follow a fairly standard format. After a brief description of the distinguishing features of the adults and other stages, ecology (in the broad sense), geographical distribution, and the division into families, where relevant, will be discussed. Within a group certain genera and species may be used as examples of particular ecological, behavioural, or zoogeographical phenomena. The number of species and their geographical distribution are generally approximate.

Suborder Zeugloptera

The Zeugloptera comprise small, metallically coloured moths, characterized by functional mandibles and the absence of a proboscis. The wings are homoneuran and covered by scales and spines (*microtrichia*). They are coupled by the jugum. In the female the vagina and the rectum open through a single cloacal aperture on the ninth to tenth sternite. In addition to three pairs of thoracic legs the larvae have eight pairs of abdominal prolegs. These are conical, lack muscles, and bear a claw. The head has long antennae. In the pupae the mandibles articulate (*decticous* pupa). This is considered the most primitive lepidopteran suborder because it retains many characters (the presence of laciniae, unspecialized galeae, functional mandibles, homoneuran wings, well-developed tegulae) that are similar to those of other panorpoid insects, such as the Mecoptera and Trichoptera. Whether the suborder Zeugloptera can be regarded as moths or not is still not settled, and certain experts would rank it as a separate order. Zeugloptera contains a single superfamily, the Micropterigoidea.

Superfamily Micropterigoidea

This superfamily is represented by just one family, the Micropterigidae.

Family Micropterigidae (Plate 28). The diagnosis of these primitive moths is exactly the same as for the suborder Zeugloptera (see above). The adults feed on the pollen of trees, shrubs, and herbs, including oaks, maples, hawthorns, and sedges with a distinct preference, it would seem, for the Ranunculaceae; significantly this group of plants is one of the most primitive of angiosperm families. The adults prefer cool, shady, places and spend a long time feeding on flowers, where they are easy to observe. At night they are quite often attracted by lights. At temperate latitudes they fly from late spring to late summer. The Micropterigidae are divided into eight genera and a total of about eighty species. They are found throughout virtually all temperate areas. Five species occur in Britain. There are a score of species of *Micropterix*, a genus with holarctic distribution, present in Italy. *Epimartyria* is another northern genus, while *Sabatinca* occurs in Australia and New Zealand. Micropterigids or similar forms dating from the Cretaceous have been found in fossil amber.

Suborder Dachnonypha

The Dachnonypha include small moths [wingspan varying between 8 and 16 mm (0.3 to 0.63 in)] that fly by day, with chewing or suctorial mouthparts. In fact in some groups the mandibles and maxillary laciniae may be present and functional. Similarly the galeae may be unmodified or have been transformed into only a short proboscis. The wings are always homoneuran and bear microtrichia, while the wing coupling is jugo-frenate. The female has a single genital aperture on the ninth to tenth abdominal sternite. The larvae are apodous. The pupae are decticous and have large articulated mandibles. Together with the Zeugloptera the Dachnonypha are the only Lepidoptera with decticous pupae. However, unlike the former, only a few members of the Dachnonypha (*Agathiphaga*) have adult chewing mouthparts. In the majority the maxillary galeae form a rudimentary proboscis. This suborder contains a single superfamily, the Eriocranioidea.

Superfamily Eriocranioidea

The Eriocranioidea are divided into five families: Eriocraniidae, Agathiphagidae, Neopseustidae, Lophocoroniidae, and Mnesarchaeidae. Apart from the Eriocraniidae these families contain very few species and are predominantly distributed in the Australian region.

Family Eriocraniidae (Plate 28). This is regarded as one of the most primitive lepidopteran families. In the adult the mouthparts are intermediate between the chewing type (mandibles are present but not functional) and the normal suctorial form. In the female the last abdominal segment is modified into an ovipositor capable of piercing plant tissues. The caterpillars are legless (apodous) and dig extensive galleries (mines) in the leaves of young food plants, especially birches (Betulaceae) and oaks and beeches (Fagaceae) but also the Pinaceae. Pupation begins inside the galleries in the leaves; when the pupae have limbs and other free appendages, they then descend to the ground, where they burrow into the soil and make a cocoon. Once development is complete, they emerge by cutting the cocoon with their free, functional mandibles. The Eriocraniidae are principally holarctic in distribution, with a few species in tropical Asia and Australia. In all a score of species have been described, all of which are small [8 to 15 mm (0.3 to 0.6 in)]. The most important genera are *Eriocrania*, *Eriocraniella*, and *Mnemonica*. In North America most eriocraniids are miners of young oak leaves, a typical example being the golden *M. auricyanea*. The remaining four families of the Dachnonypha contain small, closely related moths, which are characterized by the presence of several primitive morphological features.

Family Agathiphagidae. This comprises a single genus, *Agathiphaga*, with two species, one from Fiji and the other from Australia. The larvae feed on conifer seeds of the genus *Agathis*. Until recently *Agathiphaga* was placed in the Eriocraniidae.

Family Neopseustidae. This contains the three extremely rare species of *Neopseustis*, whose habits are virtually unknown. These moths occur in India, Burma, and Taiwan.

Family Lophocoroniidae (Plate 28). This comprises three species, all of which belong to the genus *Lophocorona*. Occurring in Australia, they were discovered and described by I. F. B. Common in 1973.

Family Mnesarchaeidae. A family of six species of small moths [8 to 12 mm (0.3 to 0.47 in)], which are restricted to New Zealand. Their habitat is virgin forest, where they rest close to the ground on ferns and shrubs, holding their antennae upwards in a characteristic V. Despite the fact that they are small, they are fast, direct flyers. The larvae live in damp, mossy places and in the soil. They are highly polyphagous, feeding indiscriminately on leaves, fern spores, fungal hyphae, and filamentous algae. After spending a long time as caterpillars (nine months or more), they form pupae characterized by the absence of functional mandibles (adecticous pupae). Their limbs, wings, and antennae are protected by a single, continuous cuticular case (obtect pupae), but they are free of the abdomen. The genus *Mnesarcha* was once placed in the Micropterigidae, but it has subsequently come to be considered a separate family. Mnesarchaeids differ greatly from the rest of the Dachnonypha in the characteristics of their larval stages. As a result some experts regard them as "primitive" Ditrysia rather than modified Dachnonypha. The taxonomic position of the Mnesarchaeidae is still highly uncertain, but it is widely thought that careful study of this group will clarify its relationships with more primitive moths (Zeugloptera and Dachnonypha) and with modified, more highly evolved ones (Monotrysia and Ditrysia).

Suborder Monotrysia

Monotrysians vary considerably in size but are all characterized by the lack of mandibles and maxillary laciniae and the presence of two or three segmented labial palps. The wings usually bear microtrichia. In the Hepialoidea they are homoneuran, but in the other two superfamilies they are heteroneuran. The female has one or two genital apertures on the ninth to tenth abdominal sternite. The former condition occurs in the Nepticuloidea and the Incurvarioidea, while the latter is typical of the Hepialoidea. Although this last group has two genital pores like the Ditrysia, it differs in other anatomical features, including the genital system. The pupae of the Monotrysia are adecticous and obtect. The suborder is a rather heterogeneous group, whose members display marked similarities in their primitive chracters, such as wings bearing numerous microtrichia. There are three superfamilies: Hepialoidea, Nepticuloidea, and Incurvarioidea.

Superfamily Hepialoidea

The Hepialoidea are homoneuran moths with genital apertures that are intermediate in type between the remainder of the Monotrysia and the Ditrysia. The adults are highly variable in size; but the antennae are always short, and the proboscis, if present, is rudimentary. There are three families, the largest of which is the Hepialidae; the Prototheoridae and the Palaeosetidae contain specialized forms and probably evolved from an ancient group of hepialids.

Family Prototheoridae (Plate 28). This is a small family of about ten small [20 to 27 mm (0.79 to 1.06 in)] species, occurring in Australia (genus *Anomoses*) and South Africa (genus *Prototheora*). Study of the few available specimens in collections has shown that this group has many primitive features. Its biology, however, is completely unknown.

Family Palaeosetidae. A family of small moths [of about 20 mm (0.79 in)] with a few species in India (genus *Genustes*), Taiwan (genus *Ogygioses*) and Australia (genus *Palaeoses*). The mouthparts of the adults are so reduced that they do not feed. The larvae and pupae are unknown.

Family Hepialidae (Plate 28). Although the family contains only some 300 species, it is worldwide in distribution. Its members, the ghost moths, vary greatly in size [20 to 200 mm (0.79 to 7.9 in)]. The adults have a rounded head, highly developed eyes, very short antennae, reduced labial palps, and the vestiges of a proboscis. The wings are homoneuran, aculeate, and have a jugate wing-coupling apparatus. The rounded abdomen is markedly elongated. The eggs are roughly spherical and give rise to long, cylindrical, tuberculate, caterpillars with few setae. The larvae are subterranean in those species which feed on roots. The pupae are primitive and elongate and have free appendages. The abdominal segments are highly mobile and bear spines, which are used by certain species to move from their burrow to the surface, where they emerge. Hepialids are pale and uniformly coloured (ocher, salmon, light blue, brown) or striking and variegated (green, black, yellow, white). They flight is agile and rapid. Crepuscular or noctural, they display interesting and rather unusual mating behaviour. Unlike most other butterflies and moths the male is courted. The sexes recognize each other and subsequently attempt to mate partly by visual signals and partly by chemical ones. The glands on the male's third pair of legs secrete a scent that is sexually highly attractive. In some species the females produce enormous numbers of eggs (more than 18,000 in *Abantiades magnificus*), which are shed in flight and on plants quite at random. Consequently the mortality among the early developmental stages is extremely high. Hepialid caterpillars are among the few to feed on ferns (the European species *Hepialus sylvina* and *H. hecta* feed on species of *Atridium* and on *Pteridium aquilinum*). However, they normally attack the cork or wood of many plants, in which they burrow, digging long galleries and spending up to four or five years in them before metamorphosing. The greatest concentration of species is in Australia, where there are more than 100, in New Guinea, and in tropical America. Many Australian forms are endemic, and they include the small species of *Fraus*, *Trictena*, and *Abantiades*, whose larvae feed on eucalyptus roots, *Aenetus*, whose members exhibit striking sexual dimorphism, the gigantic *Zelotypia staceyi* (Plate 109), and the numerous members of *Oxycanus*. *Endoclita* comes from tropical Asia, and *Leto* is endemic to South Africa; *Trichophassus giganteus* [200 mm (7.9 in)], a neotropical species, is the largest member of the Hepialidae and indeed of the Montrysia. The genus *Hepialus* has an enormous range. The ghost moth (*H. humuli*) is the commonest and largest European species and is known for the way its caterpillars attack hop roots. In coastal California *H. sequoiolus* larvae cause serious damage to tree lupines. In fact quite a few hepialids damage crops: In Australia, for example, the caterpillars of *Oncopera fasciculata* ravage pasture land.

Superfamily Nepticuloidea

These are small moths with a short or rudimentary proboscis, five-segmented maxillary palps, three-segmented labial palps, and heteroneuran wings with reduced venation and partially covered in aculei. The females are monotrysian and have a flexible, fleshy ovipositor. Together with highly primitive characteristics (monotrysian genitalia, aculeate wings, multisegmented maxillary palps) the Nepticuloidea also display such highly evolved features as unusual larval morphology, linked to their mining habit. There are two families: the Nepticulidae and the Opostegidae.

Family Nepticulidae (Plate 28). This family consists of about 300 species, most of which are leaf miners and sometimes gall forming (genus *Ectoedemia*). Some of the species, with wingspans of 2 to 4 mm (0.079 to 0.16 in), are among the smallest of all moths. They have no proboscis. The antennae are usually longer in the male and the scape (the basal antennal segment) is expanded, is covered in scales, and has a concave underside so as to form a sort of hood covering part of the eyes. At the rear of the head a tuft of backward-pointing bristles forms a collar that partially overlaps the thorax. The wings are heteroneuran, aculeate (that is, covered in microtrichia), and frenate. The forewings are covered with scales that are enormous compared with the tiny bodies of the moths. They are often beautifully coloured, ranging from iridescent golds to violets, blues, and purples, highlighted by metallic stripes and spots. The hind wings are usually colourless, but in the males they may bear tufts of brown androconia. The eggs are rather large; each species differs in its egg-laying habits, but with few exceptions the eggs are deposited on only one surface of a leaf. The larva bores directly into the plant tissue from inside the egg, and it then uses its mandibles to eat its way along. The larvae are highly modified because of their unusual habits. They appear delicate and colourless, smooth, and virtually apodous (their jointed limbs are reduced to fleshy appendages), and the head has just two ocelli. The larvae are incapable of cutting through the tough walls of the larger leaf veins, and so, as they mine, their initially straight galleries start to turn back on themselves and become more and more tortuous. The various species differ in their ability to attack the plant parenchyma; consequently the form of the mines, which is constant for a particular species, is an unmistakable trademark. Pupation usually occurs outside the galleries. The larva weaves an oval, yellowish-brown silk case to protect the pupa. There may be two or three generations a year. The family is cosmopolitan, but the greatest numbers of described species occur in the Nearctic and Palaearctic regions. There are sixty-seven British species. The more important genera include *Stigmella* (synonym of *Nepticula*), which occurs on Rosaceae, Fagaceae, Betulaceae, Ericaceae, and Juglandaceae; *Ectoedemia* lives on fewer groups of plants, including the Platanaceae and the Ulmaceae; the genera *Obrussa* and *Glaucolepis* are found on maples. Fossilized remains of Lepidoptera include a *Stigmella* mine in oak leaves from the North American middle Miocene.

Family Opostegidae. This small cosmopolitan family is more abundant in the tropics than it is in temperate regions. Very similar, overall, to the related Nepticulidae, opostegids differ from them in characters that are difficult to observe — apart from their eye cap, which is considerably larger. They are predominantly black or white. The adults are nocturnal and may be caught in light traps. Unfortunately, very little is known about the immature stages. *Opostega* is the most important genus, and it has a vast range.

Superfamily Incurvarioidea

Although this group numbers only a few hundred described species, the biology of the Incurvarioidea is most interesting. The adults are generally small with a wingspan of less than 10 mm (0.4 in), no ocelli, and a short proboscis. The labial palps have three segments, and the hind wing bears a frenulum. The female's abdomen is characterized by an elongated cloaca and a strong ovipositor capable of piercing the plant tissues in which the eggs are laid. The usually legless larvae are miners and in many instances cover themselves in a case, within which they feed. Before emerging, the pupae leave the cases. Four families are known: Incurvariidae, Prodoxidae, Heliozelidae and Tischeriidae.

Family Incurvariidae (Plate 28). This is a cosmopolitan family characterized by a markedly reduced wing venation; in this respect it is regarded as a relatively advanced group. There are two quite distinct subfamilies: Adelinae and the Incurvariinae. The head is covered with bristles, and in the Adelinae the antennae are very long — indeed longer in relation to their body length than those of any other lepidopteran, being up to four times as long as their bodies in certain species. The males have enormously developed compound eyes. Coloration ranges from metallic golden or coppery yellow in the Adelinae to the dull, striped patterning of the Incurvariinae. The larvae are miners. The adults fly both at midday and at twilight, often in great swarms. The Incurvariinae include *Incurvaria* (*I. koerneriella* is common in Italy) and *Lampronia*, whose members often lay their eggs in strawberry plants and other Rosaceae. The Adelinae comprise *Nemophora* and *Adela*, a group widely distributed in Europe and North America. Adelines are especially numerous in California where J. A. Powell has elucidated their host preferences and distributions.

Family Prodoxidae (Yucca Moths). This small, exclusively New World group is closely related to the Incurvariidae. Its members are nocturnal and are well known for their relationship with yuccas, first studied by C. V. Riley, one of the pioneers of American entomology. The principal genera are *Prodoxus* and *Tegeticula*. In the commonest true yucca moth, *Tegeticula yuccasella* (synonym *Pronuba yuccasella*, Plate 74), the female uses her modified maxillary palps to gather pollen from the yucca flower anthers and then to transfer it to the pistils, so pollinating the flower and leading to the formation of fruit and seeds. The moth then flies to another yucca and deposits one or more eggs in its ovary. Subsequently the larva will feed on some of the yucca seeds. It is a monophagous moth, and the caterpillars depend exclusively on yucca seeds. Equally, the plants are unable to reproduce without the yucca moth's intervention. The cooperation between the two species is so close that the evolutionary destiny of the one affects the fate of the other (see Plate 74, on coevolution). Bogus yucca moths (*Prodoxus*, *Parategeticula*) feed on parts of yucca plants other than the seeds; they play no role in pollinating yuccas.

Family Heliozelidae. A small, widely distributed family, in which the adults have shortened or rudimentary labial and maxillary palps and a well-developed proboscis. The wings have reduced venation and lack microtrichia. Coloration is metallic. The larvae are apodous and excavate mines of distinctive form: a short, straight, initial section that then expands to form broader patches. A hundred species in ten genera have been described. *Heliozela* and *Antispila* contain species dependent on grapes (*Vitis*), dogwood (*Cornus*), and various oaks (*Quercus*). The adults are diurnal and sometimes rest on flowers.

Family Tischeriidae (Plate 28). This family has small, predominantly holarctic species, characterized by pointed wings, reduced proboscis, and dull colours. The caterpillars are trumpet miners, and pupation takes place in the silk-lined galleries. There are fifty described North American *Tischeria* but only about a dozen in Europe. Oaks, wild lilacs (*Ceanothus*), roses and their relatives, and such composites as sunflowers are common hosts.

Suborder Ditrysia

This suborder contains the vast majority of the Lepidoptera. Its members lack mandibles and have a long coiled proboscis, reduced maxillary palps (except in certain tineids and lyonetiids), and three-segmented labial palps. The wings generally lack microtrichia and have reduced venation. The females have two genital apertures: an egg opening on the eighth abdominal segment and an opening for mating on the last abdominal segment. The pupa is always adecticous and obtect. As a result of its size this suborder displays enormous diversity in its biological forms, as will be seen from the comparative analysis of the families. In the Ditrysia a wide array of secondary features, such as the production of chemical substances or sounds that play an important part in intraspecific communication (for example, in the recognition of sexual partners), appears or becomes established. These chemicals or sounds may also be important in relationships with such other organisms as vertebrates (see the sections on mechanisms of defense against predation). The suborder is divided into eighteen superfamilies.

Superfamily Cossoidea

The Cossoidea are primitive Ditrysia; the adults vary greatly in size and have antennae that are usually either fully or partially bipectinate, a reduced proboscis, small maxillary palps, and heteroneuran and frenate wings. The robust, smooth caterpillars are mostly wood borers. Cossoid larvae, together with those of the two groups phylogenetically closest, the Tineoidea and Tortricoidea, have three prespiracular bristles on the prothorax. The superfamily Cossoidea contains four families: Cossidae, Dudgeoneidae, Metarbelidae, and Compsoctenidae. Apart from the carpenter moths (Cossidae) the systematic position of all these groups is uncertain. The Dudgeoneidae and Metarbelidae are sometimes included in the Cossidae, whereas the Compsoctenidae were initially placed in the Tineidae and subsequently in the Psychidae.

Family Cossidae (Plate 28). The carpenter moths are a cosmopolitan group containing more than 500 species varying greatly in size [20 to 240 mm (0.79 to 9.4 in)]. The larvae are generally xylophagous (wood boring). The adults are massive in appearance, with abdomens that are especially broad. They are grey, cream, or brown with a striped or spotted pattern. The wings are sturdy and narrow. The median vein, M, is well developed and forked in the discal cell, while the cubital vein, CuP, of the hind wing is usually reduced. The caterpillar has dorsal thoracic plates. The pupa is elongated, has abdominal spines, and lacks a cremaster; depending on whether it is male or female, it has four or five movable abdominal segments. The Cossidae are primitive moths similar to certain tineoids in their wing venation and to the tortricoids in the larval bristle patterns and in the male genitalia. Highly modified characters include the mouthparts, which are reduced to such an extent that the adults generally do not feed, the lack of ocelli, bipectinate antennae, and (in the Zeuzerinae) the loss of tibial spurs. There are two subfamilies, the Cossinae, in which the tibiae have spurs, and the Zeuzerinae, where they do not. About forty carpenter moths occur in North America, where *Prionoxystus robiniae* is the best known. *Cossus cossus*, the best-known cossine, occurs throughout the Old World. Its larvae attack broad-leaved trees in particular (willows, elms, ash). They take three or four years to emerge and so can cause immense harm to their host plants. The slow growth of these caterpillars is probably due to the low nutritional value of the wood. On the other hand the continuous availability and the virtually limitless extent of this resource is a considerable advantage to xylophagous insects in general. The females of *C. cossus* lay hundreds of eggs in cracks in the bark of food plants. Only some of the eggs hatch and only a small proportion of the young larvae manage to dig their own galleries in the wood, which they soften with a secretion from their mouths. After one year the larvae are already 10 cm (4 in) long, and at the end of the second year they pupate. Late in the spring of the third year the pupa twists and uses its abdominal spines as levers to drag itself out of its pupal chamber so that it can emerge. The larvae of *C. cossus* are notorious for emitting a penetrating and unpleasant goatlike odour, which is so strong that the wood still smells even when the galleries are empty. The accumulation of plant matter by the larvae and the presence of resinous drops in the foul-smelling excreta are two reasons for this unpleasant feature. Some carpenter moths have gregarious larvae. The gregarious larvae of the Australian *Macrocyttaria expressa* attack the mangrove *Excaecaria agalloche*. Other members of the Cossinae include *Chilecomadia* and *Allostylus*, which are mainly neotropical. *Zeuzera pyrina* [about 70 mm (3 in)], a beautiful white moth with numerous bluish-black spots on the forewings, is the commonest European member of the subfamily Zeuzerinae. In the males the antennae are bipectinate along the basal half. The present range of *Z. pyrina* includes the eastern United States, where it was accidentally introduced a century ago. The caterpillars of European populations feed on roses and their allies in particular; the North American ones attack more than 100 different plants. *Zeuzera coffeae*, a relative of *Z. pyrina*, attacks young coffee and cocoa plants. Its range is highly discontinuous and extends from India to Indonesia and New Guinea. Like other cossids (*Xyleutes durvillei*, for example), *Z. coffeae* is extremely prolific. The female lays large numbers of eggs, but the mortality rate is very high at this stage, and only a few larvae develop. These are usually dispersed by the wind, which catches their silk threads. Unlike many cossids the life cycle is short, lasting little more than a year. One very interesting genus of the Zeuzerinae is *Xyleutes*, which is best represented in Australia, where there are about seventy species. *Xyleutes amphiplecta* displays marked sexual dimorphism in both overall size and wing length. The male has a very short abdomen but much broader wings than does the female, which is brachypterous (reduced wings) and flightless. In *Xyleutes* the wingspan is variable: In males of *X. amphiplecta* the wingspan is 30 mm (1.18 in), whereas it is 250 mm (9.84 in) in the female of *boisduvali*. The abdominal diameter is similarly variable, the maximum being an enormous 22 mm (0.87 in) in *X. boisduvali*.

Family Dudgeoneidae (Plate 28). This is a small family that some authors place in the Cossidae; others believe it to be related to the Sesiidae. Dugeoneids are distributed in India, Australia, New Guinea, and possibly Africa. They are small moths [about 30 mm (1.18 in)] more strikingly coloured and patterned than are the Cossidae.

Family Compsoctenidae. The family numbers only a few dozen species. The adults are small [12 to 20 mm (0.47 to 0.79 in)] and have a rudimentary proboscis, well-developed labial palps, bipectinate antennae in the males, and wing venation similar to the Cossidae. The caterpillars mine stems and roots. *Compsoctena* occurs in Eurasia, in India, and in some areas of Africa.

Family Metarbelidae (Plate 28). The metarbelids (or teragriids or lepidarbelids) are small moths related to the Cossidae, from which they differ in lacking a frenulum. The males are characterized by a hairy abdomen ending in a forked process. Little is known of their biology. The family is mainly African, with important genera being *Lepidarbela*, *Otharbela*, which is dependent on Tiliaceae of the genus *Grewia*, and *Salagena*, whose caterpillars feed on Myrtaceae (*Eugenia*).

Superfamily Tortricoidea

The small moths of this large superfamily have a dense covering of scales on their heads, sometimes grouped into longer tufts, and hairy antennae, which are of medium length and pectinate or filiform. The proboscis and labial palps are variable in length. The larvae may appear delicate or robust, depending upon whether they feed inside or outside leaves, flowers, fruit, seeds, or bark. The pupa has a cremaster and differs clearly in the two sexes. In the males the last three (eighth to tenth) and in the females the last four (seventh to tenth) abdominal segments are fused. It is a fairly homogeneous superfamily and is closely related to the primitive Cossoidea and the Tineoidea. There are two families: the Tortricidae and the Phaloniidae.

Family Tortricidae (Plate 28). Tortricidae is a vast family of some 5,000 species, many of which are cosmpolitan and economically important. The principal adult characters include maxillary palps with between two and four segments and labial palps that vary in length, a pecten of scales on the CuA vein of the hind wing, as well as, in almost all cases, a CuP vein on both pairs of wings. The eggs are flattened and are laid singly or in small clusters. The larvae have prolegs bearing one to three rows of hooks, and they have the distinctive habit of rolling themselves up in leaves to pupate. Some may be miners in their early larval stages.

Two subfamilies that we shall discuss are the Tortricinae and the Olethreutinae (for a long time this was regarded as a separate family), which in turn are divided into numerous tribes, members differing widely in appearance. One common feature is that the outline of the closed wings is bell shaped (this is so in many Archipini); in other cases the forewings are oval (in the Cnephasini, for example) or rectangular (as in the Tortricinae). The basic pattern of the forewings ranges from the primitive type consisting of transverse bands to a more advanced one with longitudinal stripes or irregular spots. In certain Olethreutinae the wings are spotted and the colours metallic and iridescent, a rather advanced characteristic linked to the diurnal habits of these moths. Many cosmopolitan species display a high level of geographical variation in their coloration and wing patterns, which makes their taxonomic recognition more difficult. The codling moth, *Cydia pomonella* [synonyms *Carpocapsa pomonella* and *Laspeyresia pomonella*; 12 to 22 mm (0.47 to 0.87 in)] has a worldwide range and is possibly the best known tortricid. Its flesh-coloured larvae are frequent and unwelcome occupants of worm-eaten apples (Plate 111), causing immense economic damage out of all proportion to their insignificant appearance. In addition to *pomonella*, the genus *Cydia*, a cosmopolitan olethreutine group, contains numerous other species dependent on fruit trees. The oriental peach moth (*Cydia molesta*), which is smaller than the species mentioned above, is cosmopolitan and attacks plants of *Malus*, *Prunus*, and other Rosaceae. The grey caterpillars spin webs around the young twigs. In their early stages they are miners and cause the leaves to drop, but later on they bore directly into the fruit. In warm climates the temperature hastens the pace of development, and there may be two or three generations a year, resulting in serious damage. The Mexican jumping-bean moth (*C. dehaisiana*) is one of the insects responsible for the strange Mexican jumping beans. This moth [about 20 mm (0.79 in)] occurs in central America and the southern United States. In the latter, however, it is not beans that are affected but the seeds of *Croton* of the spurge family, which jump in response to sharp movements by the larvae inside them. Before pupating, the caterpillars bore through the walls of the seed and cover the opening with silk. Once metamorphosis is complete, the moth breaks through this fragile cover, leaving the exuvium in the seed, and flies away. The ugly-nest caterpillar [*Archips cerasivorana*; 18 to 25 mm (0.71 to 0.98 in)] is typical of the fauna of North America. It is an orange-red moth with gregarious caterpillars dependent upon *Prunus*, which are capable of spinning webs 20 to 30 m (65 to 100 ft) long. There are too many important tortricid groups to be able to discuss them exhaustively here. However, certain genera should be mentioned: for example, *Choristoneura*, which is mainly northern in distribution, *Eucosma* and *Tortrix*, which are cosmopolitan, and *Sparganothis* and *Epiblema*, which are holarctic. The spruce budworm [*Choristoneura fumiferana*; 20 to 25 mm (0.79 to 0.98 in)], a nearctic species, is infamous as in all probability it is the most harmful insect in North America, where it ravages coniferous forests. The caterpillars of *fumiferana* and several close relatives mine leaves and buds and form large nests in many conifers, particularly on species of *Abies* and *Picea*. *Tortrix viridiana* [16 to 23 mm (0.63 to 0.91 in)], on the other hand, is a widespread and common European species, which is clearly identifiable by its pale green colour. The larvae roll up and sew oak leaves. *Sparganothis pilleriana*, which is similar in size to *T. viridiana* and also inhabits Europe, lives on a variety of plants but is particularly known for the damage it causes to vines. The tropical American *Pseudatteria* moths are conspicuous tortricids, with wingspans of over 30 mm (1.18 in) and striking colours, which possibly serve as a warning. However, it is not known whether they are protected by repellant chemical substances in their tissues that make them unappetizing to predators.

Family Phaloniidae (synonym Cochylidae; Plate 28). Phaloniids are closely related to the tortricids and are particularly common in Europe and North America. The family is fairly homogeneous. Both the fore- and hind wings are narrow and triangular. At rest they surround the abdomen like a sleeve. The larvae are often miners. The type genus, *Cochylis*, has many members but is absent from Africa and Australia.

Superfamily Tineoidea

This is a large superfamily of mostly small- or medium-sized moths. The adults often have an eye cap, a naked and reduced proboscis, and up to five maxillary segments. The more primitive groups have a marked CuP vein, and in general venation is greatly reduced in those species with narrow hind wings. The larvae mine leaves or cover themselves with cases built from various types of material. The pupa usually has dorsal spines and free abdominal segments. It too is protected by a case that varies in shape, from which it emerges as an adult. Many psychids, however, are an exception since the adult females remain in the pupal case. Primitive characteristics include the limited reduction of the venation, the persistence of the median vein, M, in the discal cell, and the multisegmented maxillary palps. Seven families are known.

Family Pseudarbelidae. This family comprises a few small species [about 30 mm (1.18 in)] belonging to the genus *Pseudarbela*. Its systematic position is uncertain; only a few examples are known from the Philippines and New Guinea. Initially they were placed in the Cossidae and subsequently in the Tineoidea.

Family Arrhenophanidae (Plate 29). This is a family of medium-sized moths [40 to 70 mm (1.57 to 2.76 in)] from Central America. Little is known of the biology of this group and their classification in the Tineoidea is provisional. The species illustrated is *Arrhenophanes perspicilla*.

Family Psychidae (Plate 29). About 800 species of small- and medium-sized moths belong to this interesting family. They frequently display marked sexual dimorphism, with winged males and apterous females, whose anatomical organization is greatly altered. The adults have very hairy bodies, antennae that are bipectinate at their tips, reduced mouthparts (they do not feed and live only a short time), and the median vein, M, of the forewing forked in the discal cell. As soon as the larvae hatch from their eggs, they begin to build a case of sand, pebbles, bits of leaves, twigs, lichens, mosses, or other material, which they glue together with silk produced by their labial glands. As the larvae grow, they enlarge this case and in certain tropical species may reach 10 to 15 cm (4 to 6 in) in length. The head and thorax are free, and sometimes the larvae move, carrying the case around with them. The case itself usually has an opening at the front for access to food and one at the rear for the expulsion of excreta. The shape, thickness, and appearance of these cases is variable but typical for each species. Some are spiral and resemble snail shells — so much so that they were wrongly classified as such by early collectors. The cases are usually suspended from branches or leaves and may contain a pupa or an adult female and numerous eggs rather than a larva. Psychid caterpillars are usually highly polyphagous and are capable of dispersing over a wide area. Dispersal is aided by the fact that the case can be hermetically sealed and so can be carried by sea currents. Other adaptations to facilitate dispersal include the caterpillar's ability to feed on dead vegetation or to fast for long periods. They are also resistant to wide temperature fluctuations. The pupae, as has been mentioned, often occupy larval cases; the females may have reduced eyes and appendages. In most instances the females emerge within the pupal case, extend the top of their abdomen outside the case, and wait for the male. The latter, guided by female pheromones, finds the female and mates with her. In many species with few males parthenogenesis occurs, whereby larvae develop from unfertilized eggs. Psychids are normally a uniform, dull grey or whitish. The wings are sometimes semitransparent, and only rarely (in certain Australian *Cebysa* species, in which the female is also winged) are they patterned and coloured. The family is fairly heterogeneous, with highly modified species among the Psychinae (the females of which are apterous and have reduced appendages) and species displaying more primitive characteristics in the Taleporiinae (females winged). There are about fifty species of psychids in Italy (only a few in Britain) from quite a few genera, including *Phalacropterix*

Plate 28 **Primitive Lepidoptera from the Micropterigidae to the Phaloniidae**

Suborder Zeugloptera, *superfamily Micropterigoidea*, family Micropterigidae: (1) *Micropterix calthella* [8 mm (0.31 in)], Palaearctic, (2) *M. anderschella* [9 mm (0.35 in)], Palaearctic. **Suborder Dachnonypha, *superfamily Eriocranioidea*,** family Eriocraniidae: (3) *Eriocrania* sp. [12 mm (0.47 in)], Nearctic; family Lophocoroniidae: (4) *Lophocorona pediasia* [12 mm (0.47 in)], Australian. **Suborder Monotrysia, *superfamily Hepialoidea*,** family Prototheoridae: (5) *Prototheora petrosoma* [20 mm (0.79 in)], Ethiopian, (6) *Hepialus humuli*, female [44 to 70 mm (1.73 to 2.76 in)], Palaearctic, (7) *H. sylvina* [25 to 45 mm (0.98 to 1.77 in)], Palaearctic, (8) *Charagia daphnandra* [60 to 100 mm (2.36 to 3.94 in)], Australian. ***Superfamily Nepticuloidea*,** family Nepticulidae: (9) *Nepticula tityrella* [5 mm (0.20 in)], Palaearctic. ***Superfamily Incurvarioidea*,** family Incurvariidae: (10) *Adela reamurella* [15 mm (0.59 in)], Palaearctic; family Tischeriidae: (11) *Tischeria ekebladella* [10 mm (0.39 in)], Palaearctic. **Suborder Ditrysia, *superfamily Cossoidea*,** family Cossidae: (12) *Xyleutes eucalypti* [130 to 200 mm (5.12 to 7.9 in)], Australian; family Dudgeoneidae: (13) *Dudgeonea actinias* [42 mm (1.65 in)], Australian; family Metarbelidae: (14) *Metarbela triguttata* [22 mm (0.87 in)], Ethiopian. ***Superfamily Tortricoidea*,** family Tortricidae: (15) *Tortrix viridana* [16 to 24 mm (0.63 to 0.94 in)], Palaearctic, (16) *Epiblema foenella* [18 to 25 mm (0.71 to 0.98 in)], Palaearctic, (17) *Cerace xanthocosma* [35 to 50 mm (1.38 to 1.96 in)], Palaearctic; family Phaloniidae: (18) *Agapeta zoegana* [18 mm (0.71 in)], Palaearctic, (19) *Commophila aeneana* [14 mm (0.55 in)], Palaearctic.

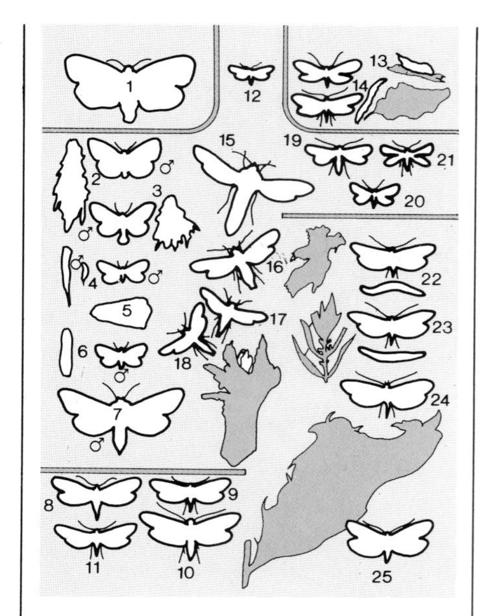

Plate 29 **Tineoidea and Yponomeutoidea**

Superfamily Tineoidea, family Arrhenophanidae: (1) *Arrhenophanes perspicilla* [40 to 67 mm (1.57 to 2.64 in)], Neotropical; family Psychidae: (2) *Pachytelia unicolor*, larval case and male [25 mm (0.98 in)], Palaearctic, (3) *Hyalina albida*, larval case and male [16 mm (0.63 in)], Palaearctic, (4) *Rebelia nudella*, larval case, female, and male [18 mm (0.71 in)], Palaearctic, (5) *Amycta quadrangularius*, larval case, Ethiopian, (6) *Scioptera plumistrella*, case and male [20 mm (0.79 in)], Palaearctic, (7) *Eumeta cervina*, male [40 mm (1.6 in)], Ethiopian; family Tineidae: (8) *Moerarchis australsiella* [30 mm (1.18 in)], Australian, (9) *Tinea fuscipunctella* [25 mm (0.98 in)], cosmopolitan, (10) *Episcardia lardatella* [30 mm (1.18 in)], Palaearctic, (11) *Monopis fenestratella* [18 mm (0.71 in)], Palaearctic; family Lyonetiidae: (12) *Lyonetia clerckella* [8 mm (0.31 in)], Palaearctic; family Gracillariidae: (13) *Caloptilia alchymiella* [10 mm (0.39 in)], (14) *Gracillaria swederella*, adult, larva, and leaf sewn by the larva [13 mm (0.51 in)], Palaearctic.
Superfamily Yponomeutoidea, family Sesiidae: (15) *Sesia apiformis* [34 to 44 mm (1.33 to 1.73 in)], Palaearctic, Nearctic, (16) *Synanthedon stomoxyphormis* [22 mm (0.87 in)], Palaearctic, (17) *Chamaesphecia chrysidiformis* [32 mm (1.26 in)], Palaearctic, (18) *C. stelidiformis*, adult and branch with pupal exuvium [35 mm (1.38 in)], Palaearctic; family Glyphipterigidae: (19) *Glyphipterix perornatella* [15 mm (0.59 in)], Palaearctic; family Douglasiidae: (20) *Tinagma perdicellum* [10 mm (0.39 in)], Palaearctic; family Heliodinidae: (21) *Heliodines roesella* [15 mm (0.59 in)], Palaearctic; family Yponomeutidae: (22) *Swammerdamia pyrella*, adult, attacked leaf, and larva [12 mm (0.47 in)], Palaearctic, (23) *Zelleria phillyrella*, attacked leaf, adult, and larva [12 mm (0.47 in)], (24) *Yponomeuta padellus* [20 mm (0.79 in)], Palaearctic, (25) *Y. cagnagella*, with larval nest [20 to 26 mm (0.79 to 1.02 in)], Palaearctic.

(*P. apiformis*). *Solenobia* (*S. triquetrella* has been extensively studied because of its parthenogenesis and colonization) is exclusively European. *Pachytella* and *Psyche* (*P. casta*, synonym *Fumea casta*, occurs in Italy) are more widely distributed in Eurasia. There are about twenty North American species. In the eastern and central United States these include *Thyridopteryx ephemeraeformis*, whose caterpillars are dependent upon conifers, especially red cedar. *Kotochelia* occurs in Africa, *Oiketicus* occurs in South America, and *Narycia*, *Lomera*, and *Clania* are among the largest Australian genera.

Family Tineidae (Plate 29). Tineids are small moths [less than 15 mm (0.59 in) on the average] that are often brightly coloured with beautiful wing patterning. Cosmopolitan, the family contains more than 2,400 species and includes such domestic pests as the clothes moths. The adults have very hairy, ruffled heads, the antennae are usually simple, and the proboscis is either absent or short. The maxillary palps vary in form, while the labial palps bear bristles on the second joint and have a thin, pointed, third joint. In many species the larvae, which have a protective case, are highly polyphagous and will eat lichens, fungi, bird and bat droppings, feathers, wool, silk, fur, grains, or other foodstuffs. The pupae bear dorsal abdominal spines and have a number of free segments. They live in the larval cases. Many tineids have become cosmopolitan as a result of the exportation from one country to another of the foodstuffs or fabrics on which the caterpillars feed. There are two principal groups of cosmopolitan genera whose members are of major economic importance: *Tinea*, *Tineola*, and *Trichophaga*, three closely related genera, belong to the first group, which is characterized by larvae that feed on plant and animal fibers; *Monopis* and *Nemapogon* are two genera typical of the second group, whose members have a broader diet, including dried plants, grain, and flour as well as dried fruit and even cured tobacco. *Tinea pellionella* [about 12 mm (0.47 in)], the case-bearing clothes moth, is common in houses and warehouses. The females lay up to 300 eggs, which hatch after a week. *Trichophaga tapetzella* about 20 mm (0.79 in)], the carpet moth, is similar in habit, as is *Tineola bisselliella*, another clothes moth, whose larvae will attack both natural and synthetic fibers. As is well known, to prevent infestation, it is usually sufficient to expose the garments to sunlight or to treat them with naphthalene or camphor.

Family Lyonetiidae (Plate 29). Lyonetiids are microlepidopterans [5 to 10 mm (0.20 to 0.39 in)] with a worldwide distribution whose mining larvae are often very harmful to crops. Their principal morphological features are narrow forewings and even narrower lanceolate hind wings with an open discal cell and reduced venation. The pupa is specialized and immobile, lacks thoracic spines, and lies inside an ellipsoidal case, which may be in a gallery or suspended by a thread of silk outside. The family is heterogeneous and contains such typical groups as *Bedellia* (*B. somnulentella* is cosmopolitan, lives on yams and other Convolvulaceae), *Bucculatrix*, *Leucoptera*, and *Lyonetia* and such atypical ones as *Opogona* (*O. glyciphaga* attacks the banana, of the Musaceae, and sugar cane, of the Poaceae), *Comodica*, and *Erechthias*. *Lyonetia*, the type genus of the family, consists mainly of holarctic species as well as species such as *Lyonetia clerckella* [7 mm (0.28 in)], which also occur in the Oriental and Ethiopian regions.

Family Phyllocnistidae. This family of about fifty species of small moths [about 10 mm (0.39 in)] has an extensive, worldwide distribution. Similar in appearance to lyonetiids and gracillariids, they differ mainly in their wing venation. The mining larvae are apodous and have mandibles that have been transformed into extremely sharp blades. They use these to cut through the walls of the leaf vessels so that they can feed on the sap. The final larval stage does not feed but instead prepares to pupate within the gallery. *Phyllocnistis saligna* [6 to 7 mm (0.24 to 0.28 in)] is the commonest European species and is dependent on willows.

The Lyonetiidae, Phyllocnistidae, and Gracillariidae (see below) are three very closely related families, whose larvae are similar in their ecology. The morphology of the caterpillars, which is also similar in all three groups, is the strongest evidence for the strong adaptive convergence that may be observed in their flattened form, in the lack or extreme reduction of the limbs, and in the form of the mouthparts and hence their way of feeding. In all three families the adults have narrow, more or less pointed wings, bordered by fringes of elongated scales, which are usually pale in colour, often with silvery or red-gold spots.

Family Gracillariidae (Plate 29). These small, delicate moths have narrow, tapering wings. The antennae are simple and long; the maxillary palps have four segments, the labial ones being slender. The legs may bear tibial spines, while on the wings the radial vein R_5 extends to the costal margin. In one subfamily, the Gracillariinae, the resting position assumed by the adult, with the body angled sharply away from the support, the head thrown up, and the first two pairs of legs extended, is highly characteristic; in the Lithocolletinae subfamily the head is appressed to the substrate. As in the Phyllocnistidae the mining larvae often undergo hypermetabolic development. Until the third instar they are flattened with atrophied limbs and reduced prolegs. Final-instar larvae are cylindrical, possess thoracic limbs and prolegs, and chew the leaf parenchyma. Pupation occurs in the gallery, and the pupa has three free abdominal segments. The mines of certain *Phyllonorycter* species (synonym *Lithocolletis*) are highly typical. The larvae stretch threads of silk inside the galleries, causing the leaf epidermis to contract and giving it a corrugated appearance from the outside. *Phyllonorycter* is a large cosmopolitan genus containing numerous tiny species [6 to 7 mm (0.24 to 0.28 in)] whose

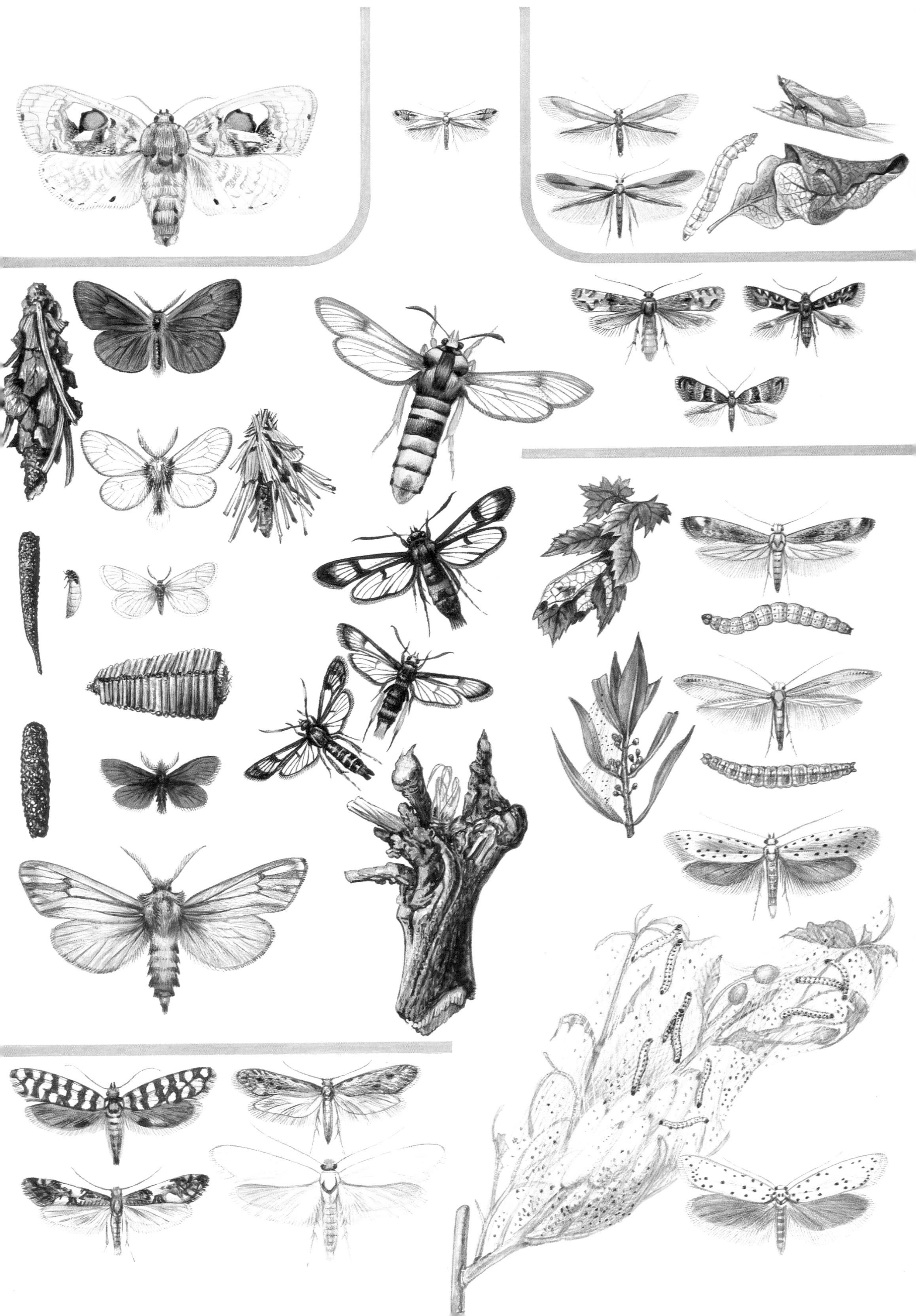

mines are species specific in terms of their host, form, and position in the leaf. *Phyllonorycter loxozona*, Ethiopian, mines along the edges of leaves of *Dombeya* (a relative of the cacao, Sterculiaceae); *P. tremuloidiella*, nearctic, causes the leaves of many ornamental plants to curl up; *P. blancardella* mines apple leaves; *Cameraria* is a genus found especially in North America, with upperside blotch mines on leaves of oaks and a few other woody plants. Other genera with many species include *Epicephala* from Africa, India, and Australia, and *Caloptilia* and *Acrocercops*, cosmopolitan. This last genus contains more than 300 species, a third of which are Australian and many of great economic importance. *Acrocercops cramerella*, for example, attacks the cacao (*Theobroma cacao*). The larvae bore into the fruit and mine the seeds. They they pupate in a case on the outside of the fruit. The pupal cases of many *Acrocercops* are covered with little frothlike bubbles, the number of which is often constant for a given species. This family contains bout 1,000 species.

Superfamily Yponomeutoidea

This medium-sized superfamily is distinguished by the great variety of its forms. The heads of the adults have rather pronounced ocelli (except in the Yponomeutidae and the Eperminiidae), bare proboscis, small maxillary palps (which vary in shape), and short labial palps. The larvae have three prespiracular bristles on the prothorax (only two in the Epermeniidae); they mine leaves, stems, and roots or build silk nests for themselves. The pupa, which has dorsal and sometimes ventral abdominal spines, may remain in the larval mine, or it may lie in a silk cocoon, which is usually abandoned before emergence. The mobile, spined pupae of the Sesiidae and the presence of prominent ocelli in the adults are regarded as primitive characters. Specialized characters — or ones that occur in a more or less highly modified condition — include the reduced wing area and the markedly reduced venation in the Sesiidae and the Douglasiidae. There are six families.

Family Sesiidae (synonym Aegeriidae; Plate 29). The principal characteristic of this family is undoubtedly its marked similarity to the Hymenoptera. The overall appearance of the Sesiidae is so close to that of wasps or hornets in some species that even expert entomologists find difficulty in distinguishing sesiids in flight from Hymenoptera. Many diurnal species are in fact potential mimics, quite harmless themselves, of certain well-protected and strikingly coloured Hymenoptera (models) and as such are examples of Batesian mimicry. The adults are small or medium sized with simple or pectinate antennae that are enlarged and curved at their tips and with or without apical bristles, atrophied maxillary palps, and erect, scale-covered labial palps. The forewings are much narrower than the hind ones as a result of the marked reduction of the anal region. The venation is also greatly reduced, and the wings are often transparent as a result of the loss, at emergence, of the scales. The frenulum is usually single, and in addition there tends to be a series of hooks on the radius of the hind wing that engages the folding of the inner margin of the forewing, so increasing the coupling of the wings. The hind legs are covered by thick tufts of bristles, increasing their resemblance to those of the Hymenoptera. In some species the abdomen ends in a fan of elongated scales. In temperate latitudes these moths fly during the summer. The length and mode of development resembles that of the Cossidae. The larvae are xylophagous and in certain cases harm crops such as peaches. The pupa has dorsal and ventral abdominal spines, which it uses to drag itself out of the larval shelter at the time of emergence. The marked tendency of the Sesiidae to Batesian mimicry involves two different kinds of adaptation: One is behavioural, and the other is morphological. Hymenopteran behaviour is mimicked by powerful, darting, somewhat intermittent flight, while the shape and coloration of the wings and abdomen are the principal morphological features involved in mimicry. Sesiids are worldwide in distribution and number about 800 species in 170 genera. Tropical Africa and the Oriental region, with about 50 genera each, are where the sesiids are most diverse; the Palaearctic, Nearctic, and Neotropical regions have little more than 20 genera apiece. The most widely distributed palaearctic species, which also occurs in the United States, is *Sesia apiformis*, whose larvae are dependent on poplars (Salicaceae). The cosmopolitan genera include *Synanthedon* and *Chamaesphecia*.

Family Glyphipterigidae (synonym Atychiidae, Plate 29). This cosmopolitan, fairly heterogeneous, group contains about 900 species and is well represented in the southern hemisphere. The adults are small [about 10 mm (0.39 in)] and are characterized by brilliant metallic or sometimes iridescent coloration, as in *Mictopsichia*. They are usually diurnal. The larvae have prolegs with hooks and mine seeds and roots. Some groups, such as *Burlacena* (synonym *Sesiomorpha*) from the Oriental region, are thought to mimic hymenopterans. The largest genera are *Glyphipterix*, with over 200 forms found particularly in the Australian region, and *Himma*, with 150 species concentrated in New Guinea. One glyphipterigid, *Tabenna bjerkandrella*, has one of the most extensive ranges of all the Lepidoptera.

Family Douglasiidae (Plate 29). This cosmopolitan family has only a few members. The adults are small [about 10 mm (0.39 in)] and have narrow, pointed wings with greatly reduced venation and an open discal cell. As in the previous family the caterpillars are miners, but they lack hooks on their prolegs.

Family Heliodinidae (synonym Schrechensteiniidae; Plate 29). The Heliodinidae contain almost 400 small species [about 20 mm (0.79 in)], which often display well-defined and

Plate 30 **Gelechioids and Copromorphoids**

Superfamily Gelechioidea, family Coleophoridae: (1) *Coleophora virgatella* [12 to 15 mm (0.47 to 0.59 in)], Palaearctic; family Agonoxenidae: (2) *Agonoxena argaula* [15 mm (0.59 in)], Australian; family Elachistidae: (3) *Elachista regificella* [9 mm (0.35 in)], Palaearctic; family Scythridae: (4) *Scythris cuspidella* [13 to 17 mm (0.51 to 0.67 in)], Palaearctic; family Stathomopodidae: (5) *Stathmopoda pedella* [10 to 14 mm (0.39 to 0.55 in)], Palaearctic; family Oecophoridae: (6) *Wingia lambartella* [37 mm (1.46 in)], Australian, (7) *Alabonia geoffrella* [18 mm (0.71 in)], Palaearctic (see also Figure 15); family Cosmopterigidae: (8) *Cosmopterix scribaiella* [9 mm (0.35 in)], Palaearctic, Oriental; family Metachandidae: (9) *Metachanda citrodesma* [7 to 12 mm (0.28 to 0.47 in)], Ethiopian; family Pterolonchidae: (10) *Pterolonche pulverentella* [20 to 30 mm (0.79 to 1.18 in)], Palaearctic; family Blastobasidae: (11) *Holocera iceryaeella* [15 mm (0.59 in)], Australian; family Xyloryctidae: (12) *Xylorycta porphyrinella* [33 mm (1.3 in)], Australian; family Stenomidae: (13) *Stenoma sequitiertia* [46 mm (1.81 in)], Neotropical; family Gelechiidae: (14) *Pthorimaea fischerella* [11 mm (0.43 in)], Palaearctic; family Oecophoridae: (15) *Thalamarchella aveola* [26 mm (1.02 in)], Australian.
Superfamily Copromorphoidea, family Alucitidae: (16) *Alucita (= Orneodes) grammodactyla* [13 mm (0.51 in)], Palaearctic; family Carposinidae: (17) *Heterogymna pardalota* (25 mm (0.98 in)], Oriental.

strikingly coloured wing patterns. The larvae have long prolegs with hooks and are normally phytophagous. However, a few species are carnivorous and prey on scale insects. *Eretmocera* and *Heliodines* are two of the principal genera. The family is widely distributed.

Family Yponomeutidae (Plate 29). This important family contains about 800 species of small- and medium-sized moths. It is virtually cosmopolitan although its greatest concentration is in the tropics. The adults almost always lack ocelli, the proboscis is bare, and the head is covered with elongated scales. The palps vary greatly in their morphology. The forewings often bear a pterostigma running along part of the costal margin. The larvae have prolegs with hooks and may be solitary or gregarious, in which case they may protect themselves by weaving webs over flowers and leaves, or they may be free living and sometimes external feeding. Some species cause considerable damage to orchards. The pupae vary greatly in appearance and are protected by a cocoon, which they leave prior to emergence. There is no typical model of yponomeutid coloration or wing pattern: Although many species are inconspicuous — *Yponomeuta*, for example, has grey wings with black spots — some tropical forms, such as the Indo-Australian *Lactura*, display exceptionally beautiful coloration. The Yponomeutidae is a heterogeneous family. It is divided into four subfamilies, which some authors rank as families in their own right. *Prays* and *Plutella* are cosmopolitan genera from the Plutellinae and contain many harmful species. *Prays oleae* [about 15 mm (0.59 in)], which has been known since antiquity, produces three generations a year: The larvae of the first generation mine the leaves of the olive, those of the second attack the flowers, and those of the third, the most harmful, feed on the fruits. The range of the species, originally restricted to the Mediterranean basin, now extends over several continents but does not yet include North America. *Plutella xylostela* [synonym *P. maculipennis*; about 15 mm (0.59 in)] is a cosmopolitan species that attacks many Brassicaceae, such as cabbages and cauliflowers. *Yponomeuta* is a large cosmopolitan genus from the subfamily Yponomeutinae and contains two well-known species, *Y. padella* and *Y. malinella*. Morphologically these are virtually indistinguishable, but they differ in their host plants. The caterpillars live on different genera of Rosaceae: The former feeds on hawthorn, plum, and sorb, whereas the latter lives on apple and pear trees. For a time the larvae are miners and then become external feeders. Many adult yponomeutids, particularly among the *Argyresthia* and *Zelleria* of the Argyresthiinae, assume a highly characteristic resting position, the head pointing downwards and the rest of the body, supported by only the first two pairs of legs, angled upwards at 45°.

Family Epermeniidae. A very widely distributed family of about 100 small species [15 mm (0.59 in)], it is fairly close to the Yponomeutidae but has some extremely unusual features compared with other families. The caterpillars have only two pairs of prespiracular bristles on the thorax, and the pupae do not leave the cocoon at emergence.

Superfamily Gelechioidea

This is one of the largest lepidopteran superfamilies, with over 12,000 species, characterized by the base of the proboscis densely covered with scales and surrounded by small maxillary palps with four segments. The median vein of the forewing is almost always external to the discal cell. The venation of the hind wings is reduced, and the posterior cubital, CuP, may be absent.

The back of the abdomen may bear spiny plates, while the wings frequently have long fringes of scales. The larvae have three prothoracic bristles and are highly diverse in their ecology. The pupae do not have dorsal spines, but they have two or three mobile abdominal segments and are surrounded by a cocoon that they do not leave prior to emergence. The number of gelechioid families varies with the system of classification. In this book the Gelechioidea will be divided into fifteen families, some of which contain very few species and a large number of which are of uncertain taxonomic validity. Furthermore, about half the superfamily forms a homogeneous group whose individual members can be distinguished only by detailed morphological analysis. Given that the species involved are extremely small, it is not easy to give a concise description of their morphology. Consequently only the major morphological features will be dealt with, and the discussion of more detailed differences will be restricted to ecology and distribution.

Family Coleophoridae (Plate 30). A cosmopolitan family, concentrated in the Holarctic region, of about 600 medium-sized [8 to 15 mm (0.31 to 0.59 in)] species. At rest the antennae are held parallel and pointing forwards in line with the axis of the body. The narrow, pointed wings have fringes of hair, which are particularly extensive on the hind pair. The larvae are initially miners and subsequently feed externally. They build sacklike cases, which they use as portable shelters. These vary considerably in shape and substance among the various species although they are uniform within a species. They may be gradually enlarged as the larva grows, or they may be abandoned for other, larger ones. Coleophorids, together with psychids, are the most accomplished builders of larval cases. *Coleophora* is the main genus and contains almost 400 holarctic species dependent on the Fabaceae, Rosaceae, Labiatae, and conifers as well as on other plants.

Family Agonoxenidae (Plate 30). It consists of the genus *Agonòxena*, which is represented by possibly just three species. It occurs in part of the Indo-Australian region and on some Pacific islands. These small moths [about 15 mm (0.59 in)] have maxillary

Plate 31 **Castnioidea and Zygaenoidea**

Superfamily Castnioidea, family Castniidae: (1) *Cyanostola diva* [75 mm (3 in)], Neotropical, (2) *Zegara zagraeoides* [95 mm (3.7 in)], Neotropical, (3) *Riechia acraeioides* [65 mm (2.6 in)], Neotropical, (4) *Neocastnia nicevillei* [80 mm (3.1 in)], Oriental, (5) *Castnia licus* [80 mm (3.1 in)], Neotropical. ***Superfamily Zygaenoidea,*** family Heterogynidae: (6) *Heterogynis penella* [22 mm (0.9 in)], Palaearctic; family Chrysopolomidae: (7) *Chrysopoloma isabellina* [40 mm (1.6 in)], Ethiopian, (8) *C. similis* [45 mm (1.8 in)], Ethiopian; family Megalopygidae: (9) *Megalopyge lanata*, with (9*a*) larva [50 mm (2 in)], Neotropical; family Cyclotornidae: (10) *Cyclotorna monocentra* [30 mm (1.2 in)], Australian; family Epipyropidae: (11) *Epipomponia nawi* [25 mm (1 in)], Palaearctic; family Zygaenidae: (12) *Pollanisus viridipulverulentus* [27 mm (1 in)], Australian, (13) *Arniocera erythropyga* [34 mm (1.3 in)], Ethiopian, (14) *Netrocera basalis* [30 mm (1.2 in)], Ethiopian, (15) *A. auriguttata* [30 mm (1.2 in)], Ethiopian, (16) *Eterusia repleta* [75 mm (3 in)], Oriental, (17) *Campylotes desgodinsi* [70 mm (2.8 in)], Palaearctic, Oriental, (18) *Himantopterus dohertyi* [25 mm (1 in)], Oriental, (19) *Ino ampelophaga* [23 mm (0.9 in)], Palaearctic, (20) *Zygaena occitanica* [27 mm (1 in)], Palaearctic, (21) *Agalope infausta* [21 mm (0.8 in)], Palaearctic, (22) *Zygaena* larva.

palps with just one segment, narrow, lanceolate hind wings, reduced venation, and radial veins separate from the median ones. The larvae feed on palms; *A. argaula*, from Fiji, lives on the coconut palm. The three members of *Agonoxena* are believed to be related to the Elachistidae.

Family Elachistidae (Plate 30). Beautiful microlepidopterans [about 10 mm (0.39 in)] that are particularly common in southern Europe and Africa. About 300 species have been described, of which 200 belong to *Elachista*, a genus with a worldwide distribution and larvae dependent on Poaceae and Cyperaceae.

Family Scythridae (Plate 30). A microlepidopteran group [12 to 20 mm (0.47 to 0.79 in)] with some 300 species in the genus *Scythris*, which is worldwide in distribution although the largest number of species occurs in southern Europe and western North America. The larvae weave webs or small silk galleries, particularly on Asteraceae, Fabaceae, and Ericaceae. *Areniscythris*, an unusual monotypic genus, with both sexes flightless, lives in sand dunes in central California.

Family Stathmopodidae (synonym Tinaegeriidae; Plate 30). Small moths [about 15 mm (0.59 in)] with fringed, narrow, and pointed hind wings with a slightly concave costal margin. The resting position of adults is highly characteristic. The body is held parallel to the support, and the first pair of limbs (with tibiae and tarsi densely covered with bristles) is raised and points upwards. In certain genera, such as *Snellenia* and *Pseudaegeria*, the coloration is very vivid, and the adults are diurnal. The largest and most widely distributed genus is *Stathmopoda*. The larvae of some of its members are miners; others prey on spider eggs and scale insects.

Family Oecophoridae (Plate 30). There are about 3,000 species, half of which are Australian, in this large cosmopolitan family. The adults usually have simple antennae with a basal pecten. The venation is almost complete, the hind wings having separate radial and median veins running almost parallel. The females may be brachypterous, and in *Macrochila rostrella* the second pair of wings may be absent altogether. The larvae mine leaves, flowers, galls, seeds, and rotting wood, and they may build protective cases. Their feeding habits are varied. Certain species, such as *Endrosis lacteella* and *Hofmannophila pseudospretella*, attack foodstuffs, biscuits, and even corks. The caterpillars of *Neossiosynoeca scatophaga* act as sweepers in the nests of certain Australian parrots. The larvae of *Depressaria*, a genus of about 260 species dependent on Asteraceae or Apiaceae, unlike those of other gelechioids, are initially external feeding but become miners before pupation. Oecophorids vary in size from 12 to 24 mm (0.35 to 0.94 in) on the average, but some species, such as the Australian *Wingia lambertella*, reach 35 mm (1.38 in) or more.

Family Ethmiidae. The family consists of beautiful microlepidopterans [20 to 30 mm (0.79 to 1.18 in)], with greenish, yellowish, or bronze-grey background wing coloration and large black spots, often with metallic highlights. The family contains many tropical taxa, but it is also a part of the European and palaearctic fauna. See the species illustrated in Plate 25, which were selected to show the principal morphological characters employed in the taxonomy of the Lepidoptera: *Ethmia dodecea* (Plate 25) shows how the second median vein, M_2, on the hind wing is closer to the first median than to the third. This fairly constant condition, together with other characters, such as the male genitalia, enables the Ethmiidae to be distinguished from the similar oecophorids. A group of early-spring, diurnal *Ethmia* is found in California, but most of the family is nocturnal or crepuscular.

Family Timyridae (synonym Lethicoceridae). A small group of moths, probably related to the oecophorids, this family is well represented in the Oriental (*Timyra*) and Australian (*Lecithocera* and *Crocanthes*) regions. Timyrids are medium-sized mircrolepidopterans [20 to 30 mm (0.39 to 1.18 in)] and are often strikingly coloured.

Family Blastobasidae (Plate 30). Like many gelechioids, blastobasids are inconspicuous grey insects with wings fringed with scales. The family is cosmopolitan, and the most interesting group is *Holcocera*, with about sixty species in North and South America. *Holcocera glandulella* is fairly common in the eastern United States, and its caterpillars live in acorns abandoned by curculionids. Although most blastobasids are seed eaters or saprophagous, there are species with predatory larvae, such as *H. pulverea* from India and the Australian *H. iceryaeella*, which feed on scale insects.

Family Xyloryctidae (Plate 30). A family of more than 1,200 species of predominantly Australian and Old World tropical distribution, it bears marked similarities to the Stenomidae and the Oecophoridae. The adults vary in size but average 20 to 30 mm (0.79 to 1.18 in) although they may be three or four times as large, as in *Cryptophasa*. They are normally pale in colour and nocturnal. However, there are notable exceptions, such as the African *Cyanocrates grandis* [60 mm (2.36 in)], whose conspicuous wing patterning may mimic the warning coloration of certain agaristids and are protected thereby. *Uzucha*, *Maroga*, and *Xylorycta* contain equally striking species. The caterpillars of many species carry the leaves on which they feed to the entrance of their shelter, whereas others (in the genus *Lichenaula*, for example) feed on lichens and build shelters by using their silk either to bind a variety of materials together or to join up several leaves. Certain species, such as *Procometis bisulcata*, dig underground tunnels, and many are miners. A wide variety of plants are attacked, from such diverse families as the Fabaceae, Moraceae, Myrtaceae (particularly eucalyptus), and Proteaceae.

Family Stenomidae (Plate 30). There are about 700 species in this family, most being neotropical. It is closely related to the Xyloryctidae, but the ranges of the two hardly

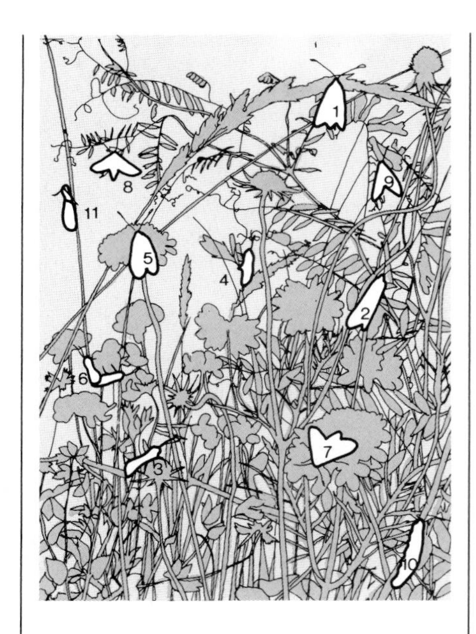

Plate 32 **Italian zygaenids,** family Zygaenidae

Zygaena contains some hundred species inhabiting the Palaearctic region. They are vividly coloured and are extremely unpalatable to birds and other predators. They are mainly tied to Fabaceae (subgenera *Zygaena* and *Agrumenia*) and Apiaceae (subgenus *Mesembrynus*).
Illustrated here are some of the twenty-eight species found in Italy: (1) *Zygaena rubicundus*, (2) *Z. erythrus*, (3) *Z. punctum*, (4) *Z. purpuralis*, (5) *Z. carniolica*, (6) *Z. loti* copulating, (7) *Z. oxytropis*, (8) *Z. romeo*, (9) *Z. lonicerae*, (10) *Z. filipendulae*, larva, (11) *Z. carniolica*, pupal exuvium.

Systematic Survey of Butterfly and Moth Families

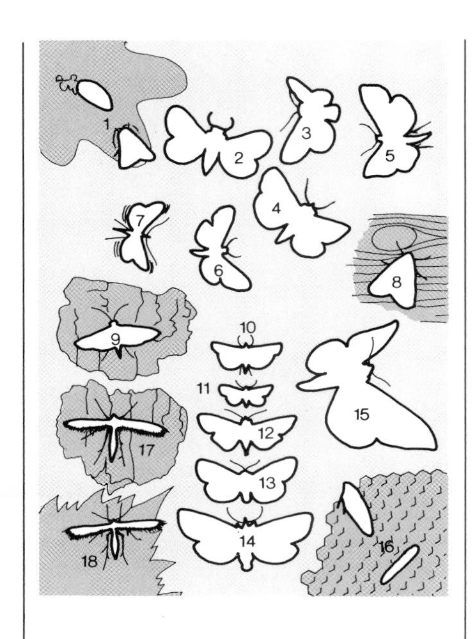

Plate 33 **Limacodidae, Pyraloidea, and Pterophoroidea**

Superfamily Zygaenoidea, family Limacodidae: (1) *Apoda limacodes*, adult and larva [20 to 30 mm (0.8 to 1.2 in)], Palaearctic, (2) *Chrysamma purpuripulcra* [32 mm (1.3 in)], Ethiopian, (3) *Coenobasis amoena* [30 mm (1.2 in)], Ethiopian.
Superfamily Pyraloidea, family Hyblaeidae: (4) *Hyblaea sanguinea* [40 mm (1.6 in)], Australian; family Thyrididae: (5) *Herdonia osacesalis* [45 mm (1.8 in)], Oriental, Australian, (6) *Rhodoneura zurisana* [33 mm (1.3 in)], Ethiopian; family Tineodidae: (7) *Tineodes adactylalis* [22 mm (0.9 in)], Australian; family Pyralidae: (8) meal moth (*Pyralis farinalis*) [24 mm (0.9 in)], Holarctic, Australian, (9) *Syngamia florella* [17 mm (0.7 in)], Nearctic, Neotropical, (10) *Hednota recurvella* [33 mm (1.3 in)], Australian, (11) *Hyalobathia miniosalis* [24 mm (0.9 in)], Australian, (12) *Margaronia agathalis* [46 mm (1.8 in)], Australian, (13) *Cardamyla carinentalis* [47 mm (1.9 in)], Australian, (14) *Hypsidia erythropsalis* [62 mm (2.4 in)], Australian, (15) *Siga liris* [75 mm (3 in)], Neotropical, (16) wax moth (*Galleria mellonella*), adult and larva [20 to 40 mm (0.8 to 1.6 in)], Holarctic, Ethiopian, Australian.
Superfamily Pterophoroidea, family Pterophoridae: (17) plume moth (*Platyptilia ochrodactyla*) [25 mm (1 in)], Palaearctic, (18) *Amblyptilia nemoralis* [20 mm (0.8 in)], Palaearctic.

overlap at all. Stenomids are generally cryptically coloured, and in some species, for example, *Antaeotricha griseana* [about 24 mm (0.94 in)], the adults resemble bird droppings. *Stenoma* (see illustration of *S. sequitiertia*) and *Antaeotricha* are large and heterogeneous genera, each with several hundred species. *Setiostoma* are oak-leaf binders in North America; they combine yellow, green, and dark iridescent colours.
Family Cosmopterigidae (Plate 30). A cosmopolitan family of about 1,200 species [10 to 12 mm (0.39 to 0.47 in) on the average], it is divided on the basis of the male genitalia into three subfamilies. The scape of the adult's antennae is long and thin, while the maxillary palps are short and the labial palps long and curved. The wings are extremely narrow; the venation on the hind one is greatly reduced. The uncus may be absent, and the male genitalia are sometimes asymmetrical. The larvae of *Cosmopterix* are miners, gall forming and ectophagous, or very occasionally predators of scales. The pupa has a cremaster. In the cosmopolitan genus *Cosmopterix* the adults are usually elegantly and brilliantly coloured, with black, orange, white, and ocher wing markings, as in the slender *C. scribaiella* [10 mm (0.39 in)]. In the past *Hyposmocoma*, a Hawaiian genus with several hundred species, and *Aphthonetus* were ranked together as a separate family, the Hyposmocomidae.
Family Gelechiidae (Plate 30). Gelechiids are small moths [less than 20 mm (0.79 in) on the average] with simple or ciliate antennae, as in *Thiotricha*, and sometimes with a basal pecten on the scape, as in *Sitotroga* or *Apatetris*. The maxillary palps, which have four segments, surround the base of the proboscis. The labial palps are curved, and the second segment bears tufts of scales. The forewings lack a pterostigma and the posterior cubital vein, CuP, and the fourth and fifth radial veins are fused. The pointed, trapezoidal, hind wings have a sinuate posterior margin. The radial veins are either fused with the first median, M_1, or are very close together at the base. The posterior cubital, CuP, is almost always absent. Coloration and patterning are fairly sober. One habit of the larvae is to sew several leaves together. They eat seeds and sometimes roots. They are only rarely miners, but when they are, the effects may be disastrous. The family, with almost 4,000 species and perhaps more than 400 genera, is cosmopolitan and displays many affinities with the oecophorids although in most cases they are easily distinguished by the shape and venation of the hing wing. In many groups of gelechiids the females have a specialized type of retinaculum, formed by a row of markedly curved bristles protruding from the proximal segment of the radial vein. Relatively few species attack crops, but those which do wreak havoc; potatoes, cotton, and cereals suffer the worst damage. One of the most notorious is the pink bollworm moth [*Platyedra* (synonym *Pectinophora*) *gossypiella*; 15 to 20 mm (0.59 to 0.79 in)], a cosmopolitan species that is extremely harmful to cotton bushes (*Gossypium*, Malvaceae). In addition to feeding on the flowers, the caterpillars bore into the capsules and destroy the seeds. *Sitotroga*, a genus that is particularly widespread in Europe and Asia, also contains the angoumois grain moth [*Sitrotroga cerealella*; 11 to 16 mm (0.43 to 0.63 in)], a species originally confined to North America but now cosmopolitan. Although it prefers maize, it will attack a wide variety of cereals in storage. The young larvae of the cosmopolitan potato tuber moth [*Phthorimaea* (synonym *Gnorimoschema*) *operculella*; 15 mm (0.59 in)] are miners, ravaging many Solanaceae, including the tomato, tobacco, and above all the potato, digging galleries even in the tubers. Harvested crops stored in warehouses may also be attacked since the females are able to identify the tubers of the food plant and lay their eggs on them. Under optimum conditions the potato tuber moth may produce five generations a year. Other important genera include the Australian *Protolechia*, with about 120 species, *Aristotelia*, with 300 species, and the type genus *Gelechia*, which is worldwide in distribution and contains about 500 species. *Chionodes* is associated with the foliage of woody plants in North America.
Family Metachandidae (Plate 30). A small family related to the Gelechiidae, it is found in tropical Africa and India. Its members are inconspicuous microlepidopterans whose habits are almost completely unknown.
Family Anomologidae. This family has been established to separate from other gelechiids two rare South African species of the genus *Anomologa*.
Family Pterolonchidae (Plate 30). A group of uncertain systematic position: The existence of one South African and one South American species of *Pterolonche*, a genus otherwise typically Mediterranean, poses a zoogeographical problem that has yet to be resolved. *Pterolonche pulverentella* is a Mediterranean species.

Superfamily Copromorphoidea

The principal characters of this superfamily are a bare proboscis, the absence of a pecten on the scape of the antenna, the tiny maxillary palps, and the presence in the larvae of only two prethoracic bristles. The wings are often divided into six or seven feathery arms. The larvae are miners, and the pupae resemble those of the Gelechoidea. There are three families: Copromorphidae, Alucitidae, and Carposinidae.
Family Copromorphidae. This family contains a small number of species from the Australian region and from South America. The adults [20 to 40 mm (0.79 to 1.57 in)] have hind wings with all the branches of the median, M, vein present; the proboscis is bare. The larvae of the Australian *Phycomorpha prasinochroa* are miners and live on fig trees (Moraceae).
Family Alucitidae (synonym Orneodidae; Plate 30). These are small, delicate moths that are unmistakable because the forewings are always divided into six feathery arms and

the hind wings into six or seven. Each arm is supported by at least one vein and is fringed at the edges. In some tropical taxa, however, the hind wings may be undivided. They are subdued in their coloration. The stout, hairy caterpillars mine flower buds and sometimes form galls. The Alucitidae are a small cosmopolitan family of about 100 small species. *Alucita* [10 to 40 mm (0.39 to 1.57 in)] is the main genus, which is also cosmopolitan and includes such widely distributed species as *A. hexadactyla* and *A. grammodactyla* from Europe and *A. montana* from North America. In *Alucita* the hind wing is always divided into six arms. Note that, although alucitids display a certain external resemblance to the pterophorids, another group of moths with feathery wings, the two families are not closely related.

Family Carposinidae (Plate 30). A small cosmopolitan family, it is concentrated in Australia and Hawaii. Carposinids differ from copromorphids in the absence of one or two branches of the median, M, vein on the hind wing, in the pointed forewing, and in the absence of ocelli. The larvae are miners, and the pupae spin a cocoon. *Carposina* is the principal genus. It is cosmopolitan and contains some seventy Hawaiian and Australian species. *Meridarchis*, with about forty species, and *Heterogymna*, with about twenty members, are Indo-Australian in distribution. *H. pardalota*, from India, is illustrated.

Plate 34 **Hesperioidea**

Superfamily Hesperioidea, family Hesperiidae: (1) *Carcharodus alceae*, adult and (1*a*) larva [26 mm (1.0 in)], Palaearctic, (2) dingy skipper (*Erynnis tages*) [27 mm (1.1 in)], Palaearctic, (3) grizzled skipper (*Pyrgus malvae*), adult and (3*a*) larva [25 mm (1 in)], Palaearctic, (4) orange skipper (*Ochlodes venatus*) [31 mm (1.2 in)], Palaearctic, (5) *Mimoniades versicolor* [50 mm (2 in)], Neotropical, (6) *Pyrrhochalcia iphis*, male above [62 mm (2.4 in)], Ethiopian, female below (72 mm (2.8 in)], (7) *Pyrrhopyge cometes* [40 mm (1.6 in)], Neotropical, (8) regent skipper (*Euschemon rafflesia*), adult and (8*a*) larva and pupa [50 mm (2 in)], Australian, (9) peacock owl (*Allora doleschalli*), in flight and (9*a*) at rest [42 mm (1.7 in)], Australian, (10) grass dart, (*Taractocéra ancilla*), adult and (10*a*) larva [35 mm (1.4 in)], Australian, (11) palm dart (*Cephrenes trichopepla*), adult and (11*a*) larva [41 mm (1.6 in)], Australian, (12) *Mysoria barcastus* [43 mm (1.7 in)], Neotropical, (13) *Urbanus simplicius* [33 mm (1.3 in)], Neotropical, (14) *Jemadia hospita* (54 mm (2.1 in)], Neotropical; family Megathymidae: (15) giant yucca skipper (*Megathymus yuccae*) [45 to 80 mm (1.8 to 3.1 in)], Nearctic, (16) Strecker's yucca skipper (*M. streckeri*) [62 mm (2.4 in)], Nearctic.

Superfamily Castnioidea

The superfamily Castnioidea is a group of moths related to the Zygaenoidea, differing in general appearance and in that they always have maxillary palps (two to four segments) and lack chaetosemata, the sensory bristles adjacent to the compound eyes. It consists of one family, the Castniidae.

Family Castniidae (Plate 31). These are essentially tropical, diurnal Lepidoptera, medium or large in size [30 to 135 mm (1.18 to 5.12 in)], which in some cases are similar in appearance to some butterflies (Papilionoidea). The wings are large and the venation is still primitive in that the median vein, M, persists in the discal cell and in addition the discal cell on the hind wing remains open. The elongated eggs have longitudinal ridges; the larvae that emerge on hatching are primitive in that they are phytophagous or, more rarely, burrowing, in which case they feed particularly on monocotyledonous roots. The pupae have dorsal abdominal spines and leave the larval shelter prior to emergence. Many species are mimics. The castniids are brightly coloured, and there are instances of sexual dimorphism. The genus *Synemon*, for example, shows sexual dimorphism in coloration, size, and the form of the abdomen, which has a long, thin ovipositor. The castniids display what is known as *flash coloration*: Very bright colours, sometimes with metallic highlights, are suddenly revealed when the moth is disturbed or when it is in flight, as in the brilliant blue *morpho* butterflies; the effect is to disconcert the approaching predator. The moths achieve this effect because their brightly coloured hind wings are hidden at rest, and the forewings are normally cryptic in their coloration and pattern. Hence, when the moth takes off, it becomes visible to the observer, whereas it disappears from view when it lands. Because of the unexpected way it appears and disappears, it is difficult for predators that rely upon sight, such as birds or indeed entomologists, to locate it.

The Castniidae are divided into two subfamilies: the Castniinae, with a well-developed proboscis, and the Tascininae, with an atrophied proboscis. The latter subfamily contains two rare species of *Tascina* from Borneo, the Philippines, and Singapore and *Neocastnia nicevillei*, a species that seems to be known from just a single specimen collected in Burma. The subfamily Castniinae contains just two genera, *Synemon* and *Castnia*, or several dozen — depending on how much weight is placed on the interspecific variability of *Castnia*. *Synemon*, a genus endemic to Australia, contains twenty-nine species, whose natural history is very little known. Most castniids live in Central and South America. The early classification of what is now referred to as *Castnia* included numerous taxa, among them *Zegara*, fifteen species, one of which, *Z. zagraeoides*, mimics certain heliconiines, *Gazera*, five species, whose members are moths with transparent wings that mimic ithomiines, and *C. licus, Cyanostola diva*, and *Riechia acraeoides*, which closely resemble acraeines of the genus *Actinote*, whose larvae are dependent on epiphytes, such as orchids and bromeliads. Most castniids live in forests, and many species form part of mimetic chains, *G. linus* (Plate 84) being one example.

Superfamily Zygaenoidea

This superfamily is highly diverse at every stage in its life cycle although a close examination of the seven families belonging to it shows that all retain such primitive characters as the persistence of the median vein, M, in the discal cell or the building of a cocoon that the pupa leaves prior to emergence. The proboscis is highly modified as are the maxillary palps, which are almost always atrophied. Tympanal organs are absent. The majority of zygaenoids are very brightly coloured, fly by day, and are often involved in mimetic associations with other butterflies and moths, beetles, wasps, and even true bugs. The caterpillars are often stout and bear verrucae. Those of the limacodids resemble slugs and display warning colorations. They are not miners but are external feeders, myrmecophils, or even, as in the Epipyropidae and the Cyclotornidae, external parasites of true bugs (Heteroptera). In certain cases they are even predators. The pupa

Systematic Survey of Butterfly and Moth Families

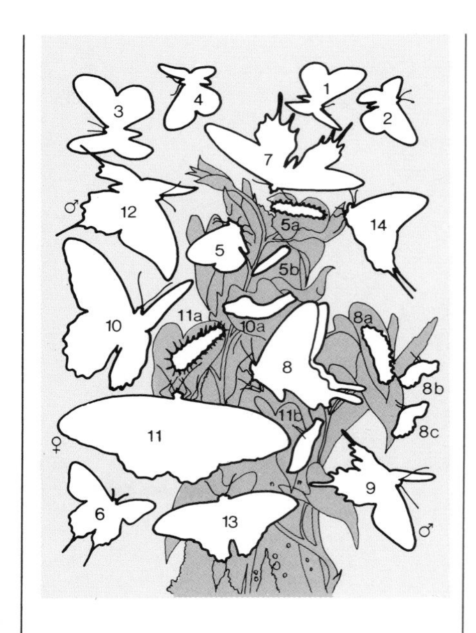

Plate 35 **Papilionids**

Superfamily Papilionoidea, family Papilionidae: (1) *Baronia brevicornis* [50 mm (2 in)], Neotropical, (2) *Hypermnestra helios* [40 mm (1.6 in)], Palaearctic, (3) *Parnassius tianschanicus* [60 mm (2.4 in)], Palaearctic, (4) false Apollo (*Archon apollinus*) [45 mm (1.8 in)], Palaearctic, (5) southern festoon (*Zerynthia polyxena*), adult, (5*a*) larva, and (5*b*) pupa [45 mm (1.8 in)], Palaearctic, (6) *Sericinus montela* [53 mm (2.1 in)], Palaearctic, (7) *Bhutanitis lidderdalei* [98 mm (3.9 in)], Palaearctic, Oriental, (8) *Atrophaneura alcinous*, adult, (8*a*) larva, (8*b*) spring-generation pupa, and (8*c*) summer-generation pupa [87 mm (3.4 in)], Oriental, (9) *Parides agavus*, male [46 mm (1.8 in)], Neotropical, (10) *Papilio polymnestor*, adult and (10*a*) larva [100 mm (3.9 in)], Oriental, (11) *Ornithoptera goliath*, female, (11*a*) larva, and (11*b*) pupa [160 mm (6.3 in)], Australian, (12) *Teinopalpus imperialis*, male [85 mm (3.3 in)], Oriental, (13) *Graphium gelon* [45 mm (1.8 in)], Australian, (14) zebra swallowtail (*Eurytides marcellus*) [58 mm (2.3 in)], Neotropical. Note that because of constraints of space the butterflies in this plate have been depicted with *Arisotolochia* although some of them, such as *Papilio polymnestor* and *Eurytides marcellus*, are not to be found on these plants in either larval or imaginal instars.

has free appendages, bears spines on its back, and lies within a rigid cocoon.

Family Heterogynidae (Plate 31). This family contains just a handful of Mediterranean species (found in southern Europe and North Africa). In certain respects these moths are intermediate between the psychids and the zygaenids, and they display marked morphological and ecological anomalies. As a rule in appearance the males resemble zygaenids of the genus *Procris*. Unlike the zygaenids, however, heterogynids have a proboscis, atrophied maxillary palps, and pectinate antennae. The two sexes differ greatly, the males being winged and having hairy legs and wings, whereas the females have atrophied mouthparts and eyes and are apterous, almost apodous, and smooth. They abandon the cocoon only to mate, subsequently returning to it to lay their eggs and die. After feeding on the secretions around the chorion, the young caterpillars eat the body of their mother. They then emerge from the cocoon and become free living and phytophagous, eating various Fabaceae, such as broom (*Genista, Sarothamnus*, and the like) and laburnum (*Cytisus*). The resistance of heterogynids and zygaenids to cyanide gas, which is often used to kill other moths, is the result of the same biochemical response, which suggests that there is a close relationship between the two families.

The Heterogynidae contain only the genus *Heterogynis*, some of whose members, such as the common *H. penella*, are diurnal.

Family Zygaenidae (Plates 31 and 32). This family differs from other zygaenoids in that it is the only one to have adults with a well-developed proboscis and chaetosemata; ocelli are normally present and are rather prominent in males. The antennae are somewhat expanded at the tips and are only bipectinate in males. The CuP vein is always present on both pairs of wings, but the hind wings are reduced in area and have three anal veins. Occasionally the wing may be unusual in shape (in *Himantopterus*, for example, the hind wings are highly tapered and elongate, like a tail). The stout, cylindrical caterpillar bears verrucae with short, tough bristles and is free living on herbaceous plants. The pupa is usually enclosed within an elongated parchmentlike cocoon attached to the stems or leaves of food plants. The diurnal adults are small or medium sized and are distinguished by their brilliant coloration. Zygaenids are well protected against birds by their unpleasant taste, which is due to the presence in their tissues of such toxic compounds as hydrocyanic acid, acetylcholine, and histamine. Such toxicity is typically associated with striking aposematic coloration, which warns any bird that has once experienced the taste. Many zygaenids choose to rest on flowers, sometimes in groups, and their flight is lazy and undulating. There are numerous cases of mimicry between zygaenids and butterflies and moths from such phylogenetically distant families as Pieridae, Nymphalidae, Arctiidae, Geometridae, and Ctenuchidae.

The Zygaenidae are a cosmopolitan family of about 800 species, mostly from the Old World and tropical Asia in particular. The Zygaeninae and the Chalcosiinae are the largest subfamilies and differ in the respective presence or absence of tibial spurs on the first pair of legs. The Zygaeninae are typical of the Palaearctic region, where about 100 species occur. The most important genus is *Zygaena*, which contains the common European forms, characterized by great intraspecific variability. There are numerous subspecies, and the number of named "forms" is absolutely vast. More than twenty species of *Zygaena* are present in Italy. *Zygaena filipendulae*, *Z. transalpina*, and *Z. carniolica* are three of the commonest, while *Z. ephialtes*, a polytypical, widely distributed European species, is somewhat rarer. Its populations may be monomorphic or polymorphic in their wing coloration and patterning, depending on the locality. The mimicry of *Z. ephialtes* is covered extensively in Plate 85. The members of *Procris* (synonym *Adscita*), a typically palaearctic group, represented by such species as *P. manni* and *P. geryon*, have metallic green wings with bluish or bronze highlights, or they are sometimes grey-brown. The intraspecific variability in size and shading is so great that their classification has to be based on more constant characters, such as the genitalia of the males. The members of *Pollanisius* and *Neoprocris* (Australia) and of *Astyloneura* (Ethiopian) are very similar to *Procris*.

The Chalcosiinae, typical of the Old World, are much more numerous. These moths have large wings [the wingspan of *Erasmia sanguiflua* reaches 80 to 100 mm (3.1 to 3.9 in)], and the hind wings often have tails (*Elcysma*, *Histia*, *Gymnautocera*) or are translucent (*Agalope*). The abdomen is slender, and the females have an ovipositor. They are brilliantly coloured and differ from other zygaenids. Mimicry is common and almost every genus contains one or more mimetic species. In the Oriental region, for example, the genus *Euploea* (Nymphalidae) is one of the models most commonly mimicked by certain *Pompelon*, *Cyclosia*, and *Erasmia*. *Campylotes* and *Chalcosia* have fewer species than does *Cyclosia* and are characterized by high-altitude species widely distributed from China to Tibet, Japan, and Vietnam. The other subfamilies are the Pompostolinae from tropical Africa (*Arniocera* and *Lamprochrisa*, for example), the Phaudinae from the Palaearctic (for example, *Pseudopsiche*), Oriental (*Phauda*), and Ethiopian (*Anomoetes*) regions, the Himantopterinae from Vietnam with the genus *Himantopterus*, which is characterized by species such as the well-known *H. fuscinervis* [30 mm (1.18 in)] from Java and Sumatra, all having slender tails and termite-eating larvae. The American zygaenids — a few species belonging to genera such as *Pyromorpha*, *Triprocris*, and *Sereda* — are often placed in a separate family, the Pyromorphidae.

Family Chrysopolomidae. A small African family with larvae similar to those of the Limacodidae, its adults are rather small [20 to 50 mm (0.79 to 1.97 in)] with no

chaetosemata or frenulum. Little work has been done on the biology of this family.

Family Megalopygidae (synonym Lagoidae; Plate 31). The number of species and the geographical distribution of this family depend on whether the North African genus *Somabrachys* is included or not. Contrary to the classification adopted here, some authors place *Somabrachys* in a separate family, the Somabrachydae. The Megalopygidae contain more than 250 species, the large majority of which are American, particularly neotropical, forms. The principal characteristics of the adults are an atrophied proboscis, short labial palps, the presence of chaetosemata, and a sturdy body densely covered with hairs. They are nocturnal. Their restrained coloration and curly, woolly appearance (see *Megalopyge lanata*, for example) enable them to be easily distinguished from other zygaenoids. The caterpillars are unsual: They are more hairy than the adults and have two extra pairs of prolegs; they withdraw their head into the first segment and in some ways resemble limacodid larvae. Beneath their covering of hair they have numerous, tough, glandular bristles with a sting that human beings can feel. A great expert on the Lepidoptera, A. Seitz, recounts how an Amazonian Indian suffered an inflamed hand, arm, and half a chest after being painfully stung by *M. orsilochus*; the Indian also suffered a feverish reaction for fifteen days. The larva weaves an oval cocoon with a circular opening and an operculum. Then, like limacodids, it enters a long diapause (prepupal stage). Only a month before emerging the prepupa becomes a pupa. In certain American species the cocoon is constructed with great care. Even after emergence the operculum, which is linked to the cocoon by a sort of hinge (as in the Limacodidae), may be made to fit the cocoon perfectly so that it appears intact. The hairy caterpillars of *Somabrachys* have an unusual feature that enables them to carry out an energetic toilet: It consists of a kind of rake with four teeth, usually concealed beneath the anal segment, which comes into play during defecation and prevents the excreta from remaining around the edge of the anus. Megalopygid larvae do not appear to be harmful. They are gregarious during the first half of their lives but solitary later. Their food plants include the Fabaceae, Rosaceae, Asteraceae, and Apiaceae.

Family Cyclotornidae (Plate 31). In this family of five endemic Australian species the adults are small, lack a proboscis and maxillary palps, and have stout bodies and sober coloration. The biology of its larvae is quite exceptional. The females of *Cyclotorna monocentra* (which will be used as an example) deposit their eggs on the bark of trees infested with nymphal leafhoppers (a group of Homoptera capable of producing sugary solutions that ants are very fond of). The eggs hatch to produce barrel-like larvae that parasitize the leafhoppers for a time and then spin a flattened case in which they shelter and moult. The next stage is that of a brilliantly coloured larva, which is flattened along its dorsoventral axis with a retractile head and antennae. At this point the larva is carried off by ants into their nest, where they feed off its anal secretions. The caterpillar in its turn lives on the ants, preying on their larvae until at maturity it leaves the ant nest and spins a cocoon.

Family Epipyropidae (Plate 31). Like the previous family the larvae of this group have also developed some highly specialized and in certain respects aberrant habits. The caterpillars are ectoparasites and in some cases predators of leafhoppers and fulgorids. As adults epipyropids [10 to 20 mm (0.39 to 0.79 in)] are characterized by the marked reduction or loss of anatomical structures (for example, proboscis, maxillary palps, or ocelli). The eggs are deposited individually or in small groups on the food plants of the host homopterans; they hatch to produce larvae with long thoracic limbs but no prolegs (these appear after the moult). The larvae then wait for a host to pass by, attach themselves to its back, and feed on the waxy and sugary secretions that it produces. Once mature, they abandon their hosts and prepare to pupate. There are numerous variations on the model described. Sometimes the larva searches actively for a host; in other instances the host is killed although it is not clear whether actual predation or parasitism takes place, or sometimes the larva merely feeds on the waxy secretions of the host without harming it. The Epipyropidae are a cosmopolitan group represented by a few dozen species, including *Epipomponia nawi* from Japan.

Family Limacodidae (synonyms Eucleidae, Heterogeneidae, Cochlidiidae; Plate 33). The Limacodidae are a group of nocturnal moths with a worldwide distribution. The adults are small or medium in size and are characterized by a reduction in their anatomical structures (the mouthparts, for example) and in the sense organs such as the ocelli and chaetosemata. The wings are rounded, particularly in the males, and are short and scaly. They have a frenulum and a retinaculum, and their venation is quite similar to that of certain microlepidopterans. They may be brightly coloured (greens, yellows, reddish browns), but often the colours are more sober, with silvery or golden stripes and spots forming primitive types of patterning. They are hairy, with tufts of hair on the head and on the limbs as well as on the wings and the short abdomens. The family takes its name from the sluglike appearance of the caterpillars, which are highly unusual and diverse. In some genera the larvae are oval, rectangular, or irregular in shape. The body is fleshy, sometimes gelatinous, and vividly coloured with stripes and spots of different colours. It is expanded ventrally with a convex back. In most species the exoskeleton bears verrucae, bristles, spines, fringed excrescences, or integuments covered with papillae or tubercles. In other cases the caterpillars are smooth and covered in a wax layer that enhances their coloration. Certain *Euclea* and *Prolimacodes* species have cryptic larvae that resemble the galls of food plants. In other genera there is a marked reduction in

Plate 36 **Pieridae**

Family Pieridae: (1) *Colotis zoe*, male [41 mm (1.6 in)], Ethiopian, (2) broad-margined grass yellow (*Eurema candida* × *anthomelaena*), male and (2*a*) larva [44 mm (1.7 in)], Australian, (3) brimstone (*Gonepteryx rhamni*), male, (3*a*) larva, and (3*b*) pupa [61 mm (2.4 in)], Palaearctic, (4) clouded yellow (*Colias crocea*), adult and (4*a*) larva [45 mm (1.8 in)], Palaearctic, (5) northern Jezabel (*Delias argenthona argenthona*), female and (5*a*) larva [67 mm (2.6 in)], Australian, (6) Australian wood white (*D. aganippe*), male and (6*a*) pupa [60 mm (2.4 in)], Australian, (7) caper white (*Anaphaeis java teutonia*), female, (7*a*) larva, and (7*b*) pupa [58 mm (2.3 in)], Australian, (8) orange tip (*Anthocharis cardamines*), male, (8*a*) larva, and (8*b*) pupa [40 mm (1.6 in)], Palaearctic, (9) *Pereute telthusa*, male [54 mm (2.1 in)], Neotropical, (10) wood white (*Leptidea sinapis*) [40 mm (1.6 in)], Palaearctic, (11) *Dismorphia nemesis*, male [51 mm (2 in)], Neotropical, (12) *D. orise* [74 mm (2.9 in)], Neotropical, (13) mothlike white (*Pseudopontia paradoxa*) [35 mm (1.4 in)], Ethiopian.

Plate 37 **Ithomiinae**

Family Nymphalidae, subfamily Ithomiinae: This is a neotropical subfamily with the exception of the Indo-Australian genus *Tellervo* (Plate 38): (1) *Hypoleria libethris* [36 mm (1.4 in)], (2) sweel oil (*Mechanitis polymmia polymmia*), male, female laying, (2*a*) larva, and (2*b*) pupa [60 mm (2.4 in)], (3) *Elzunia cassandrina* [67 mm (2.6 in)], (4) the tiger (*Tithorea harmonia pseudethra*) [66 mm (2.6 in)], (5) *Thyridia psidii* [67 mm (2.6 in)], (6) *Melinaea comma simulator* [68 mm (2.7 in)], (7) *Hyalyris avinoffi* [56 mm (2.2 in)], (8) green sweet oil (*Aeria eurimedia pacifica*) [45 mm (1.8 in)], (9) *Ithomia iphianassa* [43 mm (1.7 in)], (10) blue transparent (*Pteronymia* sp.) [32 mm (1.3 in)], (11) pair of Ithomiinae copulating, (12) *Hypothyris lycaste dionaea* [62 mm (2.4 in)], (13) *Mechanitis mantineus* [50 mm (2 in)].

segmentation often in conjunction with such marked anteroposterior symmetry as to make it difficult to tell the head from the tail. The larvae are also sluglike in their movements. The thoracic legs are rudimentary, and the caterpillars slide on their undersides, using prolegs that have been transformed into adhesive suckers. Like the Megalopygidae the larvae of the Limacodidae are highly urticating. *Doratifera vulnerans* from Australia and the American species *Sibirne stimulae* reflect this method of defense in their names; the larvae of the American genus *Phobetron* have tufts of serrated spines that can inflict painful stings or cause violent allergic reactions. Poison and aposematic coloration serve to protect the larvae from predators by sight but are no defense against parasitic Hymenoptera, which identify their hosts by scent. When they reach maturity, these larvae do not enter the pupal stage directly but like the Megalopygidae undergo an intermediate prepupal stage. The larvae are skillful at weaving a solid cocoon, rather like a nut, with a more or less smooth exterior and an operulum. As in certain *Phobetron* it is sometimes covered and protected by the larval skin. The limacodids, which include some thousand species in all, are concentrated in the tropics, with about forty North American species and as many in Eurasia. Africa has a great many native groups, including the genera *Chrysamma* and *Coenobasis*, illustrated here in the species *Chrysamma purpuripulcra* and *Coenobasis amoena*. Oriental species are also numerous, belonging to genera such as *Thosea*, *Parasa*, *Susica*, or *Narosa*. Australia has some eighty species.

Superfamily Pyraloidea

The Pyraloidea are an enormous group of small- to medium-sized Lepidoptera; they are fragile in appearance with rather long legs and conspicuous tibial spurs. The maxillary palps have between two and four segments, while the labial palps are well developed and point forwards or upwards. The median veins in the wings lie outside the discal cell. One important distinctive character occurs in the hind wing, where the group $Sc + R_1$ (the subcostal and first radial veins) is close to, or frequently anastomosed with, the radials. Tympanal organs are found only in the pyralids. The eggs are flattened. The larvae are slender and usually colourless, with two short prolegs with hooks; they are highly active and inhabit a fairly diverse range of habitats. They are usually external feeders, and some species are extremely destructive. The pupa is surrounded by a cocoon and does not protrude from it when the imago emerges; it has maxillary palps, and the fifth to seventh abdominal segments are free. The superfamily Pyraloidea is not homogeneous. I. F. B. Common recognizes five families, but other authors exclude the hyblaeids and include the alucitids and pterophorids.

Family Hyblaeidae (Plate 33). The hyblaeids occupy an uncertain position in the system of classification. In the past they were included among the Noctuoidea, but other systems, based on characters in the primitive state, relate them rather to the Sesiidae. The adults are of medium size; the head is very small and has ocelli but no chaetosemata. The antennae are simple, and the proboscis is sturdy and bare. The labial palps are prominent and form a beak, and the maxillary palps are also well developed. The wings are rectangular; the abdomen is short and strongly pointed. In overall appearance these moths resemble certain noctuids, even in coloration. The larva is of the pyraloid type and builds silken tunnels among the leaves; the pupa has a cremaster. The few species of Hyblaeidae belong to the genus *Hyblaea* and are concentrated in tropical regions of southeast Asia although a few species are present in tropical Africa and Australia. The larvae of *H. puera*, a species close to *H. sanguinea*, from Fiji, eat Verbenaceae and in east Asia attack the leaves of teak (*Tectona grandis*).

Family Thyrididae (Plate 33). This mainly tropical family consists of small- to medium-sized [12 to 66 cm (0.5 to 2.6 in)] forest moths with wing patterning resembling foliage. The phylogenetic links between the thyridids and the other families are problematic. In the past it was thought that the ancestors of modern butterflies were descended from the thyridids, which were therefore classed among the skippers (Hesperioidea). The adults resemble pyralids or some geometers; some species have regularly indented wing borders, with cryptic yellowish or reddish-brown colouring, transparent patches, and a reticular pattern. The proboscis when present is bare, chaetosemata are absent, and the maxillary palps are tiny. The hind wings have two anal veins. On the whole very little is known about the larvae, but they are of the pyraloid type, usually internal feeding and sometimes gall forming, as in the Ethiopian *Cecidothyris*. Some species are active by day, others by night. Many species settle on the underside of leaves and, like the geometers, fly off when disturbed and settle rapidly a short distance away. Other groups, such as the neotropical *Draconia rusina* and *Belenoptera*, are cryptic and are virtually indistinguishable from dead leaves. Four subfamilies have been described. Genera are fairly numerous and include the oriental *Herdonia*, illustrated here by *H. osacesalis*, and the pantropical *Banisia* and *Rhodoneura*; the last has more than 100 species, including *R. zurisana*, native to Madagascar, and the holarctic *Thyris*. The Thyrididae are more strongly represented in the tropics than they are in temperate zones. There are some 600 species, of which just over twenty are palaearctic. It has been observed that naturally occurring populations of thyridids are rather small. There are many rare species and relatively few specimens in collections.

Family Tineodidae (Plate 33). This family is closely related to the Pyralidae and consists of a few species found from India to New Guinea and Australia. One of the ten

Plate 38 **Danainae**

Family Nymphalidae, subfamily Ithomiinae: (1) Cairn's hamadryad (*Tellervo zoilus*), male [43 mm (1.7 in)], Australian.
Subfamily Danainae: (2) larva and chrysalis of the plain tiger (*Danaus chrysippus*), Ethiopian, (3) *D. hamata* [80 mm (3.1 in)], Palaearctic, (4) chrysalis of the black and white tiger (*D. affinis*), Australian, (5) *Idea durvillei*, male [115 mm (4.5 in)], Australian, (6) large tiger (*Lycorea ceres*) [75 mm (3 in)], Neotropicál, (7) oleander butterfly (*Euploea core corinna*), adult and (7*a*) pupa [75 mm (3 in)], Australian, (8) Eichhorn's crow (*E. eichhorni*), adult, (8*a*) larva, and (8*b*) pupa [63 mm (2.5 in)], Australian, (9) *E. nemertes swierstrae*, female above [62 mm (2.4 in)], male below [57 mm (2.2 in)], Australian, (10) beautiful monarch (*D. formosa*) [72 mm (2.8 in)], Ethiopian, (11) *Amauris vashti*, male [90 mm (3.5 in)], Ethiopian, (12) Friar (*A. niavius*) [80 mm (3.1 in)], Ethiopian, (13) *Ideopsis gaura* [73 mm (2.9 in)], Oriental.

Australian species, *Tineodes adactylalis*, is illustrated, showing the elongation of the antennae, labial palps, and legs characteristic of these moths.

Family Oxychirotidae. The Oxychirotidae are Indo-Australian microlepidoptera. There are probably no more than five species and these are quite rare. The wings of Oxychiotidae are very narrow with reduced venation and are either subdivided into two lobes, as in the Australian genus *Cenoloba*, or else entire, as in *Oxychirota*.

Family Pyralidae (Plate 33). With some 15,000 known species this is the third largest family of Lepidoptera. The pyralids display great morphological variation and are found in the most diverse habitats throughout the world. The greatest number of species is concentrated in hot zones. The adults are slender and of medium size, varying from 15 mm (0.6 in) to 30 mm (1.2 in); they are usually soberly coloured, but there are many exceptions in which the geometric patterns and vibrant colours of the wings are extremely striking. Most species fly at night or at dusk, only a few being exclusively diurnal. The adults may have ocelli or chaetosemata, and the antennae are morphologically variable. The base of the proboscis is usually covered with scales. The maxillary palps have four segments; the labial palps are always present and sometimes well developed and vary in form. The legs are long, and the tibiae bear clearly visible spurs. The forewings are often elongated and triangular with the third and fourth radial veins partly fused; venation is usually absent in the hind wing. It is always found in the forewing, where the group $Sc + R_1$ is typically either close to the radial veins or partially fused with them proximally to the discal cell; there are three anal veins. All pyralids have an abdominal organ with which they can perceive the ultrasounds emitted by the bats that prey on them. The caterpillars are glabrous and have prolegs bearing uni- or multiserial circles of crochets. They spin webs, build cases, and excavate galleries in roots, stalks, branches, seeds, and fruit. The larvae of some species are gall forming, whereas others are dependent on mosses; aquatic larvae and those which are destructive to agriculture and food stocks are not uncommon. The pupa is usually protected by a cocoon and lies in the larval shelter. The pyralids can be divided into ten subfamilies, which differ in appearance, habitat, and behaviour.

The Schoenobiinae are a small group of whitish-yellow moths with rather narrow wings, a rudimentary proboscis, and a tuft of anal hairs as in *Scirpophaga*. The adults are found among aquatic plants, resting on rushes and sedges, on or in which the larvae feed. The latter are well adapted to freshwater environments, where they spend all or most of their life. The most important feature of this adaptation is the ability to breathe under water, which the pyralids do in different ways. *Scirpophaga* is a typically oriental genus although it is also found in the Palaearctic and Indo-Australian regions. The larvae mostly mine Cyperaceae, such as rushes, and the Poaceae, such as rice and wheat. The adult Schoenobiinae are dimorphic, the females being larger with more pointed wings. The European *Acentropusis* are particularly interesting since, uniquely among the Lepidoptera, the adult has managed to adapt to life under water. The best known species is *A. niveus*, the larvae of which practice cutaneous breathing.

The Nymphulinae (synonym Hydrocampinae) are also tied to inland waters and have the most varied larval habitats. There are species with phytophagous or saprophagous larvae that live on dry land and species with underwater leaf- and stem-mining larvae. The connection with water may be more or less close. In some cases the caterpillars need oxygen from the air and have to come to the surface to obtain supplies: in other species, like the common palaearctic *Nymphula nympheata* [22 to 30 mm (0.9 to 1.2 in)], only the neonate larvae display cutaneous respiration. The larvae of this species live in small flattish cases constructed from pieces of *Potamogeton*. Other members of the Nymphulinae, like *Paraponyx stratiotata* [20 to 25 mm (0.8 to 1 in)], a European species of holarctic genus, have larvae that spend the entire instar underwater and are equipped with gills. Sometimes the pupa is also aquatic.

The small subfamily of Scopariinae contains moths with a small tuft of scales in the discal zone of the forewing. Very often the larvae feed on mosses and lichens. A large number of species is found in temperate latitudes, but there are also tropical species like those of the large genus *Eudonia*, which contains some hundred species in Hawaii alone. The related genus *Scoparia* has a huge geographical distribution.

The Crambinae form a very numerous group and can be distinguished from the other pyralids by their elongated wings and extremely prominent labial palps. The bright striped patterning of the Australian *Hednota recurvella* is typical of this subfamily. Crambinae are common in grasslands and bogs. Many tropical species are destructive to cereals: *Chilo suppressalis* attacks rice, and *C. zonellus* maize, while the larvae of many *Crambus*, a cosmopolitan genus with more than 400 species, are capable of destroying large expanses of grass, such as pastures or golf links. The neotropical *Diatraea saccharalis* is extremely destructive to sugar-cane plantations. Crambinae rest in a characteristic manner, with the wings tightly folded around the body; they can be mistaken for twigs if appropriately coloured. The subfamily Pyraustinae consists of nearly 3,500 species of moths that are common everywhere but particularly so in the tropics. One example is *Margaronia*, a highly heterogeneous group, with adults that have pearly wings, as in *M. unionalis*, dependent on the Oleaceae. The illustration depicts the Australian *M. agathalis*. The Pyraustinae rest with their wings outstretched, hanging on the underside of leaves. The best known species in this subfamily is the

European corn borer, *Ostrinia (=Pyrausta) nubilalis* [around 30 mm (1.2 in)], which was introduced into North America some seventy years ago. It is migratory and is now considered to be cosmopolitan. The larvae are polyphagous and attack corn, hemp, tomatoes, and fruit trees. The damage to corn can actually reach catastrophic proportions in some years. It has been calculated that in 1969, for example, the populations of this moth in the United States caused damage of nearly $190 million. Unfortunately, despite the intensive research that has been carried out on the biology of this moth, no means of controlling its exploding populations has been discovered.

It is interesting to note that the species of the small cosmopolitan subfamily Galleriinae frequently have a close relationship with other animals. Thus the wax moths, *Galleria mellonella* and *Achroia grisella*, are feared by beekeepers because their larvae feed off the wax in honeycombs and can damage a whole hive. The larvae of *Tirathaba parasitica* feed on hepialid larvae, while the adults of *Bradypolicola* and *Cryptoses* live in the coats of sloths.

The subfamily Pyralinae contains a number of very large striking moths, like the Australian *Hypsidia erythropsalis*, but mostly they are modest in appearance, like the well-known meal moth, *Pyralis farinalis*. The larvae of this cosmopolitan species eat not only stored cereals but also flour, which they soil with their excreta. Related to the Epipaschiinae is the subfamily Chrysauginae, which contains some ant-loving species such as the Javan *Wurthia myrmecophila* and *W. aurivillii*, the larvae of which raid the nests of two different species of host ants.

The final subfamily is that of the Phycitinae, a group containing many tens of thousands of species, more than 800 of which are palaearctic; the adults characteristically have elongated forewings lacking the fifth radial vein. Coloration like that of *Siga liris* from South America is relatively rare among the Phycitinae, most of which have transverse bands on their forewings; typical examples are *Plodia interpunctella*, the Indian meal moth, and *Ephestia kuehnellia*, the Mediterranean flour moth. Both species are cosmopolitan and dependent on the human habitat; they are very common in houses, food stores, bakeries, and even ships' holds since the larvae eat all kinds of foodstuffs. The larvae of *Ephestia*, a cosmopolitan genus with many very similar species, such as *E. cautella*, *E. figulilella*, or *E. elutella*, seem to have a special preference for flour although they are highly polyphagous and will eat chocolate, dried fruit, tobacco, and even paper and polyvinyl packing material. *Ephestia kuehnellia* is easy to rear and is to be found in many zoology and genetics laboratories. Most of the Phyticinae have larvae that live in silken tubes that they leave only by night to feed; in addition to phytophagous larvae some are predatory. The latter include those of the North American *Laetilia coccidivora*, which live on scale insects (Hemiptera, Homoptera), and some Ethiopian *Metoecis*, which prey on the larvae and pupae of Lymantriids and Notodontids. This subfamily also contains one of the most famous pyralids, *Cactoblastis cactorum* from South America.

Superfamily Pterophoroidea

This superfamily contains the single family Pterophoridae, which resembles the pyralids in some respects and has consequently been classified by some authors among the Pyraloidea.

Family Pterophoridae (Plate 33). This is a family of small, delicate, moths, easily distinguishable by their extremely narrow wings, which are deeply incised to form several lobes edged with fringes of hair. The forewings are normally divided into two, the hind wings always in three, with the exception of the species of *Agdistis* and related genera, which have entire wings. The adult is very soberly coloured, without ocelli or chaetosemata; it has a bare, well-developed proboscis and tiny maxillary palps. The legs are slender, long, and spurred except in *Agdistis* and *Pterophorus*. The larvae are short, cylindrical, and rather hairy and have elongated ventral prolegs with a semicircular band of crochets. They are usually external feeders on flowers and the leaves of shrubs and grasses, showing a preference for Asteraceae. The pupa is shaggy, lacks maxillary palps, and is suspended by the cremaster or else lies on the ground, protected by a fine cocoon. Pterophorids are insubstantial moths, whose flight resembles that of mosquitoes, and they are easily carried by the wind. The adults rest with their wings folded fanwise at right angles to the body, forming a T. They are typically nocturnal, and some species like the holarctic *Emmelina monodactyla* are attracted to light, but many others come out immediately after dusk. Subgroups can be distinguished on the basis of wing characters as follows: Agdistinae (wings entire), Platyptiliinae (wings fissured with one anal vein on the hind wing), and Pterophorinae (fissured wings with two anal veins). There are more than 600 species in the family, which seems to have developed along two evolutionary lines, giving rise to entire-winged species and fissure-winged species. Despite their fragility, pterophorids are found everywhere. The main cosmopolitan genera are *Platyptilia*, illustrated here by *P. ochrodactyla*, and *Trichoptilus*, containing the North American *T. parvulus*, the larvae of which feed on insectivorous sundews (*Drosera*). Another interesting platyptiliine species is *Lantanophaga pusillidactyla*, originally from Central America but now pantropical. The larvae are linked with *Lantana* (Verbenaceae), and the species is used to control this weed. Mention may also be made of the *Convolvulus*-eating *Pterophorus pentadactylus* and *Amblyptilia punctidactyla*, which feed on Labiatae.

Plate 39 **Satyrinae**

Family Nymphalidae, subfamily Satyrinae: (1) eyed brown (*Satyrodes eurydice*) [47 mm (1.9 in)], Nearctic, (2) Scotch argus (*Erebia aethiops*), adult and (2*a*) larva [45 mm (1.6 in)], Palaearctic, (3) *Cithaerias aurorina* [51 mm (2 in)], Neotropical, (4) Kershaw's brown (*Oreixenica kershawi*) [32 mm (1.3 in)], Australian, (5) palmfly (*Elymnias agondas melantho*), female above, male below [82 mm (3.2 in)], Australian, (6) *Aphysoneura pigmentaria* [40 mm 1.6 in)], Ethiopian, (7) evening brown (*Melanitis leda*) [68 mm (2.7 in)], Ethiopian, (8) lady skipper (*Pierella lena*) [65 mm (2.6 in)], Neotropical, (9) ringlet (*Aphantopus hyperantus*) [43 mm (1.7 in)], Palaearctic, (10) *Argyrophorus argenteus* [40 mm (1.6 in)], Neotropical, (11) meadow brown (*Maniola jurtina*), adult and (11*a*) larva [46 mm (1.6 in)], Palaearctic, (12) *Enodia andromacha* [43 mm (1.7 in)], Neotropical, (13) Russian heath (*Coenonympha leander*) [32 mm (1.3 in)], Palaearctic.

Systematic Survey of Butterfly and Moth Families

Plate 40 **Brassolinae**

Family Nymphalidae, subfamily Brassolinae: This consists of some eighty large, neotropical species that fly at dusk, preferably in clearings and beside woodland watercourses: (1) *Caligo uranus* [100 mm (3.9 in)], (2) *C. placidianus* [112 mm (4.4 in)], (3) *Dasyophthalma rusina* [75 mm (3 in)], (4) *D. creusa* [75 mm (3 in)], (5) *Narope sarastro* [57 mm (2.2 in)].

Superfamily Hesperioidea

This is one of the two superfamilies made up of the butterflies; its members, the skippers, are sturdy and small to medium sized with a rapid, fluttering flight. Ocelli and chaetosemata are absent, the scapes of the antennae are set far apart and bear a tuft of scales, and the flagellum is clavate and often narrowed to appear uncinate. The wings lack frenulum and retinaculum except in the males of the Australian species *Euschemon rafflesia*, which is thought to be the most archaic butterfly, and the posterior cubital vein CuP, is also absent. The larvae are glabrous and look insignificant. The pupae have a cremaster. There are two families, Hesperiidae and Megathymidae. Some systems include the megathymids among the hesperiids and place the latter in the superfamily Papilionoidea.

Family Hesperiidae (Plate 34). The true skippers are sturdy Lepidoptera of medium size, between 20 mm (0.8 in) and 80 mm (3.1 in) (*Pyrrhochalcia iphis*), with the first pair of legs bearing well-developed epiphyses, short wings often with patches of androconia in the males, broad heads, and well-separated antennae and eyes. They also have tufts of hairs around the base of the antennae, the distal portion of the club is usually uncinate, and certain species have scaleless patches on the wings. Flight is very rapid and unpredictable. Skippers are rare at high elevations and altitudes. They are not greatly sought after by collectors because the wing muscles are extremely strong, and since they fold the wings beneath the body after death, they are difficult to mount. At rest they hold their wings closed in the manner of the true butterflies although they sometimes hold them like moths. The eggs are hemispherical and finely sculpted. The larvae have large heads and are hairless to the naked eye. They live protected by leaves, which they fold up with silken threads; pupation takes place in a cocoon made of silk and bits of leaf. In some species the larva and pupa are covered with a dusty material. Their food plants consist of a huge range of angiosperms. The subfamilies Trapezitinae and Hesperiinae feed on monocotyledons with the exception of a single Asian species; all others eat dicotyledons. Some 3,000 species are known, divided into five subfamilies. The Coeliadinae are palaeotropical and comprise some seventy species, twenty of which are African, including *Phyrrhochalcia iphis* — possibly the largest hesperiid in the world — which flies very slowly and resembles an agaristid. The Pyrrhopyginae consist of some hundred neotropical species, including *Pyrrhopyge cometes*, *Mimoniades versicolor*, *Mysoria barcastus*, and *Jemadia hospita*. The Pyrginae are cosmopolitan with a few hundred species. European species include *Pyrgus malvae*, *Erynnis tages*, and *Carcharodus aldeae*. The Australian *Euschemon rafflesia* is unusual in that the male has a frenulum. America is the home of many long-tailed skippers, including the widespread *Urbanus proteus*. The Trapezitinae consist of some sixty exclusively Australian species, the larvae of which feed on monocotyledons. The Hesperiinae are a group of more than 200 cosmopolitan species, feeding on grasses, sedges, and palms. *Atalopedes campestris* is found in the most varied environments in North America.

Family Megathymidae (Plate 34). This is a small family that, as mentioned above, some authors consider a subfamily of Hesperiidae. There are some fifty yucca skippers, mostly inhabiting arid habitats in the southwestern United States and northern Mexico. The larvae live first on the leaves, then in the roots, and subsequently at the base of the leaves of such Liliaceae and Agavaceae as *Yucca*, *Agave*, and *Manfreda* (from which Mexican tequila is distilled; this is not considered genuine unless there is a caterpillar in the bottle). The egg is hemispherical and flattish with indented edges and is laid on *Agave* (*Agathymus*) or *Yucca* (*Megathymus*). The larvae are whitish or coloured as in *Aegiale* and resemble the caterpillars of some moths. They pupate in holes mined in the thickness of the leaves, lining these with silk and closing them with a door at the top until the time comes for them to emerge. The adults are on the average lager than most true skippers, with a slight resemblance to a castniid, a family to which the megathymids were once held to be related. The head is narrower than the thorax, the antennae are clavate and bend slightly back, the palps are relatively small, and the second and third pairs of legs bear claws. All known species produce only a single generation: Those of the genus *Megathymus* usually appear in spring, and those of *Agathymus* usually in autumn.

Superfamily Papilionoidea

Together with the Hesperioidea this superfamily makes up the butterflies: diurnal butterflies often regarded as the most advanced and highly evolved members of all the Lepidoptera. The adults lack ocelli and have developed chaetosemata and slender antennae with marked clubs. The proboscis is hairless, and the maxillary palps consist of a single short segment or are entirely absent. The posterior tibiae are clawless. The wings lack posterior cubital veins and a frenulum; tympanal organs are absent. The eggs are usually elongated. The larvae are often highly coloured and ornamented; when newly hatched, they have prolegs with a full ring of crochets, but in the later stages this ring is incomplete, marking a higher stage of evolution. The pupae are obtect with a cremaster and sometimes succinct; their appearance varies greatly among the different groups. The origin of the Papilionoidea is still obscure; at one time it was thought that they were descended from an ancestor common to the present-day Thyrididae, but they are now considered by be related to the castniids. There are six families in the superfamily.

Family Papilionidae (Plate 35). The swallowtails are undoubtedly the best known and

Systematic Survey of Butterfly and Moth Families

most "popular" family of butterflies. The limited number of species, their large size, and their great variety of shape and colour makes them one of the most useful animal groups in terms of differentiating zoogeographical regions and biomes. On examining a small collection of swallowtails or indeed even a single specimen, an experienced lepidopterist would be able to identify more or less accurately the geographical origin and type of environment of the species in question. The family contains some 500 cosmopolitan species with a wide distribution in the tropics, some of which are large, the females of some Ornithoptera exceeding 200 mm (8.0 in), while others are very small, such as *Parnassius simo huntingtoni* with an extended wingspan of only 30 mm (1.2 in). The adults are characterized by epiphyses on the forelegs and a single anal vein on the hind wing with the exception of the primitive genus *Baronia*, which has two. The eggs are smooth and nearly spherical and are usually laid singly. The larvae often have fleshy tubercles and typically have an osmeterium, a glandular process that secretes butyric acid. The pupa is frequently attached by means of the cremaster and a silken girdle but in other instances lies on the ground. Within the Papilionoidea the papilionids are phylogenetically most closely related to the Pieridae in terms of the complete development of their prothoracic limbs, their patagium, the structure of the head, and the insertion of the antennae. The food plants of the papilionids give an important insight into their evolution. It is thought that the most primitive swallowtails are those which feed on the Aristolochiaceae, whereas the others have gradually adapted to plants with more or less similar essential oils, culminating with the Rutaceae and Apiaceae. The papilionids are divided into three subfamilies: Baroniinae, Parnassiinae, and Papilioniinae. The Baroniinae consist of a single species, *Baronia brevicornis*, which inhabits semidesert areas of Mexico and feeds on the legume *Acacia cymbispina*. The females are dimorphic, being either similar to the males or red or black. Dating from the Oligocene and found in Colorado, a fossil swallowtail known as *Prepapilio* seems to resemble the Baroniinae. The Parnassiinae embrace some fifty species that are chiefly palaearctic although a few species extend into North America (*Parnassius clodius*). In the Parnassiini tribe the greatest variety of species of *Parnassius* are found in central Asia. *Parnassius autocrator* has marked sexual dimorphism and was much sought by lepidopterists at one time because it was known only from a single female in Leningrad museum. Thanks to the work of the Kotzsches, famous German collectors, it was rediscovered in a remote district of northern Afghanistan. Many *Parnassius* such as *P. simo*, *P. epaphus*, and *P. acco* fly at a height of more than 5,500 m (18,000 ft). Other members of this tribe are *Hypermnestra helios*, restricted to the steppes of southern and central Asia, and *Archon apollinus*, which is found in the eastern Mediterranean.

Mention must finally be made of the Zerynthiini, which include the well-known species *Zerynthia polyxena*, *Z. rumina*, *Allancastria cerisyi*, and *Luehdorfia puziloi*. Other typical members of the Zerynthiini are *Bhutanitis lidderdalii*, which is found in subtropical mountain forests from India to northern Thailand, and *Sericinus montela*, from China and Korea. The Papilioninae are subdivided into three tribes distinguished particularly by their food plants: These are the Troidini, chiefly circumtropical, whose caterpillars feed on Aristolochiaceae; the Papilionini, which are cosmopolitan and feed on a wide variety of plants, including the Rutaceae, Lauraceae, Apiaceae, Labiatae, and Magnoliaceae; and the Leptocircini, which feed on Annonaceae, Hernandiaceae, Lauraceae, Magnoliaceae, and Winteraceae. The Troidini comprise more than 130 species, which are Indo-Australian and neotropical in distribution, including *Parides gundlachianus*, native to Cuba; a few species extend into the Palaearctic region as far as Japan, for example, *Atrophaneura alcinous*. Only one Troidini is found in Africa: *Pharmacophagus antenor*, which is native to Madagascar and is tied to the only known tropical African species of *Aristolochia*, *A. acuminata*. A characteristic feature in the males of many species of Troidini is the rolling of the margin and anal area of the hind wing, which contains odoriferous hairs. In *Parides* during courtship the male flies slowly in front of the female so that she can catch these hairs and the pheromones released into the air in special receptor grooves situated in her antennae. The females of *Battus*, which are related to *Parides*, do not have these sulci, and the odoriferous hairs in the males are much reduced so that courtship is different. The north Australian genera *Atrophaneura* and *Cressida* replace the neotropical *Parides* and *Euryades*. Mention must be made of the superb birdwings (*Ornithoptera*), which are so extraordinary that even an experienced naturalist such as A. R. Wallace recorded how exciting they were to catch. The genus is found only in the Indo-Australian region and shows marked sexual dimorphism (see Plate 10). The females of *O. alexandrae* and *O. goliath* are the largest known butterflies. Finally we must mention the *Troides* and *Trogonoptera*, which are found in the Oriental region; the latter contains the species *T. brookiana*, named after Rajah Brooke.

The Papilionini, which contains more than 200 species, make up the majority of swallowtails. Many genera have been described (such as *Pterourus*, *Heraclides*, or *Achillides*), but most authors prefer to classify them in a single genus, *Papilio*, which they subdivide into groups of species since there are no real taxonomic breaks between one group and another. Only three species are known in Europe, the common swallowtail (*P. machaon*), the Corsican swallowtail (*P. hospiton*), and the southern swallowtail (*P. alexanor*); the Palaearctic region has a few species that are evidently originally Oriental and include *P. agestor*, *P. clytia*, and *P. epycides*. Most species, some ninety, are found in the Indo-Australian region; a number of them exhibits mimicry,

Plate 41 **The Genus *Morpho***

Family Nymphalidae, subfamily Morphiinae: The genus *Morpho* is found from Mexico to northern Argentina and southern Brazil and includes some of the most beautiful and popular butterflies. The glittering bright blues of the males are not a result of pigmentation but of diffraction and light interference (see page 17). *Morpho* specimens are often used for making pictures, brooches, and jewellery and particularly in the past were much hunted for these purposes. Today the governments of some South American countries have passed conservation laws to limit or forbid the practice: (1) *Morpho didius*, male [155 mm (6.1 in)], (2) *M. neoptolemus*, male [130 mm (5.1 in)], (3) *M. menelaus terrestris*, female [120 mm (4.7 in)], (4) *M. polyphemus*, male [124 mm (4.9 in)], (5, 5*a*) *M. thamyris*, male [75 mm (3 in)], (6) *M. cypris*, male [120 mm (4.7 in)], (7) *M. helena*, female [150 mm (5.9 in)], (8) *M. hecuba*, female [140 mm (5.5 in)].

Plate 42 **The Genus *Anaea***

Family Nymphalidae, subfamily Charaxiinae: The genus *Anaea* is basically neotropical in distribution, with four species extending into the United States. It includes butterflies with fairly varied coloration: The upper wing surfaces are vivid, while the lower ones are markedly cryptic, resembling leaves. They live mostly at the edges of decaying forests. They easily elude predators by exploiting the difference between their upper- and lower-wing coloration: (1) *Anaea marthesia* [56 mm (2.2 in)], (2) *A. pasibula* [72 mm (2.8 in)], (3) *A. panariste* [73 mm (2.9 in)], (4) *A. alberta* [53 mm (2 in)], (5) *A. glaucone* [45 mm (1.8 in)], (6) flamingo (*A. ryphea*) [44 mm (1.7 in)], (7) *A. electra* [56 mm (2.2 in)], (8) tiger with tails (*A. fabius*) [65 mm (2.6 in)], (9) leaf shoemaker (*A. itys*) [60 mm (2.4 in)].

such as *agestor*, *clytia*, and *laglaizei*, whereas *P. toboroi* may not. There are also the famous groups *P. memnon* with the nonmimicking *P. polymnestor* and *P. helenus*, which includes the rare *P. iswaroides* and the blue *P. ulysses*. There are only some forty-five species of Papilionini in Africa, the most notable of which include the group *demodocus*, with a heavy speciation in Madagascar (see Plate 20B), the mimicking *dardanus*, *P. antimachus* and *P. zalmoxis*, which are among the largest butterflies in the world, and the *nireus* group. Some sixty species are found in the Neotropical and Nearctic regions, some of which are extremely rare and known from only a small number of specimens. Notable groups include *anchisiades*, which is taxonomically complex, *thoas*, represented in the Antilles by a number of species, including *P. caiguanabus*, and the superspecies *zagreus* and *menatius*. *Machaon*, *glaucus*, and *troilus* are well-known groups that have been studied for speciation and competition; *P. homerus*, a native of Jamaica, is the largest of the neotropical papilionids. The scarce swallowtail (*Iphiclides podalirius*) is the only European representative of the 130 species of Leptocircini, and the zebra swallowtail (*Eurytides marcellus*) the only nearctic one; there are fewer than ten species in the Palaearctic region, the majority being found in the Neotropical and Indo-Australian regions. Africa has some thirty species. The most interesting representatives in the Neotropical region are *E. pausanias*, *E. phaon*, *E. protesilaus*, and *E. epidaus* and in Indo-Australia *Graphium milon*, *G. encelades*, *G. androcles*, and *G. sarpedon*. A number of genera are remarkable for their unusual shapes and colours, including the Indian *Teinopalpus imperialis*, the long-tailed *Lamproptera*, and the Sino-Himalayan genera *Meandrusa* and *Dabasa*; the distribution of the latter extends as far as the Sunda Islands. The Papuan *G. weiskei* has extraordinarily beautiful colouring, combining shades of blue, green, pink, and violet.

Family Pieridae (Plate 36). This family comprises more than 1,000 species found throughout the world, some of which are very numerous, for example, the genera *Phoebis*, *Eurema*, and *Pieris*; others such as *Pieris rapae* are emigratory, which accounts for their wide distribution in many different countries. There are some tiny species, as in the genus *Nathalis* [25 mm (0.9 in)], and some large ones, as in the genus *Hebomoia* [100 mm (4 in)]. Many genera and species show marked sexual dimorphism (for example, *Colotis*, *Colias*, *Dismorphia*) or seasonal variation (*Pieris napi*); some are polymorphic (*Colias*) or exhibit mimicry (*Dismorphia*, *Valeria*, *Archonias*).

White and yellow are the predominant colours of this family, owing to the presence of pteridine pigments. Some genera are quite varied in colour (*Delias* and *Dismorphia*). They are closely related to the swallowtails, but they differ in various characteristics. The forewings have three to five radial veins and the hind wings two anal veins; the forelegs are well developed in both sexes, the tarsi having five subsegments and two forked claws. The eggs are elongated and fusiform, often a brilliant orange-yellow in colour, with longitudinal grooves, and are laid either singly or in clusters. The larvae have very short hairs and are cryptically coloured with longitudinal stripes in solitary larvae; they are aposematically coloured in some gregarious caterpillars, such as *Pieris brassicae*. The pupa has a frontal prominence, which is sometimes very long, and is attached to its support by means of the cremaster and a silken girdle except for species of the genus *Zegris*, which make a cocoon. Four subfamilies are recognized: Pierinae, Coliadinae, Dismorphiinae, and Pseudopontiinae. The first two differ from the others in that the male genital valves are unfused, in addition to which the second two have different venation and antennae.

The Coliadinae are the only group with sclerotized patagia. The Pierinae and Coliadinae are cosmopolitan, while the Dismorphinae are basically neotropical, with the exception of the genus *Leptidea*, which has four or five Eurasian species. The sole species of Pseudopontiinae is the tropical African *Pseudopontia paradoxa*, which was confused in the past with the noctuids.

The food plants of *Dismorphia*, *Pseudopieris*, *Leptidea*, and many of the Coliadini are chiefly Fabaceae and Brassicaceae, which are eaten by the larvae of the Euchloini, including *Anthocharis* and *Zegris*. However, *Hebomoia*, found in Indo-Australia, feeds on the Capparidaceae. In tropical and subtropical zones this plant family is the food of many genera of Pierinae that in temperate zones feed on Brassicaceae. All these plants contain thioglucosides or mustard oils, which are stored by the larvae. The Indo-Australian *Delias*, the neotropical *Catasticta* and *Archonias*, and the tropical African *Mylothris* form a group of genera that feed on mistletoes (Loranthaceae), which do not contain glucosides. It is thought that this group became isolated and differentiated earlier than did those genera which live on the Capparidaceae and Brassicaceae.

Delius contains more than 130 species, half of which are native to New Guinea, which may be the center from which the genus has dispersed. It shows wide ecological variation, some species being restricted to small plateaux or high-altitude rain forests, whereas others are found in areas inhabited by man. The genus *Mylothris* has some thirty species and is found in the Ethiopian region, where it is thought to replace *Delias*. Various pierids are found at the extreme limits of the geographical and ecological distribution of the Lepidoptera and indeed of animal life. *Colias nastes* and *C. hecla* fly within the Arctic Circle at a latitude of 83°, while *Baltia shawi* is found at over 5,000 m (16,500 ft) in the Pamir and Himalayas. In the Neotropical region *Piercolis*, *Phulia tatochila*, and *Hypsochila* have colonized a variety of environments along the Andes chain as high as 5,000 m (16,500 ft). The resemblance between the genera *Baltia* and

Plate 43 **The Genera *Prepona* and *Agrias***

Family Nymphalidae, subfamily Charaxiinae: These two genera are also found in Central and South America, the majority of species inhabiting the hottest zones. Both genera fly swiftly and powerfully and are eagerly sought by collectors, who use crayfish, fruit, or rotting organic matter as bait. The males of *Agrias* have olfactory hair pencils on the hind wings: (1) *Prepona praeneste* [88 mm (3.5 in)], (2) *P. chromus* [80 mm (3.1 in)], (3) *P. xenagoras* [86 mm (3.4 in)], (4) *P. pheridamas* [80 mm (3.1 in)], (5) banded king shoemaker (*P. meander megabates*) [82 mm (3.2 in)], (6) *Agrias amydon, muzoensis* form [72 mm (2.8 in)], (7) *A. amydon tryphon* [71 mm (2.8 in)], (8) *A. beata, staudingeri* form [72 mm (2.8 in)], (9) *A. claudina sardanapalus* [82 mm (3.2 in)].

Phulia is probably due to adaptive convergence in similar environments. The genus *Colotis*, with some fifty Ethiopian and oriental species is to be found in the Palaearctic region, following the distribution of its food plant, *Capparis*. The genus *Colias* has more than seventy primarily holarctic species but is also represented by a few species in the Andes and on the plateaux of Africa. The genus *Appias* has some forty Indo-Australian species. *Pereute* is a neotropical genus with some ten mountain species, which are territorial; their habitat is the tops of trees more than 10 m (30 ft) high, and they tenaciously defend their tiny territories of about 1 sq m (10.5 sq ft).

Family Nymphalidae (Plates 37 to 45). The nymphalids comprise more than 5,000 species; together with the lycaenids they are the most numerous family of butterflies. It would be hard to imagine a greater variety of shapes and colours than those to be seen in this family. It contains a number of subfamilies, some of which have such marked characteristics that they are classified as separate families in many systems; this is the case with the Satyridae, Heliconiidae, Ithomiidae, Brassolidae, and others. However, despite their diversity, the Nymphalidae have certain features in common that suggest that they should be treated as a single family. The most important character is the reduction of the foreleg, which is replaced by a hairy cushion, particularly in the males. The forelegs are not used for walking in either sex. This differentiates them from the Lycaenidae, where the reduction of the foreleg is limited to the tarsus. They differ from the Libytheidae in having much shorter palps and again in the reduction of the legs, which is a feature of both sexes among the Nymphalidae. The claws are not forked except in some Acraeinae. Venation is comparatively uniform among the various categories. The egg is higher than it is wide, with vertical and horizontal ribbing, or else spherical and occasionally smooth. The larvae lack an osmeterium, have long, paired, fleshy spines covered with down, and possess a forked anal segment as well as a head capsule with elongated processes. The pupa is not girdled but is suspended by the cremaster or else lies on the ground. Wingspan varies from 25 mm (1 in) in *Dynamine* and 130 mm (5 in) in some species of *Charaxes* and *Prepona* to 150 mm (6 in) in *Zeuxidia* and 200 mm (8 in) in some *Morpho* species. Eleven subfamilies are recognized. We shall discuss these first comparatively and then singly in more detail.

The Ithomiinae and Danainae are considered to be the most primitive of the Nymphalidae and are thought to be related as a result of certain features of venation and the reduction of the tarsus in the females. The subfamily Ithomiinae consists of some 300 species found in the Neotropical regions and one in Australia. The Danainae are cosmopolitan, with some 200 species distributed over the Palaeotropical region. The Satyrinae, Amathusiinae, Brassolinae, and Morphiinae form a very homogeneous group. The males have a simple valve, the cell in the hind wings tends to be closed (in the forewings it is always closed) and the larvae have forked tails and feed on monocotyledonous plants, with the exception of certain *Morpho* species. Many species have striking ocelli on the underside of their hind wings. The Satyrinae differ from the other three subfamilies in that the base of one or two of the veins of the forewings is swollen, a feature also found in some Nymphalinae, such as *Bulboneura*, *Callicore*, and *Cystineura*. Satyrinae are cosmopolitan but are found chiefly outside the tropics; there are more than 1,000 species. The other three subfamilies include some 200 species in all, distributed over South America and Indo-Australia. The Calinaginae are represented by a single known Sino-Himalayan species, *Calinaga buddha*, and are of uncertain classification although they are related to the Satyrinae. The larvae live on Moraceae. The Charaxinae, included as a tribe of Nymphalinae by some authors, are typically circumtropical, some being found in temperate zones, and are made up of some 300 to 400 species, including such genera as *Agrias* and *Prepona*, which are much sought by collectors. The Nymphalinae are thought to be made up of some 150 genera and 3,000 cosmopolitan species of the most varied shapes, patterns, and colours. The 200 circumtropical species of Acraeinae display the greatest diversity both in terms of species and colour in Africa. The last subfamily to have evolved, the Heliconiinae, has some sixty species found only in the Neotropical region. They are recognizable by the narrow, elongated shape of their forewings. Coloration is aposematic, usually in shades of yellow, black, and red, and their flight is stately and elegant.

Ithomiinae (Plate 37). This group is interesting for a number of reasons, principally that many species are links in chains of Müllerian mimicry (see the chapter on strategies against predators and mimicry); they also have a scattered distribution, and they have been taken as typical evolutionary models for the dense forests of South America. They have long, fine antennae and reduced forelegs; the maxillary palps are almost absent; and although they have no olfactory setae on the abdomen, a similar structure is present along the costal vein of the forewing of the males. Many genera and species are almost entirely unpatterned and are distinguished by differences in venation (*Hypoleria*, *Pteronymia*, *Greta*, *Patricia*). Some ten tribes are known from South America and one, the Tellervini, from Australia. The latter is justly considered by some authors to be related to the Haeterini (Satyrinae) and has a single polytypical Papuan species, *Tellervo zoilus*, which is not found west of Wallace's line. The larvae live on *Parsonsia* (Apocynaceae). Some of the neotropical groups are extremely numerous both in terms of individuals and species; this is the case with the genus *Mechanitis*, dozens of specimens of which may be found clinging to a single sprig of boneset, *Eupatorium* (Compositae), or ten to fifteen species may be found within a few square meters (yards) of thick, humid forest. The way in which single genera occupy specific zones in the forest

is also notable (Plate 84). By contrast some species are rare and found only on the forested slopes of the Andes. It is well known that the ithomiids are poisonous, inasmuch as the larvae, which are gregarious, are found especially on various Solanaceae, which are rich in alkaloids. The larvae are smooth with alternately coloured segments. The chrysalids closely resemble those of the Danaids, with a metallic sheen, and are suspended beneath leaves. Most species are attracted by the volatile alkaloids of *Heliotropium*, a plant genus that in the Oriental region also attracts two genera of Danainae, *Danaus* and *Euploea*, which are phylogenetically related.

Danainae (Plate 38). These are all tropical or subtropical, many are migratory, and some, like the famous monarch (*Danaus plexippus*), extend into temperate zones. They are fairly constant in pattern, at least among the Danaini, being black and pale blue, black and white, or black and yellow; their flight is slow, and they are sometimes to be seen in great numbers. The males have abdominal hair pencils used for stimulating the female antennae during courtship. They may be divided into three tribes: the Lycoreini, which are neotropical and exhibit mimicry in some species (*Lycorea ceres*, *Ituna phaenarete*); the Danaini, which include the species of the oriental genus *Idea*, which inhabits mangrove swamps, and the genus *Amauris*, which is Ethiopian and exhibits mimicry in some species; and finally the Euploeini, which are basically Indo-Australian but have some species scattered over Oceania (*Euploea lewinii* in Fiji and New Caledonia) and others in Africa (*Euploea mitra*, native to the Seychelles) and Japan. They display a degree of sexual dimorphism. The larvae are smooth with rows of parallel white, yellow, and black bands and two to four pairs of fibrous, fleshy tubercles. The pupa is compact and barrel-shaped, may be a variety of colours, and is suspended by the cremaster. Almost all the danaids live on Apocynaceae and Asclepiadaceae, which are rich in alkaloids and glucosides.

Satyrinae (Plate 39). This group contains more than 1,500 species; the characteristic coloration is brown with eyespots on the ventral hind wings. They are found everywhere: the scrub of the Mediterranean (*Pararge*, *Coenonympha*), the steppes (*Pseudochazara*), the Andean *páramos* (*Pronophila*, *Argyrophorus*), oriental forests (*Lethe*, *Elymnias*), neotropical forests (*Pierella*), alpine pastures (*Erebia*), and tundra (*Oeneis*). Some display marked sexual dimorphism, such as *Elymnias*, the females of which imitate *Taenaris* (Amathusiinae) and *Papilio* (Papilioninae), whereas the males imitate various species of *Euploea* and *Danus* (Danainae). Many tribes are indigenous to specific regions, such as the neotropical Haeterini (*Cithaerius aurorina*), Biini, and Pronopholini, the Dirini in South Africa, and the single species *D. armandi* of the genus *Davidina* in Tibet, which resembles a pierid and was long thought to be one. The Erebiini are holarctic and have been the subject of important studies in microevolution (Plate 23). The genera *Oreixenica*, *Heteronympha*, and *Geitoneura* are native to Australia. Finally mention must be made of the holarctic Satyrini and Maniolini tribes and of the palaearctic Melanargini tribe.

Brassolinae (Plate 40). The position of this subfamily in the system is not certain. It is included among the satyrids when these are classed as a family, or it is considered close to the Morphiinae. The Brassolinae feed on monocotyledons, as do the Satyrinae, while *Morpho* live on dicotyledons. The larvae have forked tails; in the Morphiinae the tail is very reduced. The discal cells on both wings are closed as in the Satyrinae, whereas in many species of *Morpho* the discal cell on the hind wing is open. All the Brassolinae are neotropical; there are some eighty species, some of them large [*Caligo*, 200 mm (8 in)] and others small [*Narope*, 65 mm (2.5 in)]. They live in forests, coming out at dusk. The males have olfactory scales and hair pencils on the hind wings. They are found from sea level up to altitudes of 2,000 m (6,500 ft). They are cryptically coloured, with large threatening ocelli on the under surface of the wings (*Caligo*). The larvae are gregarious, with longitudinal stripes and sometimes short spines, and feed on banana leaves; many species are attracted by the fermenting fruit and can cause considerable damage to banana plantations. Mention must also be made of *Dynastor*, which resembles a noctuid and has an irregular flight, and *Penetes*, *Dasyophthalma*, and *Narope*.

Amathusiinae. This is a group of some 100 Indo-Australian species, some of which extend into the Palaearctic region of China. They are large, robust butterflies; some are a brilliant metallic blue reminiscent of *Caligo* and *Morpho*. They may be considered the ecological counterparts (Plate 88) of the Brassolinae; they too fly at dusk and are attracted by ripe bananas, also causing serious damage to banana and palm plantations. The males exhibit conspicuous secondary sexual characteristics, having a pouch with olfactory hairs on the hind wing between the anal veins. In *Discophora* the pouch is circular. Some species are quite small [*Faunis* measures 65 mm (2.5 in)], whereas others are very large, such as the female of *Zeuxidia aurelius*, whose wingspan reaches 150 mm (6 in). The larvae resemble those of the Satyrinae, being cylindrical, thicker in the center, and covered all over with fine hair with some longer tufts. The head capsule has two elongated processes except in *Discophora* and *Enispe*, and the anal segment is forked. The chrysalids are elongated and suspended by the cremaster. Notable genera are *Taenaris*, *Hyantis*, *Morphotenaris*, *Morphopsis*, which are mostly Papuan and comprise some thirty species, *Stichophthalma*, made up of a few oriental species, a couple of which are found in the Palaearctic region, and *Amathusa* and *Zeuxidia*, which exhibit the greatest degree of specific diversity in the Sonda area.

Morphiinae (Plate 41). These are some of the best known and most conspicuous inhabitants of the neotropical forest. The name *Morpho* is associated with the colour

Plate 44 **Charaxiinae and Nymphalinae**

Family Nymphalidae, subfamily Charaxiinae: (1) large blue charaxes (*Charaxes bohemani*) [70 mm (2.8 in)], Ethiopian.
Subfamily Nymphalinae: (2) Australian rustic (*Cupha prosope prosope*), (2*a*) larva [48 mm (1.9 in)], Australian, (3) cruiser (*Vindula arsinoe arsinoe*), male above [84 mm (3.3 in)], female below [90 mm (3.5 in)], and (3*a*) larva, Australian, (4) white nymph (*Mynes geoffroyi*), male above in flight, male below at rest, and (4*a*) pupa [54 mm (2.1 in)], Australian, (5) purple emperor (*Apatura iris*), male, (5*a*) larva, and (5*b*) pupa [65 mm (2.5 in)], Palaearctic, (6) *Nessaea ancaeus* [60 mm (2.4 in)], Neotropical, (7) forester (*Euphaedra medon*) [60 mm (2.4 in)], Ethiopian, (8) dead-leaf butterfly (*Kallima rumia*), male [70 mm (2.8 in)], Ethiopian, (9) small tortoiseshell (*Aglais urticae*), (9*a*) larva [46 mm (1.8 in)], Palaearctic, (10) cracker (*Hamadryas guatemalena*) [62 mm (2.4 in)], Neotropical, (11) map butterfly (*Cyrestis acilia fratercula*), (11*a*) larva [48 mm (1.9 in)], Australian, (12) the 89 (*Callicore cajetani*) [44 mm (1.7 in)], Neotropical, (13) *Perisama bonplandii* [46 mm (1.8 in)], Neotropical.

blue, but this is not true for all species. In the *hercules* and *hecuba* groups both sexes are brown, and some species are more or less thickly dusted with blue; the groups *aega*, *menelaus*, and *zephyritis* have dark- or pale-blue males and brown or polymorphic females. In the *deidamia*, *achilles*, and *helenor* species groups both sexes are black with horizontal blue stripes. The *rhetenor* group has a characteristic furrowing in the forewing and clear sexual dimorphism; *M. helena* from Peru and the famous *M. diana* belong to this group. Le Moult, a dealer in Lepidoptera, employed escaped or pardoned convicts from Devil's Island in Guyana to trap these insects in the pathless marshes of the Amacuro delta of the Orinoco. No specimen seems to have been caught since then, some fifty years ago. The classification of this group is fluid, and it is probable that its eighty species have been overestimated. Many species feed on dicotyledons, such as the Lauraceae, Fabaceae, and Myrtaceae. *Morpho rhetenor* inhabits the forest canopy, while *M. achilles* and *M. menelaus* frequent forest paths; they may be lured by bananas and blue objects such as dead specimens of *Morpho*.

Charaxiinae (Plates 42 to 44). These butterflies are large and robust; they are forest dwellers and often garishly coloured. The males are territorial and aggressive; some species exhibit the phenomenon known as "hilltopping," tending to concentrate on the tops of mountains and hills. The costal margin of the forewing is thickened and serrated. They fly rapidly and are difficult to catch; traps are laid for them baited with fruit, dung, and carrion. The subfamily is divided into four tribes. The Charaxini consist of some hundred species almost entirely from Africa although a few are Indo-Australian and palaearctic. Many show marked sexual dimorphism. The two-tailed pasha (*Charaxes jasius*) has a type subspecies with a Mediterranean distribution. The different groups of species of the genus *Charaxes*, such as *varanes*, *cynthia*, *tiritades*, *kahldeni*, and *etheocles*, differ widely in shape, size, and colour. The genera *Palla* and *Euxanthe* contain a few Ethiopian species that differ in terms of behaviour and habitat from *Charaxes*. *Polyura* is the Indo-Australian counterpart: a genus with twenty-six species, including *P. dehanii* from Java, *P. eudamippus* from India, and *P. jupiter*, a polytypical species from Australia itself. *Polyura narcaea* is found in China. The Prothoini comprise a small number of Indo-Australian species such as *Prothoe franck*, *P. calydonia*, and *P. ribbei*; they are cryptically coloured on the underside and rest on tree trunks with their heads pointing downwards. The larvae have a pair of processes on the head. The chrysalis has a very long cremaster, allowing it to swing from its support. In the Neotropical region we find the Preponini with the genera *Prepona* and *Agrias*. Collectors and entomologists have devoted a great deal of attention to these genera, classifying the most minute variations of each specimen. The genus *Agrias* is currently divided into four superspecies with nine semispecies; their fantastic adaptive coloration and mimicry of various species of *Callicore* and *Callithea* (Nymphalinae) seem to have been the dominating factors in their evolution in neotropical habitats during the Quaternary. The upper surface of the wings of *Prepona* is black with green horizontal bands; the underside is cryptic. Some species, such as *P. praeneste* and *P. xenagoras*, have red stripes and thus closely resemble *Agrias*. The discovery of what is probably a hybrid *Agrias* × *Prepona* (*P. sarumani*) suggests that the two genera are extremely close even though they appear so different at first sight. *Agrias* lives in the canopy of humid forests rarely at altitudes above 1,000 m (3,300 ft). These butterflies are attracted by animal secretions but otherwise are rarely seen on the forest floor. *Prepona* prefers to live some 10 to 20 meters (30 to 65 ft) above the ground and likes bananas; the maximum altitude at which it is found is more than 2,000 m (6,500 ft) above sea level. In the Americas we also find the Anaeini tribe, which includes *Anaea marthesia*, *A. panariste*, and *A. fabius*. The goatweed emperor (*A. andria*) is the tribe's common representative in the south-central United States. Charaxiinae larvae have a characteristic head capsule with two pairs of fleshy processes resembling horns; moreover, *Prepona* and *Agrias* have caudal processes. They feed on Annonaceae, Lauraceae, and Piperaceae and on many other families of plants, including the Rutaceae, Malvaceae, and Poaceae.

Nymphalinae (Plate 44). This taxon with its variety of shapes, colours, and sizes is probably the most familiar of all to even the ordinary observer. Typical of the Nymphalinae are the reduced forelegs of both sexes, antennae with two sulci separated by a central carina, the single anal vein, which is never forked, on the forewing, and the two anal veins on the hind wing. However, these distinguishing characters are not very satisfactory, and the different tribes are much better typified by their preimaginal stages. The osmeterium found in swallowtails is never present, nor are the processes found in the danaids nor the forking of the last segment as in the Satyrinae and Charaxiinae. The pupae lack the median silken girdle. Many of the species migrate, including the well-known painted lady (*Vanessa cardui*), perhaps the most widely distributed butterfly in the world, and *Phalanta phalanta*, *Cirrochroa ermalea*, *Hypolimnas bolina*, and the mimic (*H. misippus*). Batesian mimicry is a feature of some species: The female of *H. misippus* resembles the plain tiger (*Danaus chrysippus*, Danainae), and the two forms of the female of *Euripus nyctelius* are astonishingly like *Euploea diocletianus* (Danainae) and *Papilio paradoxa*, *aegialus* form (Papilionidae). Some species also exhibit polymorphism, for example, many of the Euthaliini. *Euthalia nomina* is polymorphic in the male, while this applies to both sexes of the map butterfly (*Cyrestis cocles*). The dead-leaf butterfly (*Kallima*) is famous for having cryptic coloration combined with aposematic coloration. These examples are butterflies that are found in the Oriental region. The Argynnini is a widespread but mainly holarctic tribe, which

Plate 45 **Heliconiinae and Acraeinae**

Family hymphalidae, subfamily Heliconiinae: Illustrated here are eight neotropical species of the genus *Heliconius*. The preimaginal stages are illustrated in Plate 74: (1) *Heliconius aoede* [62 mm (2.4 in)], (2) *H. numatus* [70 mm (2.8 in)], (3) blue Grecian (*H. wallacei*) [76 mm (3 in)], (4) postman (*H. melpomene aglaope*) [68 mm (2.7 in)], (5) *H. melpomene vulcanus* [72 mm (2.8 in)], (6) *H. astraeus rondonius* [80 mm (3.1 in)], (7) *H. hermatena* [72 mm (2.8 in)], (8) *H. sapho leuce* [74 mm (2.9 in)]. **Subfamily Acraeinae:** (9) *Acraea (Telchinia) violae*, adult, (9*a*) larva, and (9*b*) pupa [50 mm (2 in)], Oriental, (10) glass wing (*A. andromacha*), in flight and at rest [56 mm (2.2 in)], Australian, (11) *A. insignis* [45 mm (1.8 in)], Ethiopian, (12) *A. doubledayi* [48 mm (1.9 in)], Ethiopian, (13) *A. braesia* [55 mm (2.1 in)], Ethiopian, (14) small orange acraea (*A. eponina*), adult, (14*a*) egg, (14*b*) larva, and (14*c*) pupa [36 mm (1.4 in)], Ethiopian, (15) tiny acraea (*A. uvui*) [30 mm (1.1 in)], Ethiopian, (16) *A. bonasia alicia* [28 mm (1.1 in)], Ethiopian, (17) *Bematistes elgonense* [65 mm (2.6 in)], Ethiopian, (18) *Actinote laverna* [44 mm (1.7 in)], Neotropical.

Plate 46 **Libytheidae and Riodinidae**

Family Libytheidae: (1) Australian beak (*Libythea geoffroyi nicevillei*), male right, female left [58 mm (2.3 in)], Australian, (2) *L. geoffroyi maenia*, male [54 mm (2.1 in)], Australian. Note the difference in coloration between the males of the two races, *nicevillei* [Cape York Peninsula, Australia) and *maenia* (West Irian, Papua–New Guinea). (3) African snout (*L. labdaca*) [46 mm (1.8 in)], Ethiopian, (4) Mexican snout (*L. carinenta*) [44 mm (1.7 in)], Nearctic; family riodinidae: (5) blue nymph (*Mesosemia croesus*) [32 mm (1.3 in)], Neotropical, (6) *Dodona eugenes* [40 mm (1.6 in)], Oriental, (7) blue lasaia (*Lasaia moeros*) [28 mm (1.1 in)], Neotropical, (8) *Stalachtis calliope* [58 mm (2.3 in)], Neotropical, (9) *Calydna caieta* [36 mm (1.4 in)], Neotropical, (10) six-tailed helicopis (*Helicopis endymion*), male [35 mm (1.4 in)], Neotropical, (11) Duke of Burgundy (*Hamearis lucina*), top and side views [30 mm (1.1 in)], Palaearctic, (12) *Lypropteryx apollonia* [45 mm (1.8 in)], Neotropical, (13) white theope (*Theope pieridoides*) [29 mm (1.1 in)], Neotropical, (14) *Lymnas pixe* [40 mm (1.6 in)], Neotropical, (15) *Ancyluris aulestes* [34 mm (1.3 in)], Neotropical, (16) *A. inca* [33 mm (1.3 in)], Neotropical, (17) *A. formosissima*, ventral (above) and dorsal (below) views [40 mm (1.6 in)], Neotropical.

includes the nearctic genus *Speyeria* with fourteen species, the circumboreal genus *Euphydryas*, the palaearctic *Argynnis* and *Melitaea*, *Phyciodes* and *Chlosyne* from America, as well as a few species exhibiting mimicry. The southern Andean genus *Yramea* is related to the holarctic genus *Boloria*. The Argynnini are interesting from the phylogenetic point of view since like the Heliconiinae and Acraeinae, which are genetically very similar, their caterpillars feed on Passifloraceae, Flacourtiaceae, Violaceae, and Turneraceae, which are all closely related; but other plant families are used as well. It may thus be surmised that these tribes are all descended from a common ancestor tied to these plants. Of the other ten tribes of Nymphalinae the following are noteworthy: among the Vanessini *Vanessa cardui* and its respective Australian and Hawaiian counterparts, *V. kershawi* and *V. tameamea*; the Indo-Australian genera *Vindula* and *Cethosia*, which exhibit marked sexual dimorphism and are related to the Heliconiini since, like the latter, they feed on Passifloraceae; and the Ethiopian *Pseudacraea*, which mimics various species of *Amauris* (Danainae), *Bematistes* (Acraeinae), and *Danaus* (Danainae). The Eunicini include a number of neotropical genera, among them *Nessaea*, and one African genus, *Asterope*, with some fifteen species. Among the Limenitidini we may note the oriental and holarctic genus *Limenitis* with some fifty species, the neotropical *Adelpha* with more than 100 species, and the Ethiopian *Cymothoe*, including *C. sangaris*, which exhibits marked sexual dimorphism. *Euphaedra* is a group of over 100 African species, which are as yet poorly known. The larvae have plumed lateral evaginations, while the adults are scarcely less colourful than *Agrias*. *Neptis*, *Euthalia*, *Panacea*, *Batesia*, and *Marpesia* are genera that have affinities with the Apaturini, the males of which have iridescent blue and green coloration. Less than 100 species of Apaturini are known, and these fall mainly into the genera *Apatura*, *Dilipa*, *Apaturina*, *Helcyra*, *Hestina*, *Sasakia*, and *Doxocopa*. The emperors (*Astero campa*) are the United States representatives.

Acraeinae (Plate 45). Some 250 species are in this subfamily. They are small- to medium-sized; a few inhabit Indo-Australia, belonging to the genera *Pareba* and *Miyana*, some 200 are native to Africa, and thirty or so belong to the neotropical genus *Actinote*. Their flight is slow, and like the ithomiines they are often to be found in large numbers on flowering bushes. They are mimicked by various swallowtails, metalmarks, and other Nymphalidae. In addition to the red-brown typical of *Acraea* the coloration of some *Actinote* consists of vivid metallic reds and blues. The eggs are barrel shaped with numerous longitudinal sulci and are laid in clusters by most species or singly as in *Actinote nohara* and *Pardopsis punctatissima*. The larvae have plumed evaginations on each segment and are often gregarious. The pupa is more or less forked at the head; in *Bematistes* it has four pairs of elongated processes, whereas in *Acraea* and *Pardopsis* it is smooth. The food plants belong to various families, but *Acraea* feeds particularly off the Passifloraceae. A number of groups may be the ecological counterparts of certain groups of Heliconiinae.

Heliconiinae (Plate 45). During the last ten years more than 200 works have been published on this group of butterflies, whose importance and fascination can hardly be exaggerated. The classic works are by M. Bates; in 1950 W. Beebe undertook up-to-date biological research in Trinidad; recently J. R. S Turner has defined *Heliconius* as "the best known of the invertebrates . . . after the Drosophilidae." Because *Heliconius* can be raised and bred in captivity, it has been possible to conduct a series of genetic studies on them, in particular on the two species *H. melpomene* and *H. erato*; this has led to the formulation of new hypotheses concerning the development of Müllerian mimicry. The Heliconiinae are an essentially neotropical subfamily; it has recently been suggested that the two oriental genera *Cethosia*, which feeds off *Passiflora*, and *Vindula* may also belong to it. The genus *Heliconius* has been used as a model for geographical variation in the Neotropical region. It contains a large number of polytypical species, showing marked colour differentiation in individual areas; this is controlled by a small number of genes and maintained by a stabilizing process of selection in the different mimetic environments. Many *Heliconius* are long-lived and reproduce continuously; others are short-lived and tend to exist in large and taxonomically ill-defined populations. These butterflies are thought to have a very wide field of vision, and this accords well with the large size of head and eyes, which are the largest among the Lepidoptera in relation to body size. It has been seen that the females spend a long time examining a plant before laying their eggs, avoiding plants where there are predators or other *Heliconius* eggs or larvae: Cannibalism among larvae of this group has in fact been observed. The larvae are often cryptically coloured and armed with long spines or else are vividly coloured in imitation of other larvae. The pupae are usually cryptic with spurs and spines. The larvae feed on various passion vines (*Passiflora*), but the adults eat the pollen and nectar of the Cucurbitaceae (*Gurania*). Some sixty species belonging to a dozen genera are known. Some of these are common and plentiful and are the first butterflies that a visitor to South America would see. Others are extremely rare and even today are known from only a few specimens; examples are *H. hecalesia*, *H. luciana*, *H. nattereri*, and *H. hermathena*. A number of species are interesting from the ecological point of view; these include high-altitude forest-dwelling species (*H. clysonymus*) and others whose distribution is highly localized in tiny areas (*Neruda godmani*). Other species are of taxonomic interest: Some groups are highly intricate, such as the superspecies *cydno*; *Philaethria dido* at first sight does not appear to be a heliconiine; and others, such as *Agraulis*, *Dione*, and *Dryas*, resemble the argynnids. This group contains some

Plate 47 **Lycaenidae**

Family Lycaenidae: (1) brown hairstreak (*Thecla betulae*), female and (1*a*) larva [35 mm (1.4 in)], Palaearctic, (2) holly blue (*Celastrina argiolus*), female and (2*a*) larva [25 mm (1 in)], Palaearctic, (3) baton blue (*Pseudophilotes baton*) lava, Palaearctic, (4) chequered blue (*Scolitantides orion*), female (left), male (right), and (4*a*) larva [28 mm (1.1 in)], Palaearctic, (5) Damar copper (*Aloeides damarensis*), (5*a*) larva [28 mm (1.1 in)], Ethiopian, (6) lesser figtree blue (*Myrina dermaptera dermaptera*), female (left), male (right), and (6*a*) larva [30 mm (1.1 in)], Ethiopian, (7) *Talicada nyseus* [28 mm (1.1 in)], Oriental, (8) Victorian hairstreak (*Pseudalmenus chlorinda zephyrus*), female above, male below [26 to 30 mm (1 to 1.1 in)], Australian, (9) moth butterfly (*Liphyra brassolis major*), male (left) [76 mm (3 in)] and female (right) [72 mm (2.8 in)], Oriental, Australian, (10) *Aethiopana honorius*, male [52 mm (2 in)], Ethiopian, (11) *Evenus gabriela*, female [44 mm (1.7 in)], Neotropical, (12) *Mimeresia neavei* [30 mm (1.1 in)], Ethiopian, (13) *Hypolycaena liara*, male [28 mm (1.1 in)], Ethiopian, (14) *Cycnus battus*, upper and lower wing surfaces [32 mm (1.3 in)], Neotropical, (15) Adonis blue (*Lysandra bellargus*), male [34 mm (1.3 in)], Palaearctic, (16) Meleager's blue (*Meleageria daphnis*), male (left) [36 mm (1.4 in)], and female (right) [32 mm (1.3 in)], Palaearctic, (17) Damon blue (*Agrodiaetus damon*), male [30 mm (1.1 in)], Palaearctic, (18) silver-studded blue (*Plebejus argus*), male above, female below [22 to 28 mm (0.9 to 1.1 in)], Palaearctic, (19) small copper (*Lycaena phlaeas*), female [28 mm (1.1 in)], Holarctic, (20) lustrous copper (*L. cupreus*) [28 mm (1.1 in)], Nearctic.

widespread species in which geographical variation is a rarity with the exception of *Dryas julia*, a dozen subspecies of which are found in the Antilles. *Agraulis vanillae galapagensis* is notable in that it is one of the few species of butterfly found on the Galápagos Islands.

The heliconiine are thought to be protected by the poisons that they extract from their food plants, but this has not been proved. Preliminary research into *Heliconius* has revealed evidence of cyanogenic glycosides that have not been identified in their host plants. Protection from predators is enhanced by aposematic colouring in various combinations of yellow, red, and blue. Extremely complicated systems of mimicry exist among the different *Heliconius* and in imitation of other genera, subfamilies, and families, such as the Ithomiinae, Nymphalinae, and Arctiidae.

Family Libytheidae (Plate 46). This is a small cosmopolitan family of a dozen small species known as "snouts" [45 to 60 mm (1 to 1.5 in)], belonging to the single genus *Libythea*, which is usually divided into two subgenera *Libythea* and *Libytheana* (from the New World). At one time the libytheids were linked with the riodinids and Satyrinae, but today they are held to be a separate family related to the Nymphalinae. The adults, particularly of *Libytheana*, are characterized by extremely elongated labial palps, which form a kind of beak — hence the family's common name of snout butterflies. The wings are angled at the apex with wrinkled or serrated edges. The forelegs are short in the male with tibia and femur of equal length, but in the female they are longer and larger. The males have the nymphalid type of genitalia, relatively simple in structure and quite different from the more complex structures of the riodinids. The eggs are taller than wide, resembling the eggs of the nymphalids and pierids, with paired riblike structures joined by horizontal bands. The larva is of the pierid type with swollen first segments and covered with very short hair. The chrysalis is stumpy, brilliant green in colour, and resembles certain Satyrinae chrysalids, such as those of the meadow brown (*Maniola jurtina*); it is suspended by the cremaster. Food plants are usually various hackberries (*Celtis*, Ulmaceae) except those of the Japanese populations of *Libythea celtis*, which are tied to Rosaceae of the genus *Prunus*. Indo-Australian species are the Australian beaks (*L. narme* and *L. geoffroyi*); *L. cyniras* is native to Mauritius, *L. ancoata* to tropical Africa, and *L. lepita* and *L. narina* are oriental. Two species of the subgenus *Litytheana* are found in America: the snout butterfly (*L. bachmanii*), which extends from Canada to Mexico, and the Mexican snout (*L. carinenta*), from Arizona to Paraguay. The hosts of the North American species are hackberries. The nettle-tree butterfly (*L. celtis*) is of wide palaearctic distribution and is found in central Europe.

Family Riodinidae (Plate 46). This is the most neglected family of the Papilionoidea, and it presents a number of difficulties. Systematically the metalmark family is rather complex, and specimens are hard to catch since many species live in forested areas, where they settle under leaves, some of them with forewings raised and hind wings spread ready to fly off suddenly at the slightest disturbance. Some species fly very high among the tree tops, particularly the females. In addition most species are rare and extremely localized. A Brazilian collector is said to have spent thirty-three years collecting only seventy species, many represented by a single specimen.

More than 1,000 species are known, most of them found in the tropics with the majority found in Cental and South America. The Duke of Burgundy (*Hamearis lucina*) is the single representative in Europe, where it feeds on the Primulaceae. A number of species are native to North America, especially the Mormon metalmark (*Apodemia mormo*) and several *Calephelis*; in Africa there are a dozen species belonging to the oriental genus *Abisara* and the Malagasy genus *Saribia*. The variety of shapes and colours of these butterflies is quite incredible; in some cases their colouring is reminiscent of hummingbirds. Some species resemble moths, especially the geometers; others, such as *Ancyluris inca*, recall *Callicore*, a genus of Nymphalinae, while *Helicopis endymion*, for example, resembles a hairstreak. *Stalachtis calliope* is similar to an ithomiine or a pericopid moth, and *Mimocastnia rothschildi* resembles a species of *Castnia* (Castniidae). Some have raised metallic spots on the wing surface, like the Australian hepialid *Zelotypia staceyi*. The colour of the upper and lower wing surfaces provides a strong contrast: brown above and blue, grey and green, or brown and red below.

Morphologically the Riodinidae are characterized by the antennal club's often being pointed; the forelegs are atrophied in the male but normal in the female; the hind wing has a precostal vein; and the uncus is well developed. The last three characters distinguish riodinids from the lycaenids. The eggs are hemispherical, and the larvae are sometimes onisciform, are frequently very hairy, and vary in colour from green to red. Little information has been recorded concerning the food plants of this distinctive family. It is known that some species feed off Myrsinaceae and Primulaceae, closely related families eaten by few other butterflies.

Three subfamilies are distinguished: Euseliinae, Riodininae, and Hamearinae. The first two are solely American and typically lack the precostal vein in the forewing.

Family Lycaenidae (Plate 47). This huge family (together with the nymphalids) covers three-quarters of the Papilionoidea. Some authors have made a case for including the riodinids in the Lycaenidae, which contain some 5,000 or 6,000 species. They are usually small and highly coloured — red, green, and dark or bright blue — but the wingspan of genera such as the neotropical *Eumaeus* and the Ethiopian *Liphyra* can be as much as 70 mm (2.75 in). The eyes are often hairy, the maxillary palps are absent, the labial palps

are ascending, epiphyses are absent, and the forelegs are slightly reduced in the male but normal in the female. The radial vein in the forewing lacks one or two forks, the hind wing has no humeral vein but two anal veins, and in many species the hind wing has one or more tails. The egg is flattened and domed or disc shaped with delicate furrows and evaginations. The larva is sluglike with retractile head and sometimes is covered with dense short hairs; there is frequently a dorsal gland on the seventh abdominal segment or eversible organs on the eighth. The pupa lacks a cremaster and is attached at the caudal end by a girdle or else lies loose on the ground.

Eight subfamilies may be distinguished. The African Lipteninae (*Mimeresia neavei*, *Aethiopana honorius*) contains more than sixty species, which have no wing tails and mimic the *Acraea* and some pierids. The larvae have lateral hairy tufts. The Poritiinae resemble the Lipteninae in terms of the male genitals, but in terms of the larval characters, such as the tufts of sparse hairs on the abdominal segments, they resemble the Riodinidae (*Laxita*). The Liphyrinae include a few large species from Africa and tropical Asia, the larvae of which live in association with ants (*Liphyra brassolis*). The newly hatched adults are covered with white scales that, as they emerge, protect them from the predation of the ants and fall away when they make their first flights. The Miletinae form a small oriental and African group with a single North American species, the harvester (*Feniseca tarquinius*), whose larva is completely carnivorous and feeds on woolly aphids, spinning a web to conceal itself from the ants that tend the aphids. The larva of the southern African genus *Thestor* is also carnivorous; during the early stages it feeds on Psyllidae before entering the nests of ants of the genus *Anoplolepis* to pupate. The Theclinae form a very large cosmopolitan group, the hairstreaks, brilliantly coloured and with long tails, containing some fifteen tribes and a few hundred species. The grey hairstreak (*Strymon melinus*) is the most widespread American species. This subfamily includes the palaearctic brown hairstreak (*Thecla betulae*), the neotropical *Evenus gabriela*, *Hypolycaena liara* from the Congo, which has three long tails, *Pseudalmenus chlorinda zephyrus* from Australia, and the Ethiopian *Myrina dermaptera*. The Polyommatinae, or blues, consist of a huge number of cosmopolitan genera and species, most of which are holarctic. Blues are commonest in boreal habitats, and in North America the mountains throng with such as the greenish blue (*Plabejus saepiolus*), the silvery blue (*Glaucopoyche lygdamus*), and the square-dotted blue (*Euphilotes battoides*). We may also mention the holly blue (*Celastrina argiolus*), the baton blue (*Pseudophilotes baton*), the Adonis blue (*Lysandra bellargus*), Meleager's blue (*Meleageria daphnis*), the Damon blue (*Agrodiaetus damon*), and the silver-studded blue (*Plabejus argus*) — all palaearctic species — and *Tylicada nyseus* from India as well as the tiny-sized genera *Zizeeria*, *Cupido*, *Neolucia*, *Freyeria*, and *Brephidium*. The Lycaeninae are chiefly holarctic but include the genus *Melanolycaena*, the two species of which, *M. altimontana* and *M. thecloides*, are found on the plateaux of Papua. A few species of *Lycaena* are found in New Zealand and two in South Africa; a few subspecies of the small copper (*L. phalaeas*) are found in the mountains of Ethiopia and Uganda. *Lycaena phlaeas* and the North American lustrous copper (*L. cupreus*) are illustrated. Finally we may mention the subfamily Styginae, containing the single species *Styx infernalis*, found above 2,000 m (6,500 ft) in the Peruvian Andes, which some authors relate to the pierids and others to the riodinids because of the shape of the palps and mesothoracic anepisternum. The structure of the forelegs in the male of this strange butterfly resembles that of the genus *Thestor*, but it nevertheless differs markedly from the Miletinae and other groups in certain characters, such as the two veins in the discal cell of the forewing and the asymmetry of the veins in the hind wing.

The huge size of this family means that as yet we have little information about its food plants. As has been said, the Lipteninae feed on lichens; the Liphyrinae are predators and seem not to be phytophagous; and the Lycaeninae feed on a whole range of plants. *Nacaduba* eat various genera of Myrsinaceae, and *Agriades* eat Primulaceae, the same groups of plants as are eaten by the Hamaearini and the riodinid genus *Abisara*. *Philotes* and *Scolitantides* feed partly on Crassulaceae. The genus *Jamides* is tied to the Fabaceae, but *J. alecto* feeds on gingers (Zingiberaceae). Within the genus *Callophrys* the subgenera *Callophrys* and *Incisalia* live on angiosperms (Leguminosae, Polygonaceae, Rosaceae, Ericaceae) although some *Incisalia* are tied to the gymnosperms (Pinaceae); *Mitoura* feeds on Cupressaceae and mistletoes (Loranthaceae) which are epiphytic on conifers. Various Theclinae are tied to the Fagaceae and Oleaceae, but the genus *Shirozua* has become a predator on aphids. The coppers (*Lycaena*) feed chiefly on docks, smartweeds, and buckwheats (Polygonaceae). However, much of the information given here is based solely on single reports, and it is known that the holly blue (*Celastrina argiolus*) feeds on at least fourteen different families of dicotyledons.

Superfamily Geometroidea

The Geometroidea form a large and important superfamily. The adults usually lack ocelli but generally have chaetosemata; the maxillary palps are reduced or absent; and the proboscis is smooth. The wings, which are usually very broad, lack the posterior cubital vein and almost always have a frenulum. At rest they are held open, parallel to the surface on which the moth is resting, or they are frequently held close above the body in the manner of butterflies. Cryptic coloration predominates. The abdomen is slender with a pair of tympanal organs except in the sematurids. The eggs are dorsoventrally flattened. The larva is usually slender, the foreprolegs bearing two

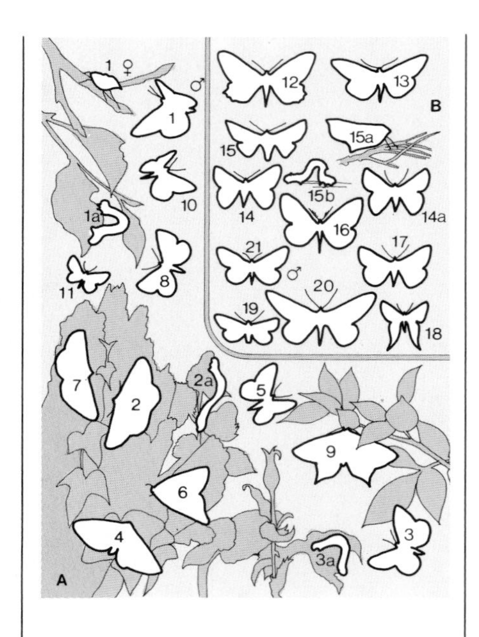

Plate 48 **Geometers**

Superfamily Geometroidea, family Geometridae, (A) European species: (1) *Erannis defoliaria*, apterous female (left), male (right) [42 mm (1.7 in)], and (1*a*) larva, (2) *Boarmia selenaria*, adult and (2*a*) larva [37 mm (1.5 in)], (3) *Anticlea badiata*, (3*a*) larva, (4) *Abraxas glossulariata* [42 mm (1.7 in)], (5) *Euchloris sardinica* [25 mm (1 in)], (6) *Aplocera praeformata* [43 mm (1.7 in)], (7) *Menophra nycthemeraria* [37 mm (1.5 in)], (8) *Pseudopantherina macularia* [26 mm (1 in)], (9) *Ourapteryx sambucaria* [48 mm (1.9 in)], (10) *Rhodostropha calabra* [31 mm (1.2 in)], (11) *Eucrostes indigenata* [12 mm (0.5 in)]; (B) Non-European species: (12) *Erebomorpha fulguritia* [70 mm (2.8 in)], Palaearctic, (13) *Percnia felinaria* [80 mm (3.1 in)], Oriental, (14) *Crypsiphona occultaria*, above and (14*a*) below [36 mm (1.4 in)], Australian, (15) *Oenochroma vinaria*, above and (15*a*) below and (15*b*) larva [50 mm (2 in)], Australian, (16) *Cartaletis libyssa*, *monteironis* form [44 mm (1.7 in)], Ethiopian, (17) *Catocalopsis medinae* [54 mm (2.1 in)], Neotropical, (18) *Erateina julia* [35 mm (1.4 in)], Neotropical, (19) *Praesos rotundata* [52 mm (2 in)], Australian, (20) *Dysphania fenestrata* [86 mm (3.4 in)], Australian, (21) *Rhodophthitus simplex roseus*, male [38 mm (1.5 in)], Ethiopian.

incomplete series of crochets; the ventral pair is reduced or absent in many cases. The classification adopted here recognizes seven families: Drepanidae, Thyatiridae, Geometridae, Uraniidae, Epiplemidae, Axiidae, and Sematuriidae. Other systems, such as that adopted by J. P. Brock are more comprehensive and perhaps correctly associate the callidulids and pterothysanids with the geometroids. Three main groups can be broadly distinguished within the Geometroidea: Geometridae, Uraniidae (including the *Epicopeia* genera), and Drepanidae-Thyatiridae, to which the Axiidae are appended.

Family Drepanidae (Plate 50). These are medium to small moths, and they resemble some of the geometrids; they are often characterized by the markedly curved apex of the forewing. The adults have short, usually bipectinate, antennae, a large head, small palps (although in the Cyclidiinae they are longer and thickened), and a proboscis that is poorly developed (Drepaninae) or vestigial (Oretinae). The wings are large, and the forewings have characteristic venation: M_2 and M_3 are approximated or identical at the base. In the males the frenulum may be present (Drepaninae) or absent (Oretinae). The tympanal organs are dorsal. The larvae lack anal prolegs, have ventral prolegs with crochets, and vary in appearance. They are generally hairless, slender, and elongated, but certain *Drepana* have flattened, thickened, first segments with short, fleshy, protuberances. In many species the anal segments are reduced to form a tail. They feed mainly on broad-leaved woody plants, both shrubs and trees. Pupation takes place on the ground, and the tiny pupae are frequently covered with a bluish wax. Drepanids are ocher and pink with a pattern of faint stripes, but *Macrauzata* from tropical Asia and Japan has wings with broad, pearly, scaleless windows between the veins. The Drepanidae have some 800 species concentrated in the Old World tropics; they are not found in the Neotropical region, and few are known in the Holarctic or Indo-Australian regions (there are five Australian species). The subfamilies are Drepaninae, Oretinae, and Cyclidiinae; the latter includes the genera *Cyclidia* and *Mimozetes* from the Oriental region and Japan, and some authors consider it a distinct family. *Drepana binaria*, *D. falcataria*, and *D. cultraria*, a moth inhabiting beech woods, are common in Europe; they belong to a mainly palaearctic genus, which has a few species in India and one in North America (*D. armata*). *Cilix* has a similar distribution; its dozen species resemble *C. glaucata* and have rounded wings. Species native to the Ethiopian region include *Epicamptoptera*, some of which are related to the Indo-Australian *Cyclura*, and the polytypical species *Negera natalensis* from central southern Africa. Mainly Indo-Australian groups include *Tridrepana*, *T. flava*, *Oreta*, and *O. erminea* from Australia, *O. rosea* from North America, and *Macrauzata*, a small genus distributed over Japan, the Philippines, China, India, and the whole of Southeast Asia.

Family Thyatiridae (= Cymatophoridae; Plate 50). The thyatirids are related to the drepanids but may be easily distinguished from them by the appearance of their wings and the clear differences in the venation of the forewing. Unlike the drepanids the M_2 and M_3 veins are distinct at their base. Other family differences are evident in the larvae. The thyatirid larvae have anal prolegs and a semicircular arrangement of crochets on the ventral prolegs, whereas the drepanid larvae lack anal prolegs, and the ventral ones have a circular arrangement of crochets. The tympanal organs are dorsal. The appearance of the adults is very like that of certain noctuids. Brown and olive-green tones predominate, enlivened by pink spots and wavy bands of lighter colour. The larvae vary more in appearance than the adults do. To simplify, two main types may be distinguished: The first represented by genera like *Habrosyne* and *Thyatira*, and the second by genera such as *Polyploca* and *Tethea* (*Palimpsestis*). The caterpillars of the first group live in the open and have tubercules and forked protuberances of the type seen among the notodontids; the second tend to live concealed between two leaves, which they sew together with silk, and are more ordinary in appearance, without ornaments. As an adaptation to their rather restricted quarters they are somewhat flattened. They are active during the summer and autumn, wintering in pupal form in a cocoon sheltered by leaves. The family contains more than 100 species and is found almost throughout the world with the exception of the Ethiopian and Indo-Australian regions. Species like *Habrosyne pyritoides* and *Thyatira batis* are palaearctic in distribution and belong to genera also found elsewhere. *Habrosyne* (17 species) is holarctic, stretching as far as Java; the vast habitat of *Thyatira* (12 species) includes the Americas, the Palaearctic region, and much of the Oriental region. In contrast the related genus *Euthyatira* (4 species) inhabits the Nearctic region. Such widely distributed species as *T. batis* are found at high altitudes in tropical zones and at lower altitudes farther north.

Family Geometridae (Plate 48). With some 15,000 known species the Geometridae are the second largest family of Lepidoptera, representatives of which are to be found in different environments throughout the world. The adults are slender bodied and broad winged; they vary greatly in size and almost invariably lack ocelli, whereas chaetosemata are always present; they have simple or pectinate antennae, short, monosegmental maxillary palps, and long labial palps. In certain species of *Alsophila*, *Apocheima*, *Erannis*, *Operophtera*, and so on, the females are brachypterous or apterous. The hind wing bears a frenulum, and in certain South American species of *Erateina*, such as *E. julia*, it has long tails. *Ourapteryx* also exhibits a number of tails and indentations; this genus is concentrated in the Oriental region. The family is represented in Eurasia by *O. sambucaria*, *Selenia*, and *Boarmia* and in India and China by *Erebomorpha fulguritia*. All the adults have tympanal organs. The ventral prolegs of the larvae are either reduced

Plate 49 **Uraniidae**

Family Uraniidae: (1) *Cyphura pardata* [35 mm (1.4 in)], Australian, (2) *Strophidia fasciata* [50 mm (2 in)], Oriental, (3) *Urania sloanus* [60 mm (2.4 in)], Neotropical, (4) *U. leilus* [60 to 80 mm (2.4 to 3 in)], Neotropical, (5) *Alcides zodiaca* [95 mm (3.7 in)], Australian, (6) *A. aurora* [78 mm (3.1 in)], Australian, (7) *A. agathyrsus* [80 to 100 mm (3.1 to 3.9 in)], Australian, (8) sunset moth (*Chrysiridia ripheus*) [80 to 100 mm (3.1 to 3.9 in)], Ethiopian, (9) *Lyssa patroclus*, Solomon Islands race [130 mm (5.1 in)], (10) *L. patroclus*, Queensland race [120 mm (4.7 in)].

Systematic Survey of Butterfly and Moth Families

Plate 50 **Other Geometroidea and Calliduloidea**

Superfamily Geometroidea, family Drepanidae: (1) *Drepana falcataria* [30 mm (1.2 in)], Palaearctic, (2) *D. cultraria*, adult and (2*a*) larva [25 mm (1 in)], Palaearctic, (3) *Cilix glaucata* [17 mm (0.7 in)], Palaearctic, (4) *Negera natalensis* [43 mm (1.7 in)], Ethiopian, (5) *Tridrepana flava* [42 to 66 mm (1.7 to 2.6 in)], Oriental, (6) *Oreta erminea* [40 mm (1.6 in)], Australian, (7) *O. rosea* [25 to 35 mm (1 to 1.4 in)], Nearctic; family Thyatiridae: (8) *Thyatira batis*, adult and (8*a*) larva [37 mm (1.5 in)], Palaearctic, (9) *Habrosyne pyritoides* [38 mm (1.5 in)], Palaearctic; family Axiidae: (10) *Axia margarita* [22 mm (0.9 in)], Palaearctic, (11) *Epicimelia theresiae* [40 mm (1.6 in)], Palaearctic; family Epiplemidae: (12) *Epiplema coeruleotincta* [27 mm (1.1 in)], Australian, (13) *E. himla* [25 mm (1 in)], Palaearctic; family Sematuriidae: (14) *Coronidia orithea*, female [42 to 45 mm (1.7 to 1.8 in)], Neotropical.
Superfamily Calliduloidea, family Callidulidae: (15) *Pterodecta felderi* [25 to 50 mm (1 to 2 in)], Palaearctic; family Pterothysanidae: (16) *Hibrildes venosa*, male (left), female (right) [52 mm (2 in)], Ethiopian, (17) *Pterothysanus noblei* [40 to 60 mm (1.6 to 2.4 in)], Oriental.

(Brephinae) or absent, and the caterpillars, sometimes called "measuring worms" or "inchworms," exhibit a characteristic looping method of walking (Plate 6). They are generally cryptically coloured and often resemble twigs or leaf veins. The pupa has a cremaster. The wing patterning of the adults is usually cryptic, but notable exceptions are common, as in *Milionia* (Plate 9), *Catocalopsis medinae* from Chile, the diurnal, north Australian *Dysphania fenestrata*, and *Praesos rotundata* from Queensland. Other species, such as the north Indian *Percnia felinaria*, the palaearctic *Abraxas glossulariata*, and the South African *Zerenopsis*, have unusual speckled markings. The most important defoliating species of Geometer are the winter moth [*Operophtera brumata*; 25 mm (1 in)], which was originally palaearctic but has spread to Canada and attacks oaks, hawthorns, and apples, the holarctic genus *Erannis*, which includes the well-known *E. defoliaria*, which damages birches, oaks, and beeches, and *Bupalus piniarius* from Europe, which causes serious damage to pines and larches. Geometers form a cosmopolitan family with more than 1,000 species in North America and some 3,000 palaearctic species. Six subfamilies are recognized, distinguished by differences in wing veining.

The Oenochrominae (over 600 species) are predominantly Indo-Australian, oriental, and neotropical and form a rather heterogeneous subfamily that includes some primitive forms such as *Oenochroma*; the Australian *O. vinaria* is illustrated. The Brephinae form a tiny holarctic subfamily or, according to some authors, a distinct family, whose species may be distinguished by their oval eyes. *Brephos parthenias*, common in central Europe, belongs to this group. The cosmopolitan subfamily Hemiteinae (Geometrinae) has many species in Australia and the Old World tropics. Their predominant colour is green, and many species are cryptic with barklike or lichenlike patterning. *Geometra papilionaria* and *Euchloris smaragdaria* belong to this group, which comprises some 1,600 species somewhat related to the Sterrhinae. The latter (also known as Acidaliinae) form a cosmopolitan subfamily containing such huge genera as *Scopula*, *Sterrha*, and *Cosymbia*, the species of which are small and usually light in colour with striped or dotted patterning. The common *Rhodostropha calabra*, which is found from central southern Europe to the Caucasus, is illustrated. The Sterrhinae are also plentiful in the tropics, and some species of the Indian and Malayan genera *Zythos* and *Antitrygodes* are notorious for their habit of sucking the tears and sweat of domestic animals and even of human beings. Most of the geometers in temperate and cold zones are members of the Larentiinae as, for example, the small *Eupithecia* (a large and cosmopolitan genus), *Operophtera*, *Cidaria*, *Chloroclystis*, and *Anaitis praeformata* and *Earophila badiata*, two species from the western Palaearctic region, illustrated in Plate 48. The largest subfamily is the Ennominae (Boarminae), a cosmopolitan and heterogeneous group that includes some geometers notable for a variety of reasons: the sexual dimorphism of the wings of, for example, *Nyssia florentina* (Plate 10) and their industrial melanism, a classic example of which is *Biston betularius* (Plate 16). They also include some of the few lepidopterans to fly in winter, such as *Phigalia pedaria*, *Apocheima hispidaria*, and *Lycia hirtaria*, while *Gnophos* lives in mountain environments. Other members of this subfamily are *Abraxas glossulariata*, a Eurasian species that has long been used for genetic research into the heredity of sex-linked characters, the beautiful *Pseudopanthera macularia* of palaearctic distribution, and *Rhodophthitus simplex* from southern Africa.

Family Uraniidae (Plate 49). The hundred-odd species of uraniids are restricted to the tropics and are found in America (the single native genus is *Urania*), Africa (for example, *Chrysiridia*), India, Malaya, and Australia (*Alcides*, *Lyssa*, *Strophidia*). The adults resemble the geometers, but the frenulum is reduced or absent. The hind wing has a very wide humeral angle, a single anal vein, and a more or less developed tail in the region of the third median vein.

The tympanal organs are abdominal and differ in the two sexes. Wing coloration varies from white to brown and polychrome tints. The coloration of some species, such as *Urania leilus*, *U. sloanus* from Jamaica, *Alcides agathyrsus*, *A. accrora*, and *A. zodiaca* from New Guinea, and the Madagascan sunset moth (*Chrysiridia ripheus*), is unusually iridescent; these species belong to the subfamily Uraniinae, which contains large and very colourful diurnal species, usually with tails. The other subfamily in the Microniinae. The species of this group are less striking, with light wings patterned with dark stripes and without tails; they are smaller and fly by night or at dusk. They include *Cyphura pardata* from New Guinea, the Indo-Malayan *Strophidia fasciata*, and the Indo-Australian and Ethiopian genera *Micronia* and *Acropteris*. Some of the best known uraniids are the large *A. agathyrsus*, a forest moth of New Guinea that is protected against predators and mimicked by the swallowtail *Papilio laglaizei*, and the incredible *Epicopeia polydora* found in India, Burma, and Tibet, which mimics *Parides philoxenus*, an aposematically coloured swallowtail that is also protected against predators. Other species of *Epicopeia*, a genus found from India to Japan and Sumatra, mimic various species of *Parides*.

Family Epiplemidae (Plate 50). The epiplemids form a cosmopolitan group of more than 550 species, many of them very rare and most of them tropical. For a long time they were considered a subfamily of the uraniids. The adults have small heads with large eyes, lack ocelli, and have tympanal organs like those of the uraniids. When at rest, they hold their forewings away from their hind wings. Usually the forewings are more or less rolled up to form a tube or else are spread out and thrown forward while the hind wings cover the abdomen. Certain species, such as *Epiplemia birostrata* from tropical America, have a

curious formation on the abdominal part of the hind wing: a fold concealing tufts of long and very fine hairs of unusual softness, 100 times smaller in diameter than a human hair. The overall appearance of the epiplemids recalls that of various geometers. Their coloration is usually neutral, but some species, such as the Himalayan *Epiplema himala*, are unusually light in colour, and more commonly others are dark and cryptic as in the northeast Australian *E. coeruleotincta*. It has been observed that many cryptically coloured species like to remain concealed beneath leaves and are loath to take off even when the bushes they are resting on are violently shaken. In their first stage the larvae are gregarious and spin a communal web.

Family Axiidae (Plate 50). This family contains less than a dozen species grouped into two genera with strong affinities with the thyatirids. The lovely *Axia margarita* and *Epicimelia theresiae*, like the other axiid species, have tympanal organs on the seventh segment and are broadly Mediterranean in distribution, covering central Europe, Turkey, and north Africa.

Family Sematuriidae (Plate 50). This is a small tropical family containing some thirty species that are medium to large in size and are concentrated in Central and South America. Sematuriids have striking coloration often with a more or less extensive grid pattern broken by transverse bands. In *Sematura* (= *Nothus*) the hind wings end in long tails with ocelli. The species illustrated, *Coronidia orithea*, is found from Mexico to Paraguay. The largest genus is *Homidiana*, containing some twenty species similar in appearance to *Coronidia*. Despite the absence of tympanal organs, the sematuriids are sometimes thought to be related to the uraniids.

Superfamily Calliduloidea

This is a small superfamily related to the preceding one and containing delicate moths similar in coloration to the butterflies. The adults lack maxillary palps, but the labial palps are well developed and often pointed. They usually have a proboscis but no tympanal organs or posterior cubital wing veins. Nothing is known of the larva or pupa. There are two families, Callidulidae and Pterothysanidae.

Family Callidulidae (Plate 50). These are medium-sized moths with delicate bodies and broad wings. Like the butterflies they fly by day and also resemble them in the position of their wings at rest. Some 100 species are known in all, most of which are from the Oriental region and Papua. Very few are palaearctic. *Callidula* with some fifty species is the most important genus; *Pterodecta felderi*, which is found from Kashmir to Japan, is illustrated.

Family Pterothysanidae (Plate 50). The pterothysanids are thought to be related to primitive geometers although in coloration they are more reminiscent of the lymantriids or even certain nymphalids. Some forms of *Hibrildes venosa* from southern and eastern Africa are vaguely similar to the plain tiger (*Danaus chrysippus*). The *Pterothysanus* from India, China, and Thailand, however, recall the appearance of the geometers and have a fringe of very long hairs on the abdominal edge of the hind wing, a feature clearly visible in the Burmese *P. noblei*.

Superfamily Bombycoidea

This is an important cosmopolitan superfamily containing moths that are medium to large in size and characterized by the reduction or less of many structures, such as proboscis, maxillary palps, ocelli, chaetosemata, frenulum, retinaculum, and tympanal organs. The adults have stout, hairy bodies. The wings are broad, and the loss of the frenulum is compensated for by the dilation of the humeral area of the hind wing, which makes amplexiform wing coupling possible. The antennae are pectinate, complex, and in the male rich in sensory apparatus. The eggs are usually smooth and flattened. The larvae are external feeding and active by day as well as by night. They vary in appearance: They may be hairy (and often sting) or hairless and covered with verrucae. The lavae have tubercles, fleshy processes, and erectile organs and usually change in appearance from one instar to the next.

This superfamily contains the species of silkworms. The pupa is stumpy and usually lies inside a cocoon. There are thirteen families. Some systems of classification include the sphingids, treated here as a separate superfamily, while others exclude the Mimallonidae and Saturniidae.

Family Endromidae. The only species of this family is the Eurasian *Endromis versicolora* [50 to 90 mm (2 to 3.5 in)]. The male and female differ in size, wing coloration, and antenna shape as well as in flight, which occurs only by night. After mating — which may last an entire day — the female begins to lay the eggs in clusters of ten or twenty on young twigs of various broad-leaved trees, such as lime, birch, or willow. The caterpillars are hairless and pale green; they are gregarious during their first stadia, becoming solitary as they mature. Pupation takes place in June and July; the light cocoon is woven on the ground and covered with vegetable remains and dead leaves. The moth overwinters in pupal form and may emerge only after two or three years if climatic conditions are adverse.

Family Lasiocampidae (Plate 51). These are medium to large hairy moths with bipectinate antennae, a rudimentary proboscis, and developed labial palps; they lack retinaculum, frenulum, or tympanal organs. The family is relatively homogeneous and cosmopolitan with more than 1,200 species, largely tropical in distribution. The imago exhibits strong sexual dimorphism. Coloration is usually sober and often cryptic as in the

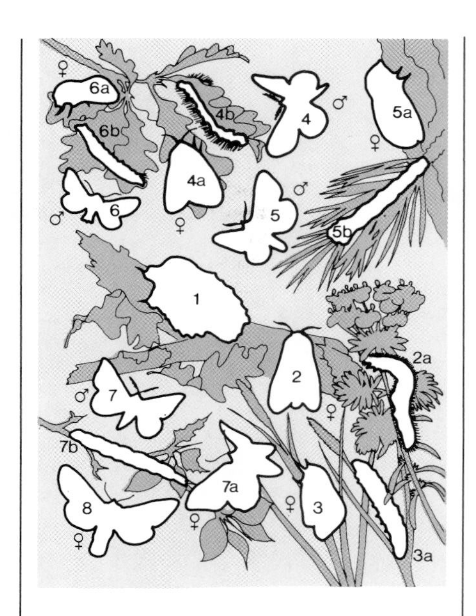

Plate 51 **Lasiocampidae**

Superfamily Bombycoidea, family Lasiocampidae: (1) *Gastropacha quercifolia* [40 to 80 mm (1.6 to 3.1 in)], Palaearctic, (2) *Malacosoma castrensis*, female and (2*a*) larva [35 mm (1.4 in)], Palaearctic, (3) *Philudoria potatoria*, female and (3*a*) larva [50 mm (2 in)], Palaearctic, (4) *Lasiocampa quercus*, male [50 mm (2 in)], Palaearctic, (4*a*) female [70 mm (2.8 in)], and (4*b*) larva, (5) *Dendrolimus pini*, male [55mm (2.2 in)], Palaearctic, (5*a*) female [75 mm (3 in)], and (5*b*) larva, (6) *Eriogaster catax*, male [40 mm (1.6 in)], Palaearctic, (6*a*) female [45 mm (1.8 in)], and (6*b*) larva, (7) *Odonestis pruni*, male [50 mm (2 in)], Palaearctic, (7*a*) female [60 mm (2.4 in)], and (7*b*) larva, (8) *Porela notabilis*, female [60 mm (2.4 in)], Australian.

palaearctic *Gastropacha quercifolia*, which at rest resembles a pile of dead leaves. In some species, for example, the palaearctic *Lasiocampa quercus*, *Eriogaster catax*, and *Philudoria potatoria*, the forewings have a white disc that may be surrounded by a small curved band. The females are typically very prolific. Shortly after copulation they lay their eggs either singly or in regular clusters on leaves or as in *Malacosoma neustria* (Plate 5) in rings around the stems of the food plant. The larvae are hairy with setae of different lengths, and they may often sting badly. The caterpillars are external feeding, polyphagous, active by day, and gregarious during their first stadia. Those of the palaearctic *Eriogaster* and the holarctic tent caterpillars (*Malacosoma*) spin large communal webs. Pupation takes place on the ground or on the food plant, and the cocoons often contain the larval stinging hairs. The pupal stage lasts three to five weeks. In temperate latitudes only one generation is produced a year, and the larvae usually hibernate. Although the females are, as has been said, very prolific, the larvae and pupae are subject to attacks by fungi (*Isaria*, *Botrytis*) and parasitic insects (tachinid flies and ichneumonid wasps), which greatly reduce the number of adults that emerge. The adult is short-lived and does not feed; it flies by night and is attracted by light traps. The females are less mobile, especially when still unfertilized, and may be brachypterous (as in *Lasiocampa staudingeri*) or apterous (as in the Ethiopian genus *Mesoscelis*). Some species are subject to irregular population explosions, which can damage crops and forests. They include *Odonestis pruni*, a Eurasian species linked with several kinds of fruit tree, *Malacosoma neustria*, related to *M. castrensis*, and *Dendrolimius pini*, which is responsible for the destruction of the Scotch pines in central and eastern Europe. Other exotic taxa are *Labeda*, *Paralabeda*, and *Trabala* from the Oriental region, the Ethiopian *Nadiasa* and *Chilena*, and the Australasian genera *Crexa*, *Pinaria*, *Isostigena*, and *Porela*; the species *P. notabilis* is illustrated. Like the palaearctic species these genera include some highly polytypical species. No lasiocampids are found in New Zealand.

Family Anthelidae (Plate 52). This family lies between the lasiocampids and eupterotids and consists of some 100 species native to Australia and New Guinea. The adults lack many anatomical structures but have retinaculum and frenulum — at least in the males. The larvae and frequently the cocoons are protected by stinging hairs. The food plants of the caterpillars include species of *Acacia* and *Eucalyptus*. The main genus is *Anthela* (fifty-one species in Australia), which shows considerable intra- and interspecific variation. As can be seen from the plate, *A. neuropasta* is completely different in coloration from *A. denticulata* or *A. nicothoe*. Sexual dimorphism also contributes to the apparent diversity of these moths, being particularly marked in *Nataxa flavescens*, for example. The larger anthelids fall into the genus *Chelepteryx*, and the females of *C. collesi* have a wingspan of 180 mm (7 in).

Family Eupterotidae (Plate 52). The eupterotids are medium to large moths found in the tropical zones of Asia, Africa, and the Australasian region. The adults have thick, bipectinate antennae and are hairy; the proboscis, maxillary palps, and frenulum are much reduced or absent. They are typically forest-dwelling moths, which fly by night, with broad wings and sober coloration, often enlivened by brown, silver, and black stripes and blotches. In *Phiala*, an Ethiopian genus with some fifty species, the antennae and abdomen are a beautiful golden yellow, and the wings are white with brown stripes. *Phiala cunina* from west Africa has a pattern that is quite unusual compared to that of other congeners and exemplifies the great variation found in the eupterotids. The larvae have long, stinging hairs, are polyphagous, and feed *inter alia* on Fabaceae, Bignoniaceae, Oleaceae, Loganiaceae, and Acanthaceae. The larvae of *Panacela lewinae* from Australia are gregarious and defoliate various species of *Eucalyptus* (Myrtaceae) and *Exocarpus* (Santalaceae). Marked sexual dimorphism is evident in the Australian species *Cotana serranotata*. The family comprises more than 300 species.

Family Mimallonidae (= Lacosomidae). This is a small family of some 200 species, consisting of small to medium robust moths [30 to 60 mm (1 to 2.5 in)] with broad, often forked wings, characterized by a number of primitive morphological characters. In the past the family was assumed to be close to the psychids on the basis of larval habits and to the drepanids in view of adult morphology, but today it is thought to be related to the bombycids and saturnids. Exceptionally among the higher Ditrysia the caterpillars of this family build movable cases in which they hide. The cases always have two openings but vary in shape and solidity; that of *Cicinnus despecta* from South America is curved and formed of a tough mixture of silk and larval excrement. The mimallonids are restricted to the Americas and are chiefly neotropical, only four species being known from the United States. The chief genera are *Lacosoma* and *Cicinnus*.

Family Bombycidae (Plate 53). This family is related to the saturnids and is characterized by the adult lack of mouth parts and by bipectinate antennae in both sexes. They are of medium size and have rather plump furry bodies, the wings are frequently falcate and relatively large and short, and the frenulum is rudimentary or absent. The caterpillars are elongated and covered with a thick pile and are armed with a dorsal horn on the eighth urite. Most species are to be found on the leaves of nettles and allies (Urticaceae). The pupa is enclosed in a silken cocoon; that of some species, such as *Bombyx mori*, the silkworm, is of great commercial value. The family is almost cosmopolitan in temperate and tropical zones but is concentrated in the Oriental region. Some 300 species belong to it. The best known genus is *Bombyx*, which has six species found from India to Japan. *Bombyx mori* is today exclusively domestic and seems to have derived from a common

Plate 52 **Anthelidae, Carthaeidae, and Eupterotidae**

Family Anthelidae: This family of Bombycoidea is restricted to Australia and New Guinea: (1) *Anthela nicothoe*, male and (1*a*) cocoon [75 mm (3 in)], (2) *Nataxa flavescens*, female (left) [35 mm (1.4 in)], male (right) [25 mm (1 in)], (3) *A. neuropasta*, male [70 mm (2.8 in)], (4) *A. denticulata*, male [45 mm (1.8 in)], (5) *Chelepteryx felderi*, male [80 mm (3.1 in)]. Family Carthaeidae: This family is also native to Australia and consists of the single species illustrated here: (6) *Carthaea saturnioides*, male at rest and with wings spread [80 mm (3.1 in)]. Family Eupterotidae: (7) *Cotana serranotata*, female (left) [62 mm (2.4 in)], male (right) [36 mm (1.4 in)], Australian, (8) *Phiala cunina*, male [64 to 82 mm (2.5 to 3.2 in)], Ethiopian, (9) *Panacela lewinae*, male (left) [28 mm (1.1 in)], female (right) [34 mm (1.3 in)], Australian.

Plate 53 **Bombycidae, Lemoniidae, and Brahmaeidae**

Family Bombycidae: (1) *Bombyx mandarina*, male [34 to 47 mm (1.3 to 1.9 in)]; family Lemoniidae: (2) *Lemonia dumi*, male [45 mm (1.8 in)], Palaearctic, (3) *L. philopalus*, male [45 mm (1.8 in)], Palaearctic, (4) *Spiramiopsis comma*, male [50 mm (2 in)], Ethiopian, (5) *Sabalia barnsi*, male [60 mm (2.4 in)], Ethiopian, (6) *L. taraxaci* [40 mm (1.6 in)], Palaearctic; family Brahmaeidae: (7) *Dactyloceras widenmanni*, male [110 mm (4.3 in)], Ethiopian, (8) *Acanthobrahmaea europaea*, male, (8*a*) larva, and (8*b*) pupa [60 mm (2.4 in)], Palaearctic, (9) *Brahmaea wallichii*, male [130 mm (5.1 in)], Palaearctic, Oriental, (10) *Calliprogonos miraculosa*, male [60 mm (2.4 in)], Palaearctic.

ancestor of *B. mandarina*, found in China, Korea, and Japan. The scientific literature concerning the silkworm is vast, and it is undoubtedly the most intensively studied lepidopteran. Briefly, its biological cycle is as follows: The virgin female attracts the male by emitting a specific pheromone, known as *bombykol*. Courtship is short and followed by mating, which lasts one or two days. The fertilized female then lays around 500 eggs measuring 1.5 mm in diameter. The eggs do not develop at once but enter diapause, the larvae emerging only at the end of winter. There are three or four larval stadia. The larvae are initially 3 to 4 mm (0.12 to 0.16 in) in length, dark grey, and quite hairy. They are reared in an ambient temperature of about 22 to 24°C and feed on a constant supply of mulberry leaves (*Morus alba*). Some thirty to forty days after hatching they reach 80 mm (3 in) in length, when their bodies are almost glabrous and lighter in colour. Ten to twelve days after the final moult the larvae stop feeding and prepare to spin their cocoon, a process that requires at least two days of uninterrupted work. The cocoon measures some 35 mm (1.3 in) in length and consists of a single strand of silk, which varies in thickness, colour, and length with the different breeds of silkworm. The extremes of length are 300 m (1,000 ft) and 2,000 m (6,500 ft), and usually more than 3,000 cocoons are needed to produce 1 kg (2 lb) of silk. Inside the cocoon the larva metamorphoses into the pupa, and some three weeks later the imago emerges, secreting a corrosive alkaline liquid that eats through the silken wall of the cocoon. Silk producers do not allow this to happen but kill the pupae by immersing them in boiling water or by drying them in ovens. Other noteworthy genera of bombycids include *Ocinara*, which is Ethiopian and oriental, and the oriental *Trilocha* and *Theophila*. The latter contains the species *T. huttoni* (India, Malaysia, Borneo, and Sumatra), which is raised for commercial purposes like the silkworm.

Family Lemoniidae (Plate 53). This is a small and heterogeneous family, containing moths that are often confused with the lasiocampids and eupterotids. There are two genera, the palaearctic *Lemonia* and the Ethiopian *Sabalia*, or three if the southern African species *Spiramiopsis comma* is included; but this is alternatively classed among the eupterotids or brahmaeids. There are twelve species of *Lemonia*, medium-sized moths without ocelli or proboscis and with tibial spurs on the forelegs; the latter is an exceptional character among lepidopterans. The larvae have no silk gland, and hence the pupa is not surrounded by a cocoon. *Lemonia* usually produces only one generation annually, and the females emerge before the males, which is rare among lepidopterans. In some species the adults live only two or three days, and the males fly by day as well as by night, even in bad weather. *Lemonia dumi* is found from western Europe to the Urals, and *L. taraxaci* is a polytypical species found from central Europe and the Mediterranean to the steppes of Russia. The third species illustrated here is *L. philopalus*, which is found in Spain and north Africa. *Sabalia*, illustrated here by the Congolese species *S. barnsi*, has not been definitively classified, and in some systems it is given the status of a separate family.

Family Brahmaeidae (Plate 53). These moths are easily distinguished from the other Bombycoidea because of their peculiar wing coloration, which consists of a mosaic of submarginal ocelli and a series of close, wavy lines. There are some twenty species, which are Ethiopian, south European, and oriental in distribution. The adults are medium sized or large with bipectinate antennae in both sexes; the proboscis is present though nonfunctional; and the labial palps are large. The abdomen is stout, and the wings broad. The larvae are curious in appearance, being glabrous and brightly coloured with fleshy appendages on the thorax and abdomen that decrease in size with age. Four genera are known. The type genus *Brahmaea* contains about half the species and is found in the Palaearctic region (an example being *B. ledereri* from the Near East) and in the Oriental region; for example, the large *B. wallichii* is found in the mountain forests of India, China, and Japan. The genus *Acanthobrahmaea* contains a single species, *A. europaea*, discovered in southern Italy in 1963 by the lepidopterist F. Hartig. This lovely moth inhabits only a small wooded area of Monte Vulture, near the lakes of Monticchio, Basilicata, and is the only European representative of the family. The larvae of Hartig's brahmaea are gregarious, at least in the first stadium. The food plant is not known although it is suspected to be ash (*Fraxinus excelsior*) or possibly privet (*Ligustrum vulgare*), two species of Oleaceae to which certain other non-European species of brahmaeid are tied. The adults emerge between March and April and live about three weeks. They fly by night and are attracted by light. The genus *Dactyloceras*, illustrated by the east African *D. widenmanni*, contains only Ethiopian species. From the Tapai Mountains in central and northern China *Calliprogonos miraculosa* is an extremely rare species that differs from other brahmaeids in coloration, as can be seen from the illustration.

Family Carthaeidae (Plate 52). This family was established in 1966 by I. F. B. Common and contains a single southeastern Australian species, *Carthaea saturnioides*, which, alone among the Bombycoidea, has primitive three-segmented maxillary palps. It has ocelli on the hind wings that become visible when the insect is disturbed. The caterpillars also have eyespots and a dorsal horn on the eighth abdominal segment. They feed on Proteaceae (genus *Dryandra*).

Family Oxytenidae. The oxytenids consist of some forty neotropical species reminiscent of the uraniids and eupterotids in appearance but phylogenetically close to the saturniids. The genera are *Oxytenis*, medium-sized moths with forewings hooked at the apex and hind wings with tails, and *Asthenidia*.

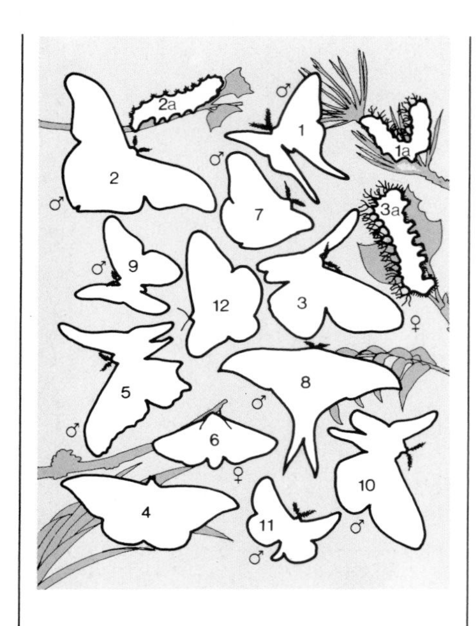

Plate 54 **Saturniidae**

Family Saturniidae: (1) *Graellsia isabellae*, male and (1*a*) larva [85 mm (3.3 in)], Palaearctic, (2) Cynthia moth (*Samia cynthia*), male and (2*a*) larva [110 mm (4.3 in)]. Palaearctic, (3) *Saturnia pyri*, female and (3*a*) larva [100 to 150 mm (4 to 5.9 in)], Palaearctic, (4) *Automeris* sp. (compare *crudelis*), Neotropical, (5) *Athletes steindachneri*, male [140 mm (5.5 in)], Ethiopian, (6) *Eochroa trimeni*, female [65 mm (2.6 in)], Ethiopian, (7) ceanothus silk moth (*Hyalophora euryalus*), male [85 mm (3.3 in)], Nearctic, (8) *Actias selene*, male [110 mm (4.3 in)], Neotropical, (9) promethea moth (*Callosamia promethea*), male [90 mm (3.5 in)], Nearctic, (10) splendid walnut moth (*Citheronia splendens*), male [110 mm (4.3 in)], Nearctic, (11) Nevada buck moth (*Hemileuca nevadensis*), male [50 mm (2 in)], Nearctic, (12) unidentified species.

Family Cercophanidae. The ten species of this family used to be classified among the oxytenids. They are natives of the Andes, extending from Colombia to Chile.

Family Saturniidae (= Attacidae; Plate 54). This family contains some of the largest moths in the world. Some enormous females of the Atlas moth (*Attacus atlas*) and *Coscinocera hercules* from Indo-Australia have wingspans of up to 300 mm (12 in). Although most are large or very large, there are also small- and medium-sized saturniids. Some 1,100 species exist in all, grouped into 125 genera. The family is cosmopolitan but is considerably more abundant in tropical and subtropical zones. The adults are hairy with small heads; they lack ocelli and have reduced mouthparts. The antennae may be simple — though only in the females — or more frequently bi- or quadripectinate and feathery; in the males they have sensitive cells capable of picking up the smell of the females even if they are a few kilometers away. The forewing is long, narrow, and occasionally falcate at the apex. It lacks the posterior cubital vein, with A_1 and A_2 forked at the base. The hind wing often has long tails, as in *Graellsia* (Europe), *Argema* (Africa and tropical Asia), *Eudaemonia* (Africa), and *Actias* (tropical Asia and North America), of which the oriental *Actias selene* is illustrated here. The luna moth (*Actias luna*) is another striking example. The wing coupling is amplexiform. Tympanal organs are absent. The egg is large, flattened, and unsually smooth. Newly hatched larvae have a dorsal horn on the eighth urite and striking tubercles often arranged in several branches with abundant setae and spines. The length of these colourful structures, known as *scoli*, generally varies in inverse proportion to the age of the larva with the exception of the Hemileucinae, where even the mature larvae have spectacularly well-developed scoli. Hemileucinae and the Saturniinae both have poisonous scoli. The pupa is glabrous, with or without cremaster. Only the two above-mentioned subfamilies have cocoons, and that of certain Asian species of *Antheraea* has commercial value. The pupa of the other five subfamilies lacks a cocoon and is buried in a cell in the earth. Most saturniids fly by night and are easily attracted by light although many species are diurnal in the male and nocturnal in the female, the nuptial flight taking place by day. It is interesting to note that diurnal forms have relatively reduced eyes and are remarkably and efficiently adapted to escape predators by such means as suddenly exposing the ocelli on their hind wings (a behaviour typical of this family; see Plate 80). Their caterpillars also have painful, stinging setae. One of the finest examples of colour variation and pattern is illustrated in Plate 54.

An interesting feature of the reproductive cycle is that the female often lays unfertilized eggs, which develop into either all males or, as in the subfamily Ludiinae, all females. Many subfamilies of saturniids have a distribution typified by geographic isolation. Three subfamilies are natives of the Old World: Agliinae, that is, *Aglia tau* and *A. japonica*, are palaearctic; the Ludiinae, which include various small saturniids such as *Ludia*, *Holocerina*, and *Carnegia*, are Ethiopian; and the Salassiinae are typical of tropical Asia. A further three subfamilies inhabit the New World. Of these the Rhescynthiinae are neotropical, whereas the Citheroniinae, sometimes classed as a separate family, and the Hemileucinae are also found in the Nearctic region; examples of these are the imperial moth (*Eacles imperialis*) and the splendid walnut moth (*Citheronia splendens*) from the United States and Mexico and the Nevada buck moth (*Hemileuca nevadensis*) from the western United States. The last subfamily, Saturniinae, is cosmopolitan in distribution. It includes the Atlas moth (*Attacus*) and *Coscinocera hercules* and many other large species well known to collectors as, for example, *Graellsia isabellae* from Spain and the French Alps, the ailanthus silkworm (*Samia cynthia*), which is native to tropical Asia but is raised commercially in various places in North America and southern Europe, and *Saturnia pyri*, the largest European moth. Of the American Satuniinae, the promethea moth (*Callosamia promethea*) from the eastern United States and the ceanothus silk moth (*Hyalophora euryalus*) from the western United States and Mexico are illustrated here. The superb *Eochroa trimeni* and *Athletes gigas* belong to genera native to tropical Africa.

Family Ratardidae. This small family is found in the Oriental region from India to Borneo and Japan. The moths are small [40 to 50 mm (1.5 to 2 in)] and soberly coloured. The wings are narrow, elongated, and rounded at the apex. Some dozen species are known; the most important genus is *Ratarda*.

Superfamily Sphingoidea

The only family in this group is the Sphingidae. Not all specialists accept the superfamily ranking of the sphingids, and in the past they were held to be related to the superfamily Bombycoidea, a notion that has recently been revived in classifications such as J. P. Brock's.

Family Sphingidae (Plate 55). This family contains common moths that may be easily distinguished by their peculiarities of shape and behaviour in both adult and larval instars. Their size varies enormously from 10 mm (less than 0.5 in) in the Madagascan species *Sphingonaepiopsis obscurus* to 200 mm (8 in) in the females of the neotropical *Cocytius antaeus*. The imago has a large head with well-developed eyes, probably adapted to its crepuscular habits. The antennae are relatively short, clubbed and hooked at the apex, and often partly pectinate in the male. The proboscis is very sturdy and varies in length from short and nonfunctional in the holarctic *Smerinthus* and African *Polyptychus* to — more commonly — long or very long, reaching as much as 250 mm (10 in) in *C. antaeus* and over 280 mm (11 in) in the tropical American *Amphimoea*

Plate 55 **European Sphingidae**

Superfamily Sphingoidea, family Sphingidae: The family is structurally very homogeneous and contains some 1,000 species found throughout the world. Illustrated here are some common European species in larval and adult instars. Other members of the Sphingidae in various stages are illustrated in Plates, 6, 7, 21, 27, 70, 73, 75, 79, 80, and 108: (1) lime hawkmoth (*Mimas tiliae*), adult and (1*a*) larva [65 mm (2.6 in)], (2) pine hawkmoth (*Hyloicus pinastri*), adult and (2*a*) larva [75 to 90 mm (3 to 3.5 in)], (3) privet hawkmoth (*Sphinx ligustri*), adult and (3*a*) larva [80 to 110 mm (3 to 4.3 in)], (4) spurge hawkmoth (*Hyles euphorbiae*), adult and (4*a*) larva [85 mm (3.3 in)], (5) broad-bordered bee hawkmoth (*Hemaris fuciformis*), adult and (5*a*) larva [40 mm (1.6 in)].

walkeri. The long proboscis allows certain species to feed on the nectar of flowers with deep corollas, a source of nourishment not available to other lepidopterans. The mouthparts lack maxillary palps. The legs are sturdy, are armed with sharp spurs, and sometimes bear coxal olfactory organs. The wings are triangular, the forewings being long and narrow and lacking the posterior cubital vein, while the hind wings are small in surface area with the subcosta joining the first radial vein. The frenulum is highly developed. In some cases the wings may be semitransparent as in *Hemaris* from the northern hemisphere and *Cephenodes* from Australia. Some groups like *Smerinthus* and related genera have characteristic ocelli on the hind wings, which they expose when they are disturbed (Plate 80). The hawkmoths include some of the heaviest Lepidoptera, but although their abdomens are thick, they are tapered and aerodynamically shaped. These features make it hard for them to glide as other lepidopterans commonly do but give them a powerful and rapid flight, which can reach 40 to 50 km/hr (25 to 30 mi/hr). The wing beat is not very deep, that of *Acherontia atropos* being estimated at 75° in comparison with the 150° of *Vanessa atalanta*, but it is correspondingly rapid — indeed the fastest among the lepidopterans — at twenty-two beats per second in *A. atropos* compared with ten in *Vanessa atalanta* and over forty in other sphingids. This rapid wing beat can be heard in some species and can raise the body temperature by as much as 8 to 10°C. This motion also allows hawkmoths to hover and thus, as with hummingbirds, to explore and exploit flower nectar. Some sphingids, such as *Macroglossum stellatarum* from the Palaearctic region, are diurnal, but most species fly by night or at dusk and are attracted by light. The females hover to lay their eggs, usually singly. The larvae are cylindrical, glabrous, and wrinkled, with the characteristic dorsal horn on the eighth urite, which decreases in length with age. The larvae are usually green or brown, often with striking stripes or eyespots. Some are cryptic and soberly coloured; others are bright but nevertheless cryptic as a result of somatolysis, stripes and spots breaking up the shape. The caterpillars generally stand with the thorax and the first two or three urites raised in the air like Egyptian sphinxes, an attitude that has given the family its name. When disturbed the larvae sway their heads rhythmically from side to side. Their food plants are varied, ranging from grasses to trees, but they show a certain preference for Euphorbiaceae, Solanaceae, and Apocynaceae. The larvae are solitary and on the whole cause negligible damage with the exception of species such as the palaearctic *Hyloicus pinastri* or the North American five-spotted hawkmoth (*Manduca guinguemaculata*), which can cause serious losses. The pupae have a cremaster and are cigar shaped and only rarely protected by a cocoon. They are usually buried in the earth or lie in the ground litter. Some pupae have a highly developed proboscis, which is folded back ventrally into a handle-shaped bar. Hibernation occurs during either the pupal or the adult instar, and occasionally there may be a long diapause, the adults emerging only after two or three years. In temperate zones there is typically only one generation a year, but occasionally a second incomplete one may be found, as in *Laothoe populi*. Two or more subfamilies are distinguished, depending on the system of classification. In this work the subdivision into Sphinginae and Macroglossinae is adopted, the species being differentiated principally by the absence or presence, respectively, of a tuft of sensorial setae on the first segment of the labial palp. The Sphinginae, also known as Asemanophorae, are characterized by a different type of male genital armature and by the reduction of the proboscis and frenulum; the Macroglossinae, also known as Semanophorae, have no significant structural modifications, have a well-developed proboscis, and are on the average smaller. The sphingids form a cosmopolitan family of some 1,000 species, which are mainly tropical in distribution. Cosmopolitan species such as the white-lined sphinx (*Hyles lineata*) are polytypical and exhibit easily distinguished geographical races. Other species too inhabit enormous areas. Thus the death's head hawkmoth (*Acherontia atropos*) is found from the Lofoten Islands to South Africa and from the Azores to Iran, while the oleander hawkmoth (*Daphnis nerii*) inhabits Europe, Africa, and India and the convolvulus hawkmoth (*Herse convolvuli*) is a native of central and southern Europe, Africa, and Indo-Australia. Various sphingids are migratory (Plate 70).

Superfamily Notodontoidea

The Notodontoidea are closely related to the Noctuoidea — so much so that they are sometimes included among them, resembling them principally in having thoracic tympanal organs. This is sufficient to distinguish them from the Geometroidea and Pyraloidea, which have abdominal tympana. The major differences between the Notodontoidea and the Noctuoidea lie in the structure of their auditory organs, which in the former are relatively simple and open downwards but in the latter are much more complex with the opening pointing backwards. The wing venation is another distinguishing feature. The vein M_2 of the forewing originates midway between the posterior cubital vein and the radial veins in the Notodontoidea, whereas in the Noctuoidea it starts nearer the anterior cubital vein. Other differences may be observed at the level of larval bristle patterns. The group is divided into three families, the Notodontidae, which is the largest, the Dioptidae, and the Thyretidae.

Family Notodontidae (Plate 56). There are some 2,600 species in this cosmopolitan family of sturdy moths, also known as the "prominents," which are medium to large in size and have large heads usually bearing ocelli but without chaetosemata and with bipectinate or occasionally in the female filiform antennae. The proboscis and labial palps are present

Plate 56 **Notodontidae**

Superfamily Notodontoidea, family Notodontidae: (1) lobster moth (*Stauropus fagi*), adult and (1*a*) larva [62 mm (2.4 in)], Palaearctic, (2) *Leucodonta bicoloria* [32 mm (1.2 in)], Palaearctic, (3) puss moth (*Cerura vinula*) [65 to 80 mm (2.6 to 3 in)], Palaearctic, (4) buff tip (*Phalera bucephala*), at rest and in flight and (4*a*) larva [54 mm (2.1 in)], Palaearctic, (5) pale prominent (*Pterostoma palpina*), adult and (5*a*) larva [45 mm (1.8 in)], Palaearctic, (6) *Notodonta ziczac*, adult and (6*a*) larva [35 mm (1.4 in)], Palaearctic, (7) *Epicoma melanospila*, male (left) [52 mm (2 in)], female (right) [62 mm (2.4 in)], Australian, (8) *Danima banksiae* [80 mm (3 in)], Australian, (9) *Tarsolepis sommeri* [50 to 78 mm (2 to 3 in)], Oriental, (10) *Moresa magniplaga* [52 mm (2 in)], Neotropical, (11) *Chliara croesus* [50 mm (2 in)], Neotropical.

and generally well developed (Notodontinae) or highly reduced (Thaumatopoeinae); the maxillary palps are tiny and multisegmented or absent. The head and thorax are often covered by a crest of hard, sclerotized scales; the limbs are hairy. The forewings are elongated and lack the posterior cubital vein, while M_2 is either almost parallel to M_3 of else close to M_1. In many genera of Notodontinae (as, for example, in *Notodonta*, illustrated here in the palaearctic species *N. ziczac*, and in *Ochrostigma*, *Lophopteryx*, and *Pterostoma*) the posterior margin of the wing bears a lobe formed of scales that resembles a dorsal tooth projecting from the middle of the insect's body when it is seen from the side, and the family name is derived from this. The hind wings have a frenulum and lack the posterior cubital vein. All prominents have metathoracic tympanal organs. In many species the female abdomen terminates in a thick tuft of setae. The eggs are almost spherical. The larvae of the two subfamilies differ in shape and behaviour, those of the Notodontinae being arboreal, glabrous, and covered with bumps, spines, and tubercles, which are especially prominent in the early stadia. They are generally cryptically coloured, but strange-looking, brilliantly coloured forms are not uncommon. Many species have much reduced or absent anal prolegs and stand with the last two or three urites raised in the air. In certain groups, such as the two cosmopolitan genera *Cerura* and *Dicranura*, the anal prolegs are transformed into a pair of long, hollow tubercles concealing delicate bright red filaments that secrete a pungent-smelling substance; they can be evaginated and moved when the larvae are aroused. The highly coloured larvae of *C. vinula* and other Notodontinae also have a remarkable chemical defense mechanism against predators: Their prothoracic gland secretes a corrosive liquid containing formic acid, which can be sprayed as far away as 15 to 20 cm (6 to 10 in). The larva of *Stauropus fagi* is equally remarkable: Its meso- and metathoracic legs are hyperdeveloped and its body warty with a forked appendage on its dilated final urites. When disturbed, this larva assumes a highly effective threatening attitude, raising its head and tail and cleaving the air with its long, thin legs. The larvae of the subfamily Thaumatopoeinae always have the full complement of prolegs and are warty, colourful, and covered with short, stinging hairs that can cause severe rashes even in man. They dwell mainly in trees and can prove harmful to forest trees; many species show distinct presocial behaviour. The gregarious larval instinct is well illustrated by *Thaumatopoea pityocampa*, the processionary caterpillar moth. At a certain point in their life cycle the caterpillars of this species spin large silken nests at the tips of pine branches, in which great numbers of individuals live. They search for food by night, moving in a long procession of up to hundreds of individuals, which can measure up to a dozen meters (yards) in length. On both the outward and return journey the procession has a leader followed by the other larvae either in single file or in several parallel lines.

Family Dioptidae (Plate 57). This is a family of some 400 species that are related to the Notodontidae, from which they can be distinguished by the formation of their tympanal organs and wing venation. They are American in distribution, most species being found in the Neotropical region. Only the California oak moth (*Phryganidia california*) is found in the United States. The adults of this group are diurnal and are often brightly coloured, frequently mimicking geometers, nymphalids, riodinids, hypsids, arctiids, and ctenuchids and thus varying greatly between species. Two neotropical species are illustrated, *Dioptis egla* and *Josia lativicta*.

Family Thyretidae (Plate 57). This family is difficult to classify; it consists of a few hundred tropical African species that were once placed in the Ctenuchidae. They differ from the latter chiefly in the typically notodontid structure of their tympanal organs. They are mostly small but sturdy with a reduced proboscis, pectinate antennae, narrow, elongated forewings, and rather small hind wings with reduced venation. They are sober in colour, and the wings often have transparent windows. Some females are brachypterous, including those of *Automolis meteus*, the male of which is illustrated. The primitive or modified state of many of their morphological characters has led to the conclusion that the thyretids and notodontids are of more recent origin than the dioptids.

Superfamily Noctuoidea

The last superfamily of ditrysian lepidopterans contains the greatest number of species as well as the largest family, the Noctuidae. The Noctuoidea form a large, homogeneous group, the adults of which vary greatly in size and may or may not possess ocelli but almost always lack chaetosemata. The proboscis is usually present and highly developed, and the maxillary palps are small. The second median vein, M_2, may be absent in the forewing, the posterior cubital vein is always absent in both wings, and the hind wing has two anal veins and the frenulum. The tympanal organs are metathoracic and usually well developed. The larvae are frequently very hairy and have prolegs with a single series of crochets. Some authors include the Notodontoidea in this superfamily; in any case the number of families recognized varies. Seven are listed here, of which the ctenuchids and hypsids are very closely related to the arctiids, whereas the nolids, lymantriids, and agaristids are closer to the noctuids.

Family Ctenuchidae (= Syntomidae, Euchromiidae, Amatidae; Plate 57). This cosmopolitan family contains some 3,000 species, which are concentrated in the tropics. They are small- to medium-sized moths with antennae of varying shape, a proboscis, and small palps. The forewings are narrow and often lack the retinaculum; the hind wings are reduced in surface area and have the vein group Sc + R_1 fused with the radials. The

tympanal organs may be rudimentary or absent. Overall the caterpillars resemble those of arctiids: They are short and cylindrical with warts bearing tufts of hairs, setae, or hair pencils. They feed mainly on herbaceous plants and lichens. The pupa is protected by an oval cocoon. Ctenuchids are closely related to the arctiids and are treated as a subfamily (Ctenuchinae) of the latter in some systems of classification. In contrast there is no real affinity between the ctenuchids and the zygaenids, the similarities observable between various species of these groups being due to convergent adaptation through a process of mimicry. Most ctenuchids are diurnal. They fly slowly and rest for long periods on flowers, and the adults of many species are gregarious in habit; together with their brilliant aposematic coloration they are thus particularly conspicuous. They are protected by a toxic substance in the hemolymph that makes them unappetizing to many predators, as has frequently been demonstrated in experiments. Mimicry is extremely common and involves coloration, morphology, and behaviour. Large numbers of species, particularly those of the Neotropical region, which is the richest in ctenuchids, mimic wasps. Practically every species of aculeate Hymenoptera in tropical America is mimicked by a species of Ctenuchid. Varied patterns of mimicry are found: Thus there are Ctenuchid genera with uniform blue-black metallic wings and abdomen; genera with semitransparent wings, partially or completely lacking in scales, and with alternating black and yellow or red and black bands on their abdomen; genera with black wings and white spots or transparent windows or polychrome iridescent metallic spots; and genera with polychrome abdominal bands. The colour variations are almost infinite and can be extremely spectacular, as in *Euchromia lethe* from the Congo. Some species even exhibit a remarkable narrowing between thorax and abdomen, which is so like a wasp waist that it deceives predatory birds and lepidopterists alike. The males of the neotropical *Trichura* are extremely unusual in that their eighth abdominal segment extends into a process reminiscent of the ovipositor of the parasitic ichneumonid wasps (see Plate 57). Antennae and legs may also be coloured. Of the many neotropical mimicking ctenuchids mention may be made of *Phaeosphecia opaca* and the dozens of forms found in the genus *Pseudosphex* that mimic the sphecid wasps, *Pompilopsis tarsalis*, which mimics the pompilids, the wasp-mimicking *Horama*, and the genera *Cosmosoma* and *Dasysphinx*. Ctenuchids mimic not only wasps but also lycid beetles (which are equally inedible) and a number of lepidopteran families, including the zygaenids. A well-known example of mimicry is that exhibited by certain species of *Amata* (= *Syntomis*), a large ctenuchid genus from the Old World, and populations of *Zygaena ephialtes* in central Italy; this is illustrated in Plate 85. Adaptation to avoid predators seems to have played a decisive part in the evolution of the ctenuchids, as is illustrated not only by the high incidence of apoesematic colorations and mimicry but also by the not-infrequent occurrence of other types of defensive mechanism, such as the production of repellent secretions and the emission of high-frequency sound waves capable of confusing the radar systems of bats. Finally mention must be made of other groups in this interesting family: the type genus *Ctenucha* with some fifty species and *Syntomeida* with some ten species, both chiefly neotropical in distribution but also well represented in the southern United States, the lovely *Euchromia* species, the small Indo-Australian and Ethiopian *Ceryx*, and the chiefly European *Disauxes*. The Palaearctic region is poor in ctenuchids: Various *Amata*, *A. phegea*, *A. ragazzii*, and *A. marjana*, are found in Europe as well as *Disauxes ancilla* and *D. punctata*.

Family Hypsidae (= Aganaidae, Pericopidae; Plate 57). These are medium to large moths with a large head, ocelli, extremely prominent eyes, antennae that are ciliate in the male, a well-developed proboscis, single-segmented maxillary palps, and long labial palps, the apical segment of which is pointed and usually glabrous, while the interemediate segment is frequently hairy and highly coloured. The retinaculum may be absent. The eggs are laid in clusters. The larvae resemble those of the noctuids and are sparsely covered with long hairs; they are often brightly coloured and gregarious during early stages. The pupa is protected by a light silken cocoon. Like the ctenuchids the hypsids are often treated as part of the Arctiidae or as a single subfamily under the name of Hypsinae or as two subfamilies, the Aganainae, which include the Old World genera, and the Pericopinae, which include the American genera. Other classificatory systems raise one or another of these two subfamilies to the rank of family (Aganaidae and Pericopidae). The hypsids are found throughout the tropics and are particularly numerous in Central and South America, some genera, such as *Gnophaea* and *Composia* (illustrated by the Antillean species *C. credula*), extending as far as the southern United States. Sexual dimorphism is common, and the males often have prominent olfactory organs, the shape and size of which cause noticeable modifications in wing structure. Thus in *Euplocia membliaria*, which is found from southern China to Java and the Philippines, the basal part of the costal margin of the forewing arches upwards to accommodate a large olfactory pouch protected by a fold of wing membrane. In males of the oriental species *Aganais orbicularis* the whole surface of the forewing is covered with olfactory hairs. Mimicry is another very common character in this family, which, together with the degree of sexual dimorphism displayed, gives rise to considerable variation in appearance among the adults. Most genera containing mimetic species are neotropical; they include *Dysschema* (= *Pericopis*) with approximately fifty species, some of which mimic swallowtails and nymphalids, *Hypocrita* (some twenty-five species), which mimics nymphalids and riodinids, *Darna* and *Sagaropsis* (some twenty species), which mimic riodinids and dioptids, and *Chetone* (about ten

Plate 57 **Other Notodontoidea and some Noctuoidea**

Superfamily Notodontoidea, family Dioptidae: (1) *Dioptis egla* [35 mm (1.4 in)], Neotropical, (2) *Josia lativicta* [30 mm (1.1 in)], Neotropical; family Thyretidae: (3) *Automolis meteus* [38 mm (1.5 in)], Ethiopian.
Superfamily Noctuoidea, family Ctenuchidae: (4) *Horama oedippus* [28 to 40 mm (1.1 to 1.6 in)], Neotropical, (5) *Euchromia lethe* [48 mm (1.9 in)], Ethiopian, (6) *Trichura cerberus*, male [30 mm (1.1 in)], Neotropical, (7) unclassified Antillean species, (8) *Amata mestralii* and (8*a*) *Amata* sp., larva [35 mm (1.4 in)], Palaearctic, (9) unclassified Guatemalan species; family hypsidae: (10) *Chetone histrio* [70 to 100 mm (2.8 to 3.9 in)], Neotropical, (11) *Composia credula* [50 mm (2 in)], Neotropical, (12) *Eucyane arcaei* [56 mm (2.2 in)], Neotropical, (13) *Darna colorata* [38 mm (1.5 in)], Neotropical, (14) *Aganais orbicularis* [75 mm (3 in)], Oriental; family Nolidae: (15) *Nola chlamitulalis* [15 mm (0.6 in)], Palaearctic; family Arctiidae: (16) *Amsacta lactinea* [52 mm (2 in)], Oriental, (17) *Viviennea moma* [35 mm (1.4 in)], Neotropical, (18) *Rhodogastria crokeri* [68 mm (2.7 in)], Australian, (19) *Diacrisia breteaudeaui*, at rest and with spread wings [24 to 37 mm (0.9 to 1.5 in)], Palaearctic, (20) *Panaxia dominula*, larva, Palaearctic.

species), which mimics castniids, papilionids, pierids, and nymphalids. Perhaps the most famous of the neotropical hypsid mimics is *Anthomyza heliconides*, one of a complex group of mimicking lepidopterans with transparent wings and black bands that includes the butterfly genera *Dismorphia*, *Ituna*, and *Philaethria* and the castniid *Gazera linus*. Outside America we may note the hypsid genera *Aganais* and *Argina*, which are Ethiopian and oriental, the Ethiopian *Phaegorista*, and the forty-odd species of *Asota*, which cover a huge area from northern Australia to the Oriental region, China, and Japan.

Family Nolidae (Plate 57). This small family used to be classed among the arctiids although it is probably in fact related to the sarrothripine noctuids; it contains smallish moths cosmopolitan in distribution. The antennae are often pectinate or ciliate in the male and simple in the female; ocelli are absent, the proboscis is short, and the labial palps are well developed. The imago is soberly coloured and differs from the noctuids in that it has tufts of prominent scales arranged in several rows on the forewing, which give the wing surface a wrinkled appearance. The larvae resemble those of the arctiids but lack the first pair of prolegs; they feed on lichens. Several hundred species have been described. The type genus is *Nola*, which is cosmopolitan and contains over 300 species. *Nola* (= *Celama*) *chlamitulalis* from Asia Minor and south and southeastern Europe is illustrated.

Family Arctiidae (Plates 57 and 58). These moths derive their name from the woolly-bear larvae. The family is large and cosmopolitan with some 10,000 species concentrated in the tropics. The adults are small to medium in size with antennae that are bipectinate in the male only, a reduced proboscis, and short palps. In the forewing the vein M_2 is close to M_3 at the base; in the hind wing the group Sc + R_1 begins at different points in the radial sector, depending upon the genus. The tympanal organs are metathoracic and open outwards towards the base of the abdomen. The eggs are hemispherical and are usually laid in clusters. The larvae are covered with dense tufts of long hair and have verrucae on the thorax. Some aposematic species are gregarious. They feed on lichens (Lithosiinae) and various families of angiosperms, such as Asteraceae, Boraginaceae, and Convolvulaceae (Arctiinae), some species causing damage to crops. The pupa is glabrous and often without a cremaster. It lies in a silken cocoon interwoven with larval hairs. The majority of both larvae and adults are brightly and frequently aposematically coloured. It has been shown that in many species the larvae absorb substances (histamines and alkaloids) from the food plant that make them poisonous or unpalatable, but a number of larvae form these substances themselves. Aposematic coloration is a highly important feature of this family — the cinnabar moth (*Tyria jacobaeae*) is just one example — as are both Müllerian and Batesian mimicry. A classic study of this phenomenon has been carried out by M. Rothschild on two palaearctic species of *Spilosoma*, one of which, *S. lubricipeda*, is unpalatable and acts as model for the other, *S. lutea*, which is quite palatable. *Nyctemera* also exhibits internal mimicry; species of *Lycomorpha* mimic the inedible lycid beetles.

The defense mechanisms of this family include certain types of behaviour that involve exposing various highly coloured body parts. Thus many species roll on their sides to reveal their bright abdomens. Other species (for example, *Rhodogastria* and *Utetheisa*) secrete an evil-smelling acidic froth from their prothoracic glands or play dead. Many nocturnal species are able to protect themselves because with their tympanal organs they can pick up the radar waves emitted by bats and thus escape with erratic flight or by dropping to the ground. Other unpalatable species emit ultrasonic waves that confuse bats that have not come across them and warn off those which have. In the case of the North American *Pyrrharctia isabella* the existence of a form of acoustic mimicry has been demonstrated; these edible moths emit ultrasounds similar to those of the inedible moths. Another characteristic of this interesting family is the intraspecific variation in patterning and colouring. Certain species of *Arctia*, *Callimorphia*, and *Parasemia* are striking examples.

The number of subfamilies varies in the different classifications. A. Seitz's work (1900 to 1934) distinguishes seven or eight, some of which (Callimorphinae, Spilosominae, Micrarctinae, Nyctemerinae) have been included in the subfamily Arctiinae and others (Pericopinae and Aganainae) among the hypsids. Today two subfamilies, Lithosiinae and Arctiinae, are recognized. The Lithosiinae have slender bodies with large heads lacking ocelli. The wings are often unicoloured and elongated and cover the abdomen at rest. The type genus is *Lithosia* with some twenty Eurasian and African species; the palaearctic genera include *Pelosia* and *Eilema* (*E. sororcula*, *E. griseola*, and *E. caniola* are found in Europe) as well as *Miltochrista* with some hundred species in Europe and Indo-Australia. Most genera are tropical. In the Old World we find *Cyana*, *Asura*, *Chionaema* (with over 200 species), and *Damias* (with approximately fifty species), while neotropical groups include the glittering forms of *Chrysochlorosia* and the little *Callisthenia* species. The best represented subfamily, the Arctiinae, has ocelli, broad forewings of various colours, and a sturdy body; aposematic coloration predominates. Its chief genera are the tiger moths (*Apantesis*), *Arctia* with a dozen holarctic species, *Diacrisia* and *Spilosoma* from the tropics of the Old World, the small cosmopolitan genus *Utetheisa*, the palaearctic *Callimorpha*, and the large genus *Nyctemera* from the Old World tropics. In the Americas we find the neotropical *Rhipha* (20 species), *Viviennea*, *Halisidota* (120 species), *Amaxia* (some 20 species), and *Bertholdia*, as well as the North American *Arachnis*. Other genera include the palaearctic *Ochnogyna* and

Plate 58 **European Arctiidae**

Family Arctiidae: (1) *Chelis maculosa* [34 mm (1.3 in)], (2) ruby tiger (*Phragmatobia fuliginosa*), adult and (2*a*) larva [30 mm (1.2 in)], (3) buff ermine (*Spilosoma lutea*), adult and (3*a*) larva [35 mm (1.4 in)], (4) white ermine (*S. lubricipeda*) [40 mm (1.6 in)], (5) *Rhyparia purpurata*, adult and (5*a*) larva [48 mm (1.9 in)], (6) *Arctia hebe* [54 mm (2.1 in)], (7) cream-spot tiger (*A. villica*) [55 mm (2.1 in)], (8) tiger (*A. caja*), adult and (8*a*) larva [60 mm (2.4 in)], (9) *Callimorpha quadripunctaria* [54 mm (2.1 in)], (10) cinnabar moth (*Tyria jacobaeae*), adult and (10*a*) larva [40 mm (1.6 in)].

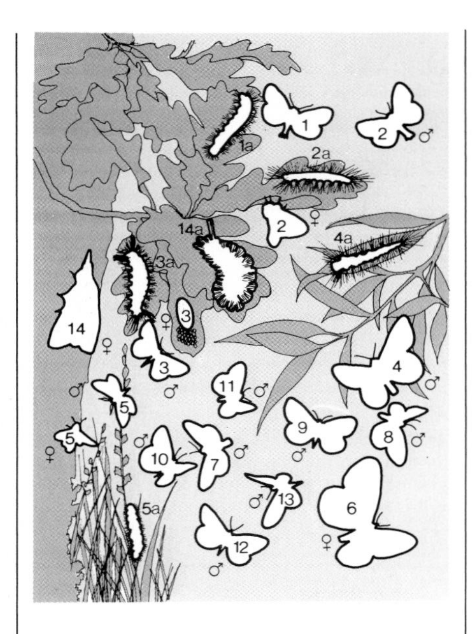

Plate 59 **Lymantriidae**

Family Lymantriidae: (1) brown tail (*Euproctis chrysorrhoea*), adult and (1*a*) larva [36 mm (1.4 in)], Holarctic, (2) yellow tail (*E. similis*), male above, female below, and (2*a*) larva [35 mm (1.4 in)], Palaearctic, (3) vapourer (*Urgyia recens*), male below [26 mm (1 in)], apterous female with egg above, and (3*a*) larva, Palaearctic, (4) black v moth (*Actornis l-nigrum*), male and (4*a*) larva [38 mm (1.5 in)], Palaearctic, (5) *Hypogymna morio*, male above [25 mm (1 in)], female below, and (5*a*) larva, Palaearctic, (6) *Cozola collenettei*, female [35 to 55 mm (1.4 to 2.1 in)], Oriental, (7) *E. limbalis*, male [40 mm (1.6 in)], Australian, (8) *E. edwardsi*, male [40 mm (1.6 in)], Australian, (9) *Porthesia lutea*, male [34 mm (1.3 in)], Australian, (10) *P. leucomelas*, male [35 mm (1.4 in)], Australian, (11) *P. melanosoma*, male [32 mm (1.3 in)], Australian, (12) *Euproctoides acrisia*, male [38 mm (1.5 in)], Ethiopian, (13) *Perina nuda*, male [30 mm (1.2 in)], Oriental, (14) pale tussock (*Dasychira pudibunda*), female and (14*a*) larva [52 mm (2 in)].

Phragmatobia, which is almost cosmopolitan — being absent only in Australia — as well as *Pericallia*, palaearctic and Ethiopian, with approximately 50 species, and *Rhodogastria*, *Asota*, and *Agape*, three chiefly Australian genera.

Family Lymantriidae (Plate 59). This is a family of some 2,500 species found chiefly in the Old World tropics. They are nocturnal and hairy and resemble the noctuids; they are medium sized, broad winged, and rather faded in colour with the exception of genera like the Ethiopian *Euproctoides* or *Cozola* from the Celebes, which resembles the butterfly *Euploea*. They lack ocelli and proboscis. The males have large, bipectinate antennae; the females have a thick tuft of hair at the tip of the abdomen that acts as a shield when they lay their eggs. They have vestigial wings and cannot fly (*Orgyia*) and are sometimes also legless (*O. dubia*). The caterpillars are very hairy and often urticating, causing serious rashes. Some species (*Euproctis similis*) have thick dorsal tufts of setae on the first to fourth abdominal segments and long, slender, hair pencils on the eighth urite. Sometimes the sixth and seventh urites bear evaginated organs like the osmeterium of swallowtails, which are connected to glands secreting aromatic substances that sometimes harden the hairs. Many of these caterpillars are brightly coloured and are not usually attacked by predators. They are arboreal and often gregarious, the larvae of some species such as the gypsy moth (*Lymantria dispar*), *L. monacha*, *Euproctis chrysorrhoea*, and some *Orgyia* causing serious damage. The pupa is enveloped in a cocoon interwoven with larval hairs. Marked sexual dimorphism is a common feature of lymantriids, in terms of both size — the females being as much as twice the size of the males — and coloration. One unusual type of dimorphism is seen in *Perina nuda*, found from India to Taiwan: The female has the appearance of a normal lymantriid, whereas the male resembles a psychid. Some of the most important genera are *Dasychira* (over 400 species), absent only from the Neotropical region, *Euproctis* (some 600 species) from the Old World and the Indo-Australian region, *Orgyia* (some 60 chiefly holarctic species), and finally two Old World genera, *Lymantria* (150 species) and *Leucoma* (some 40 species).

Family Noctuidae (Plate 60). These are robust moths of varying size, generally with ocelli and filiform antennae, which are pectinate and sometimes ciliate at the apex; the proboscis is well developed, the maxillary palps are single segmented, and the labial palps are developed and often ascending. The legs have epiphyses and spurs on the meso- or metathoracic tibiae and sometimes claws on tibiae and tarsi. Wing size is variable; the forewing is long and narrow, often with an extra cell, and the group A_1 and A_2 forked at the base. In the hind wing the base of the group $Sc + R_1$ is fused with the radial stem, and M_2 is lacking or poorly developed with its base closer to M_1 than to M_3 (in which case the wing is said to be trifid); or else M_2 is developed with the base nearer M_3 than M_1 (quadrifid). The frenulum is well developed. The tympana are metathoracic and open backwards. The males may have olfactory glands, for example, in *Phlogophora* (Amphipyrinae), and stridulating organs. The eggs are usually hemispherical with a longitudinally sculpted chorion and are laid singly or in clusters (Plate 5). The larvae are normally cylindrical, fleshy, and glabrous although there are some rare species with hairy larvae (Acronyctinae and Pantheinae). The prolegs have a single series of crochets; in some groups (Acontinae, Plusiinae, Catocalinae, Ophiderinae, Hypeninae) from one to three pairs of prolegs, usually the first and second, are lacking, and the caterpillars keep the first abdominal segments raised and are able to loop (for example, *Catocala*). The larvae of many noctuids are polyphagous and feed chiefly on leaves but more rarely on lichens (*Cryphia*, Acronyctinae) and fungi (*Parascotia*, Ophiderinae); or they may occasionally be predatory (*Cosmia*, Amphipyrinae, *Eublemma*, Acontiinae) or mining (*Sesamia*, Amphipyrinae). Some larvae are brightly coloured and ornamented (Plates 6 and 79). The pupa is cylindrical with a developed cremaster and lies on the ground in a silken cocoon (in the case of Acronyctinae, Plusiinae, and Catocalinae) or in a hole in the ground.

The noctuids are the largest lepidopteran family, consisting of more than 25,000 mostly nocturnal species, and vary in size [from 5 to 10 mm to more than 300 mm (0.2 to 0.4 to over 12 in)], colouring, and shape. However, the commonest species are cryptic and measure between 30 mm (1 in) and 60 mm (2 in). The noctuids inhabit the widest range of different environments and regions and have adapted in extraordinary ways to their ecological conditions, from the ability to use their proboscis to perforate the skin of fruit or the hides of certain mammals to the ability to perceive the ultrasounds emitted by bats and the ability to emit both ultrasounds and sounds audible to the human ear. Some species have flash colouring, but as the majority of these moths are nocturnal, cryptic coloration prevails. Many species are attracted by light or scent traps. Although certain characters vary tremendously, the noctuid family is relatively homogeneous; it is divided into approximately twenty subfamilies, some of which, such as the Ophiderinae, are very rich in species and contain more than 100 genera. The characters usually used in distinguishing the species are wing pattern and form of the external genitalia. Two groups of subfamilies are distinguished by the venation of the hind wing: In one the second median vein, M_2, is well developed (Quadriphinae), and in the other M_2 is tiny or absent (Triphinae). Other characters found in the adult that are useful in distinguishing the subfamilies are wing size, the presence or absence of cilia and hairs on the eyes, the conformation of the labial palps, and the presence or absence of tibial claws. The scheme followed here for the division of subfamilies is that proposed by W. B. Nye in the first volume of his *The Generic Names of Moths of the World* (1975).

Systematic Survey of Butterfly and Moth Families

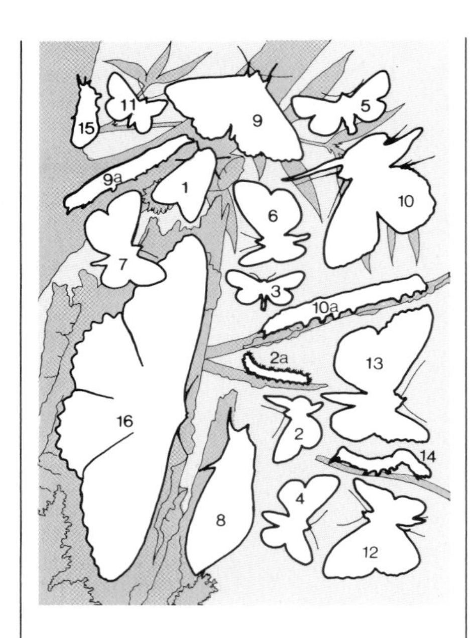

Plate 60 **Noctuidae**

Family Noctuidae, subfamily Pantheinae: (1) *Panthea coenobita* [45 mm (1.8 in)], Palaearctic.
Subfamily Acronyctinae: (2) *Arsilonche albovenosa*, adult and (2*a*) larva [30 mm (1.2 in)], Palaearctic, (3) *Mazuca amoena* [30 mm (1.2 in)], Ethiopian.
Subfamily Hadeninae: (4) *Diaphone eumela mossambicensis* (44 mm (1.7 in)], Ethiopian.
Subfamily Cucullinae: (5) *Calotaenia celsia* [40 mm (1.6 in)], Palaearctic.
Subfamily Chloephorinae: (6) *Eligma laetipicta* [62 mm (2.4 in)], Ethiopian, (7) *Egybolis vaillantina* [52 mm (2 in)], Ethiopian, (8) *Phyllodes floralis* [154 mm (6 in)], Oriental, (9) *Catocala nupta*, adult and (9*a*) larva [75 mm (3 in)], Palaearctic, (10) *C. fraxini*, adult and (10*a*) larva [85 mm (3.3 in)], Palaearctic, (11) *Grammodes stolida* [28 mm (1.1 in)], Palaearctic, Oriental, (12) *Cyligramma latona* [60 mm (2.3 in)], Ethiopian.
Subfamily Ophiderinae: (13) *Othreis fullonia* [90 mm (3.5 in)], Australian, (14) *O. ancilla*, larva, Oriental, (15) *Phytometra, argenteum* variety [37 mm (1.5 in)], Palaearctic, (16) *Thysania agrippina* [230 to 300 mm (9 to 11.8 in)], Neotropical.
Species 1 to 5 fall into the Triphinae group, and species 7 to 16 into the Quadriphinae group.

Triphinae group (corresponding names by Seitz in parentheses).
Noctuinae (= Agrotinae). These are medium-sized moths with antennae of various shapes, glabrous eyes, and the hind tibiae always bearing claws. The larvae are nocturnal and often live underground; many species are destructive, including *Agrotis segetum*, the black cutworm (*A. ipsilon*, cosmopolitan), and the variegated cutworm (*Peridroma saucia*, almost cosmopolitan). Large genera with worldwide distribution are *Euxoa*, *Agrotis*, *Xestia*, *Diarsia*, and *Ochropleura*, with more than 100 holarctic species; among mainly holarctic genera we may note *Schinia*, *Rhyacia*, and the little genus *Noctua*, which includes the common *N. comes*, *N. orbona*, and *N. janthina*.
Heliothinae (Melicleptriinae). These are often diurnal moths with hind wings lacking CuA_1. This subfamily includes the large cosmopolitan genus *Heliothis*, which contains many species of economic importance, such as *H. armigera* and *H. peltigera* from the Old World and the American corn earworms (*H. zea* and *H. virescens*). Many *Heliothis* are emigratory.
Hadeninae. The cosmopolitan subfamily contains many species that resemble the Acronyctinae, differing only in that the eyes are hairy. The tibiae are smooth. The caterpillars feed in the open and often cause damage (*Orthosia*). Some of the main genera of the Holarctic region are *Hadena*, *Mythimna* (notably the common *M. albipuncta* and *M. unipuncta*), and *Lacanobia*. *Mamestra* is a small but widely distributed genus that includes the common *M. brassicae*, *Orthosia* is rich in palaearctic species, and *Brithys* is predominantly oriental and African although *B. crini*, for example, is also common along the coasts of the Mediterranean.
Cucullinae (= Cucullianae). The eyes of these moths are hairless but have cilia; the antennae vary in length; and the thorax is covered by a cap of hair. The tibiae are usually clawless; the forewings are variable in size, often narrow and cryptically coloured; the pattern of the hind wings not infrequently exhibits sexual dimorphism. The type genus *Cucullia*, with some 200 species, is cosmopolitan but is more abundantly represented in hot zones; it includes *C. absinthii* and *C. artemisia*. Also of wide distribution are *Xanthia* (fifty to sixty species, mostly holarctic), *Agrochola*, *Lithophane*, and *Conistra*, all of which are rich in holarctic species. *Colophasia* is diurnal and concentrated around the Mediterranean. The common palaearctic species *Xylocampa areola* belongs to this subfamily.
Acronyctinae (idem + Bryophilinae + Metachrostinae). This is a small cosmopolitan group rich in species from the temperate zones. The eyes of the adults are without hairs or cilia, and the legs are smooth. The wing patterning is sometimes unusual, for example, in *Mazuca* (Ethiopian region) and *Apsarasa* (Oriental region). Some species have aquatic or semiaquatic larvae, among them the American *Bellura* and the Old World *Nonagria*. The richest genera are *Acronycta* with many holarctic species [the American dagger moth (*A. americana*) and *A. interrupta* are noteworthy nearctic species] and *Cryphia*.

Amphipyrinae (idem + Zenobiinae). The moths of this subfamily vary greatly in coloration and size; they differ from the related Acronyctinae in overall wing colouring. The main genera are northern in distribution: *Amphipyra* (some fifty species), *Apamea*, and *Cosmia* are mainly holarctic, *Archanara* is also tropical, and *Trachea* is cosmopolitan although concentrated in the New World. The largest European amphipyrine is *Mormo maura* [approximately 70 mm (2.75 in)], which is the only European species of a typically East Asian and tropical genus. Many species of *Spodoptera* are destructive; this is a small, mostly tropical genus that includes some palaearctic and emigratory species such as the beet army worm (*S. exigua*) and *S. littoralis*.
Quadriphinae group
Acontiinae (Erastrianae). These are small, slender noctuids with broad, rounded wings, hairless eyes, and smooth tibiae. Some species, such as *Acontia*, resemble bird droppings. The group is mainly tropical but contains such large and cosmopolitan genera as *Eustrotia*, *Eublemma* (represented in the Mediterranean by *E. scitula*), and *Acontia* (some 150 species) and genera of more limited distribution, such as *Stenobola*, which is oriental, and the mainly holarctic *Lithacodia*.
Sarrothripinae. This is a small subfamily of small- to medium-sized moths, often characterized by tufts of scales on the forewings. They are not generally colourful, with the exception of the palaeotropical species of *Eligma*, and are chiefly tropical in distribution. The type genus is *Nycteola*, which is not found in the Neotropical region.
Chloephorinae (Acontianae). A small group of tropical noctuids related to the Sarrothripinae, it is distinguished from the latter by characteristic green colouring and the absence of scale tufts on the forewings. The costal margin of the wings is often arched. The M_2 vein may sometimes be poorly developed. Among the chief genera are *Carea* (Oriental region) and *Earias* with some fifty mainly palaeotropical species, which include such destructive species as *E. biplaga* from tropical Africa and *E. insulana* from the west of the Palaearctic region. The two common species of *Bena* are also palaearctic in distribution: *B. phagana* and *B. prasinana*. The southeast Asian *Lobocraspis griseifusa* (Plate 73) also belongs to this subfamily.
Pantheinae (= part of the Acronyctinae). A small cosmopolitan subfamily concentrated in the nearctic, its adults resemble the Acontiinae but have hairy eyes. The well-known palaearctic species *Colocasia coryli* belongs to this group.
Plusiinae (Phytometrinae). This is an important cosmopolitan subfamily containing many

Plate 61 **Agaristidae**

Family Agaristidae: (1) eight-spotted forester (*Alypia octomaculata*) [30 mm (1.2 in)], Nearctic, (2) *Crameria charilina* [37 mm (1.5 in)], Ethiopian, (3) *Erocha leucodisca* [50 mm (2 in)], Neotropical, (4) *Musurgina laeta*, male [32 mm (1.3 in)], Ethiopian. (the male of this species produces a characteristic sound by rubbing the tops of its hind legs against a membranous area that projects from the lower surface of the forewings), (5) *Weymeria athene* [50 mm (2 in)], Ethiopian (6) *Choerapais jucunda* [40 mm (1.6 in)], Ethiopian, (7) *Pemphegostola synemonistes* [45 mm (1.8 in)], Ethiopian, (8) *Copidryas gloveri* [42 mm (1.7 in)], Nearctic, (9) *Phalaenoides glycine* [48 mm (1.9 in)], Australian, (10) *Cocytia durvillii* [70 mm (2.8 in)], Australian, (11) *Agarista agricola* [60 mm (2.4 in)], Australian, (12) *Hecatesia thyridion* [32 mm (1.3 in)], Australian, (13) *H. fenestrata* [28 mm (1.1 in)], Australian (the species of this genus produce a sound with a membranous projection — not visible in the illustration — on their wings), (14) *Coenotoca subaspersa* [30 mm (1.2 in)], Australian, (15) *Argyrolepida coeruleotincta* [49 mm (1.9 in)], Australian, (16) *Periopta ardescens* [38 mm (1.5 in)], Australian, (17) *Burgena varia* [48 mm (1.9 in)], Australian, (18) *Cruria donovani* [42 mm (1.7 in)], Australian, (19) *Platagarista tetrapleura* [40 mm (1.6 in)], Australian (the males of this species are also able to produce a sound, using a mechanism different from those of *Musurgina* and *Hecatesia*; they rub the rough surface of their forelegs against a membranous zone that stands out from the undersurface of the hind wings).

temperate species that are widely distributed and in many cases destructive, such as the silver Y (*Autographa gamma*) and the cabbage looper (*Trichoplusia ni*). These noctuids have triangular wings, which are often highly coloured and have brilliant metallic spots. The back is covered in tufts of scales that give a serrated effect. Tropical forms include *Ctenoplusia* and *Diachrysia* although the latter is also palaearctic; those concentrated in the Palaearctic region include *Euchalcia* and *Macdunnoghia*; *Plusia* is mainly holarctic, and *Autographa* nearctic. *Abrostola* and *Syngrapha*, which is rich in nearctic species, are large cosmopolitan genera.

Catocalinae. One of the most important subfamilies — the underwings — it includes medium to large moths, mostly tropical in distribution, with broad, cryptically coloured forewings and often strikingly coloured hind wings (for example, *Catacala* or *Phyllodes*). The eyes are glabrous, the proboscis highly developed, and the median tibiae prominent and clawed. Exclusively or mainly tropical genera include *Phyllodes* (Oriental region and New Guinea), *Achaea*, *Grammodes* and *Mocis*, concentrated in the palaeotropics, and *Erebus*; *E. macrops* is the largest African noctuid and resembles *Cyligramma* in overall coloration. Temperate genera include *Catocala*, half of whose 200 species are nearctic, *Clytie*, and *Minucia*.

Ophiderinae (Noctuinae). These are medium, large, or very large (*Thysania agrippina*) moths, mainly tropical and close to the Catocalinae but distinct from them in that they have smooth tibiae. The eyes are glabrous, the proboscis highly developed, and the palps prominent. The subfamily includes genera such as *Ophideres*, *Serrodes*, *Scoliopteryx*, *Gonodonta*, *Othreis*, which is able to pierce the skin of fruits to suck the juice, and *Calpe*, some species of which pierce the hide of certain mammals (Plate 73). Some species are diurnal. Important genera include *Gonodonta* (some forty species), *Thysania* and *Diphtera*, which are mainly neotropical, *Calpe* (some fifty species), palaeotropical, with one European species (*C. thalictri*), and the pantropical *Othreis* (about twenty species). *Anomis*, *Rivula*, and *Lygephila* are large cosmopolitan species.

Hypeninae (= part of the Noctuinae). This is a large and cosmopolitan group of moths of small to medium size, easily distinguished by their highly elongated labial palps (often as long as the antennae), which point forwards like a beak. The wings are broad, triangular, and dull in colour; the legs are long and slender and sometimes bear anterior olfactory organs in the male. The main genus is *Hypena*, which is found in most continents but is absent in South America and Australia. Genera rich in palaearctic species include *Scrankia* and *Herminia*, the latter being also well represented in the Oriental region. Other noctuid subfamilies are the Euteliinae, with narrow-winged, short-abdomened forms like *Eutelia*, and the Stictopterinae, which have exceptionally long antennae in such tropical forms as *Odontodes*.

Family Agaristidae (Plate 61). This is a small family of about 300 species that are chiefly tropical and have peculiarly patterned and vividly coloured wings. They are mostly diurnal and do not seem to be subject to attack from predators perhaps because their bright colouring is combined with some type of chemical defense that makes them unappetizing. The function of the stridulation emitted during flight by the males of certain genera such as *Hecatesia*, *Musurgina*, and *Pemphegostola* seems to be a mystery. It may be that it is part of courtship, but it is also possible that it plays a role similar to that of the ultrasounds emitted by the arctiids. In appearance the agaristids resemble noctuids, although they differ not only in colouring but also in having larger tympana and slender antennae or antennae with hooks at the apex. The larvae, which feed in the open, are also aposematic and appear to be fearless. Most agaristids are highly active and rapid in flight and settle readily on flowers during the day. They prefer forest environments. The Palaearctic region is almost completely lacking in agaristids, but there are many tropical genera, including the foresters (*Alypia*) with about fifteen species and *Erocha* with some ten species, both from the New World. *Brephos* and *Heraclia* come from tropical Africa, and examples of Australian species may be found in the caption to Plate 61.

Behaviour

Instinct and Learning

To the superficial observer of nature few things can appear as casual and random as the dainty fluttering of butterflies as they alight on flowers in a meadow. However, in reality few phenomena are as rigorously controlled as the type and sequence of activities undertaken by an adult butterfly or moth when flying, when resting in order to warm up or feed, or when searching for a mate. The sequence of its actions – its behaviour — is strictly determined by environmental stimuli, whether such physical or chemical factors as temperature or humidity, the products of other organisms (plants, predators, and so on), or those of the same or different species of butterfly or moth. Obviously, as with any other kind of animal, the behaviour of a butterfly is highly dependent upon the physiology of its nervous system, but it is equally clear that behaviour always has an adaptive significance. Hence, when studying behaviour, one must consider not only the physiological mechanisms that determine it but also the environmental and evolutionary factors that influence its development and manifestation.

In general an animal adapts its behaviour to its environment, through either instinct or learning. In the case of the former the animal possesses innate responses built into its nervous system, which form part of its inheritance. In the case of the latter the animal has the inherited ability to modify its behaviour as a result of experience gained during growth. Instinct and learning work together to produce adaptive behaviour patterns, but in the insects (including butterflies and moths), unlike the higher vertebrates, behaviour is largely controlled by instinct.

In the following chapters butterfly behaviour will be discussed in relation to migration, dietary habits, interactions with other organisms, and in particular defense against predators. The present chapter will deal simply with certain aspects such as flight behaviour, gregariousness, territorial behaviour, and courtship behaviour.

Regulation of Body Temperature in Flight

To fly a "cold-blooded" (*poikilothermous*) butterfly or moth must be able to warm its muscles to sufficient temperature. A butterfly *basking* in the sun with open wings is not indulging in idleness but is keeping its thoracic muscles warm for the next flight. Like reptiles — but unlike most moths — butterflies store thermal energy from the sun, warming the surfaces of the wings and the veins through which haemolymph circulates and then passes to the body tissues. The efficiency of heat absorption obviously depends upon wing coloration, the size of the butterfly, and above all its behaviour, which varies with environmental temperature. This last factor has a limiting effect on the activity of butterflies, which in temperate regions are present only from spring to autumn and which become progressively fewer the higher the latitude or altitude. Heat absorption is increased by the animal's spreading the wings, angling the exposed surface, and making a more direct contact with the substrate, when it is warmer than the butterfly (*dorsal basking*). Other butterflies orient with wings clasped above their bodies and with the ventral wing surface perpendicular to the sun's rays; this is termed *lateral basking*. In many butterflies the critical muscle temperature for flight is about 25 to 26°C, a fact that has been proved experimentally by the use of electronic microthermometers inserted into the thorax. When solar radiation falls beneath the levels needed to ensure sufficient heating (for example, at dusk or during cloud cover), butterflies immediately seek shelter or a perch. In the American black swallowtail (*Papilio polyxenes*), whose behaviour has been observed in detail, the search for the nightly roost, which is generally a stalk or the tip of a herbaceous plant, takes the form of frantic flights. Once it has found a perch that seems suitable, the butterfly rests for a few minutes, its abdomen raised, its wings open at right angles to the evening sun, ready to change its perch. Subsequently it closes its wings and lowers its abdomen into the position in which it will stay all night. In the morning, about two hours after sunrise, the butterfly again assumes its open-wing position but changes its direction to catch the sun's rays.

In the tropics flight activity lasts longer; the oppressive heat allows butterflies to become active before sunrise and to continue after dusk. The ithomiine *Mechanitis isthmia* flies in forest clearings between 5:30 and 8:30 A.M., after which, just as most other butterflies are becoming active, it takes refuge in the forest depths, searching out shade and reemerging only at dusk. Many brassolines and amathusines prefer to fly at twilight and in the evening are sometimes attracted to lights.

Unlike butterflies most moths tend to autoregulate their thoracic temperature, actively preheating the thorax by vibrating their wings. During such vibration, which resembles the shivering used by birds and mammals for thermoregulation, the dorsoventral and the longitudinal thoracic muscles contract almost simultaneously under nervous control, whereas in flight they do not. The muscle temperature needed for flight varies from family to family and is positively correlated with body weight and wing load; unlike butterflies it is independent of ambient temperature. By a series of careful measurements two California entomologists, G. Bartholomew and B. Heinrich, found that muscle temperature could vary from about 16 or 17°C in small geometrids and ctenuchids to more than 40°C in large sphingids and saturniids. In the latter groups a long period of preheating is needed before flight, especially since the external temperature is usually much lower than that of the animals. The American sphingid *Manduca sexta* takes twelve minutes to warm up if the external temperature is 15°C, whereas one minute of wing vibration is enough at 30°C. An interesting point is that in sphingids and other large moths the dissipation of heat from the thoracic muscles is restricted by numerous insulating scales and hairs on the integument of the thorax and especially by such structures as diaphragms and air sacs, which isolate the thorax from the abdomen, the latter remaining at a lower temperature.

Flower Visitation

It may seem to the casual observer that many butterflies and moths will visit any flower and that no particular limitations or preference prevail. If one makes careful quantitative observations, as have been carried out by P. A. Opler, one finds that there are both behavioural and physical determinants regulating flower visitation and thereby nectar use. Nectar of course is the energy source that keeps many adult Lepidoptera on the wing. The lengths of Lepidoptera proboscises vary, usually as a function of the adult's size, although those that feed on nonfloral resources have very short mouthparts. Obviously a butterfly or moth cannot feed on nectar from flowers whose corolla tubes are longer than the lepidopteran proboscises. Measurement and observations document this very nicely. At the other extreme very large Lepidoptera tend not to visit very tiny flowers unless they are massed because it is not energetically profitable to do so. For a large species the energy expended to visit a tiny flower might be more than that gained.

Flower colour, probably that manifested in the ultraviolet, is a cue for many species. For example, skippers tend to visit flowers that are in the white, violet, or purple range. Flower position on the plants may also be

Plate 62 **Clustering on damp ground**

Many male Lepidoptera gather to drink on damp ground and sandy riverbanks. This is particularly common in the tropics, where swarms may consist of one or more species. The butterfly and moth gathering is usually triggered by one or more individuals' settling; this acts as a visual stimulus to the others. Illustrated are some situations observed in Mexico and Guatemala.

A: Cluster of Nymphalids: the rusty-tipped page (*Metamorpha epaphus,* 1) and eighty-eights (*Diaethria clymena,* 2).

B: Julia (*Dryas iulia,* 3) and the patch (*Chlosyne janais,* 4), Nymphalidae.

C: Monospecific group of the dog's head (*Colias cesonia,* 5), Pieridae.

important; for most Lepidoptera will visit only those flowers that are directed upwards or laterally, and only a few will visit flowers that are directed towards the ground. The European cabbage white (*Pieris rapae*), for example, almost always visits flowers directed laterally; so composites are rarely visited. Hawkmoths (Sphingidae) are among the few groups that frequently visit flowers directed downwards. The height of the flowers from the ground is also critical as Lepidoptera fly at certain heights, most flying from about 0.5 to 1.5 m above the ground, where most flowers are located. A few butterflies fly within only a few centimeters of the ground and will visit only small flowers there. Similarly flowers more than 2 m above the ground, at the end of long stems, or on trees or shrubs are visited by only a small array of species. Exceptions to this rule are to be found in the tropics, where many species of Lepidoptera are to be found in the canopy levels.

Gregarious Behaviour

A prelude to primitive forms of social behaviour, gregarious behaviour patterns are relatively common among caterpillars. The larvae of various species of lasiocampids, thaumatopoeids, and lymantriids build silken communal nests for protection, and the well-known "processionary caterpillars" have developed unusual behavioural adaptations to suit a gregarious life style. These generally involve strategies to limit predation and will be discussed in more detail later, in the chapter on mimicry and strategies for avoiding predation. Most adult butterflies and moths tend to be solitary, but there are occasional instances of gregarious behaviour. These essentially fall into three categories: crowding together of migratory butterflies and moths in winter quarters, clusters of individuals on damp ground, and communal nocturnal roosts. Migrations will be discussed later. Instances of clustering to absorb water from damp earth occur in all parts of the world. European springs, particularly in the mountains, are visited by entomologists because they are readily able to collect large numbers of interesting blue butterflies, which crowd together on the wet earth. In the tropics the sandy banks of watercourses or puddles in cart tracks are favoured spots, and it is possible to find dozens or even hundreds of sulfurs and swallowtails crowded together on what may be only a few square centimeters of ground. Plate 62 shows examples observed by one of the authors in Mexico and Guatemala. The groups may be made up of just one species, as is often the case with certain pierids and papilionids (for example, *Colias cesonia* and *Eurytides epidaus*), or of several species. The reason for the apparent flocking is that the individuals have all chosen the same spot, a spot that meets their individual needs (a substrate with the right particle size, salt concentrations, and so on). Normally only freshly emerged males are found at moist spots. Flocking probably also helps to limit the toll taken by predators: A lone butterfly or moth resting on the ground to drink is statistically more vulnerable than it would be in a group. A predatory bird or reptile, unlike an entomologist armed with a net, would probably be successful in hunting a series of single butterflies rather than a group: When a predator attacks one of a group, all the other butterflies take flight simultaneously and confuse the predator.

K. Arms and P. Feeny have elegantly demonstrated that tiger swallowtails (*Papilio glaucus*) use concentrations of sodium ions as cues for places for moisture imbibition. Most species frequently found in such concentrations are freshly emerged males of what we may refer to as "patrolling" species. Studies by J. A. Scott have shown that the males of these species fly routes through likely environments in search of receptive females. Males of species with a "perching" mate-location strategy (see

Plate 63 **Roosting**

Roosting is the congregation of individual butterflies for nocturnal resting. This practice has frequently been observed in *Heliconius* and seems to be associated with a whole series of rather sophisticated behavioural features typical of these butterflies. Thus groups of the zebra (*H. charitonius*), like that illustrated in Figure A, consist mostly of individuals that return regularly to the same roost for months on end. Butterflies like *Heliconius* are protected from predators by their unpleasant smell, and roosting is a means of increasing this signal. The same may be true of roosts of other species like the black-veined white (*Aporia crataegi*, B). Nocturnal roosts have been more frequently observed in the Middle East and North Africa, even among lycaenids, which are not thought to have any protection. In Lebanon the Greek mazarine blue (*Cyaniris antiochena = helena*) has been observed to form quite large roosts (C).

below) only rarely visit moisture, and then they may be homing in on other food substances. It has not been clearly demonstrated, but moisture use by these freshly emerged patrolling males may serve to initiate an active-transport pump, which could help with temperature regulation (primarily cooling) for these butterflies.

Some groups of tropical butterflies, such as papilionids (genus *Parides*), danaines, ithomiines, heliconiines and acraeines, which are protected from predators by their unappetizing taste, have been observed in conspicuous roosting groups. These roosts are generally very stable, and their members return to the same places every evening; in heliconiines the returns may occur over a period as long as six months. In the morning the roosting butterflies each goes its own way, scattering through the forest and clearings in search of nectar or pollen (heliconiine butterflies are unique in their pollen collecting). However, at dusk they all return home, as has been proved by marking individuals. The roosts generally hold members of the same species resting in close contact with one another. Here again such groups confer protection against predators, particularly reptiles, birds, and small mammals. The scent emitted by an entire roost is a much stronger warning signal to a predator than is that given off by a single individual. Furthermore, the roost's fixed position, coupled with the unpleasant taste of the butterflies, helps the local predators to learn to avoid the butterflies and to concentrate on other prey. Plate 63 shows a roost of zebra butterflies (*Heliconius charithonius*) observed in the Sierra del Abra in the state of Tamaulipas, Mexico, and two unusual roosts observed by one of the authors in Lebanon, involving the black-veined white (*Aporia crataegi*) and the Greek mazarine blue (*Cyaniris antiochena = helena*). The first of these two species may, like other pierids, be distasteful to predators and may possibly use scents to signal its presence to nocturnal predators. On the other hand the second, which has been observed at dusk in roosts of hundreds on a few herbaceous plants on the mountain at Cedars of Barhouk, belongs to a group that tends not to be characterized by special chemical defenses, and the significance of the roost is still unclear.

Territory

When biologists use the term *territory*, they mean an area actively defended by an individual or a group of animals. Such a defense involves displays that drive away an intruder, generally a member of the same sex and species as the holder. Territorial behaviour has been extensively studied in birds and mammals, but it is also common in other groups of animals, having developed in the course of evolution as an adaptation that offers an increased chance of success in the search for a partner and also in ensuring that the offspring receive both food and protection. In butterflies, territoriality is one of two primary mate-location strategies. The other strategy is termed *patrolling*. Males with this behaviour fly purposefully through the proper habitat in search of receptive females.

Entomologists have only recently noticed that territorial behaviour is apparently widespread among butterflies even though observations made at the beginning of the century clearly described the phenomenon. As early as 1902 N. H. Joy noted that males of the purple emperor (*Apatura iris*) tended to maintain a post, known as the "throne," on the branch of an oak and that they fought other males that approached. If the holder was removed, the throne would rapidly be occupied by the next male flying into the area. It has subsequently been found in other butterflies that these posts are usually sites controlling an area that may be as much as 1,000 sq m (10,759 sq ft).

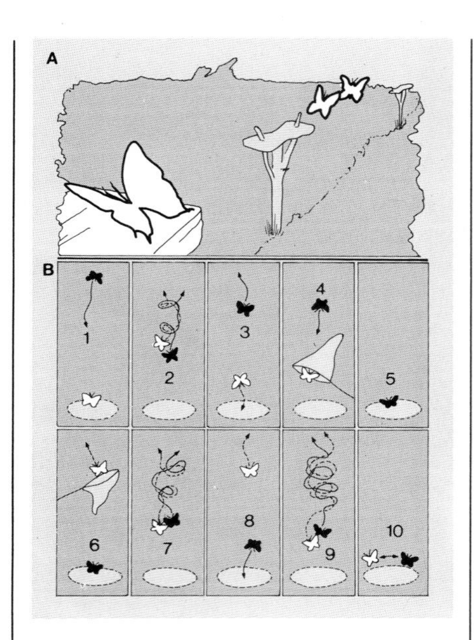

Plate 64 **Territorial behaviour**

Territorial behaviour, though more common among birds, mammals, and other animals, is also exhibited by certain Lepidoptera. It consists of a series of actions intended to defend an area, especially one used for breeding or feeding. These actions take the form of warning signals to other individuals, who are usually of the same species and of the same sex.

The males of the two-tailed pasha (*Charaxes jasius,* A) exhibit strong territorial activity during the heat of the day. In the Mediterranean maquis near Capalbio in Tuscany *Charaxes* use the wooden seats made by boar hunters as stands from which to survey their territory. Each male alternately rests on his stand for a while and makes flights patrolling the area, firmly chasing off intruders. In Figure B we see the chief stages of the territorial behaviour of the speckled wood (*Pararge aegeria,* Nymphalidae, Satyrinae), which was studied by N. B. Davies (see the text). Territorial competition in this species centers around possession of patches of sunlight on the forest floor. The owner of the territory is shown in white, the intruder in black. When an intruder approaches (1), the owner confronts it, chases it upwards in a spiralling flight (2), and then returns to its place (3). If the owner is removed (4), the patch of sunlight is quickly occupied by an intruder (5). When the former owner is returned, whether after a short or a long time (6), further spiralling confrontations take place; they may be more (9) or less (7) extended. The final winner may be either the new owner (8) or whichever of the two (10) considers itself to be the owner.

One of the authors has observed marked territorial behaviour in a southern Italian population of the two-tailed pasha (*Charaxes jasius*), the favoured posts of which were the regularly spaced wooden stools about 2 m (6 ft 6 in) high used in boar hunting (Plate 64). The male *Charaxes* defended their territories tenaciously, attacking and pursuing any intruder, whether males of the same species, other butterflies, or indeed birds. A *Charaxes* was observed to chase a hoopoe for several seconds when it flew across the defended area, while another repeatedly hit the forehead of an entomologist with its massive body. However, in *Charaxes*, as in *Asterocampa*, interactions with intruders belonging to different species are considerably shorter than are ones with males of the same species, which are pursued to the edge of the territory. Not all the guard posts are high.

So far as is known to date, territories among butterflies are purely reproductive in purpose. The males defend areas where they have a greater chance of courting females or where the best sites for oviposition lie. Territorial behaviour, however, accounts for only a small portion of the butterfly's overall activity (four hours in *N. io* and *N. urticae*, longer in *Asterocampa* and apparently in *C. jasius*). Time is also spent in warming up in the sun and in feeding, actions that precede territorial activity.

The holder of a territory generally wins clashes — if they can be called clashes. The extended flight with which the holder pursues an intruder may simply be a specific signal that tells its rival to move away. In the speckled wood (*Pararge aegeria*), a woodland species studied by N. B. Davies, this flight starts low and spirals upwards (Plate 64); the defended post is a sunlit patch on the forest floor. Such patches are relatively scarce, and there are always wandering males flying around beneath the foliage. If a holder is removed, as an experiment, another male will quickly occupy the sunlit area, which it will take for its own if it remains undisturbed for a few seconds. If this is the case, the new holder will successfully drive away other males, including the previous holder when the latter is reintroduced. If both butterflies are equally "certain" of ownership (because each has occupied the post for a short time without being seen by the other), before one or the other emerges as the victor, a protracted spiral flight will result, lasting, on the average, ten times the duration of the normal one.

In mountainous or hilly terrain males of territorial or "perching" species often select hilltops or ridgetops as sites for awaiting the arrival of receptive females. In North America this phenomenon is termed "hilltopping" by such workers as J. A. Scott and O. A. Shields.

Perching species tend to be characteristically rare or widely dispersed in the environment as opposed to the patrolling strategy mentioned earlier; patrolling butterflies tend to be found in fairly dense local colonies or are widespread lowland species. Perching males tend to locate in places that females can orient towards. These places — which may be prominent hilltops or the uppermost sunlit branches of large trees — are the chosen mating arenas for these butterflies. Virgin females may fly to these areas for mating and then, so as to begin egg laying, depart for the habitat where caterpillar host plants are found. Males may remain at the mating sites for most of their life, leaving only periodically to visit nectar sources; J. Burns has shown that males of many perching species may mate often during their short lives. Thus the ability to locate and retain a favourable perching location will mean greater selective fitness.

Sexual Behaviour and Courtship

To find a partner for breeding and to persuade it to mate, a butterfly or moth has to overcome a host of environmental difficulties.

The sexual signal, whatever form it takes, must be strong enough to be

Plate 65 **Sexual behaviour in the grayling (*Hipparchia semele*)**

The sequence of the phases of courtship has been studied in detail by N. Tinbergen and his colleagues for the grayling (*H. semele*, Nymphalidae, Satyrinae).

A: Stages of courtship: (1) The male (left) flies after the female. (2) When they settle on the ground, the male perches in front of the female, rotating his antennae and vibrating his wings. (3) the male clasps the female's antennae with his forewings, touching them with the androconial scales, which emit pheromones. This is illustrated from above and from the side. (4) If the female is receptive, the male describes a semicircle around her and makes contact with her abdomen. (5) Copulation. (6) Female laying her eggs.

B: Lures used by Tinbergen to test the discriminatory capacity of males of *H. semele* (see the text). Cardboard models varying in shape (7), size (8), and type of movement (9) were used. The diagram (10) shows that the critical distance of response of a male to a lure pulled along on a string was greater (that is, the stimulus was more effective) when the motion imparted was fluttering or rotating than when it was linear. (Adapted from Tinbergen.)

perceived at a distance. Furthermore sexual communication must take place through stimuli and responses that involve only members of the same species to the exclusion of all others; otherwise there will be wasteful confusion. Finally, to attract a partner, a butterfly has to put itself "on show" in some way, and this may put it at risk from predators. Evolution has found various ways to meet these selective pressures and has consequently produced a surprising variety of behavioural, physiological, and biochemical mechanisms. In recent years these have been studied by numerous investigators.

Sexual behaviour is yet another area in which butterflies and moths basically differ. A male butterfly actively searches for the female, using mainly visual stimuli. Olfactory stimuli may also be important, but they come into play secondarily and are almost always produced by the male. In moths, on the other hand, the male is guided to the female by olfactory stimuli in the form of her chemical pheromones, which are capable of acting over long distances. However, in addition to this general distinction there are substantial differences even among members of the same family, and it is worthwhile to refer to some specific cases.

The first butterfly to be examined in detail in this respect was the grayling (*Hipparchia semele*), a patrolling species studied by Niko Tinbergen and his colleagues in Holland in 1942 (Plate 65). In this species courtship begins with the male's flying after the female. This impulse is triggered by visual stimuli and is not very specific in that male graylings will readily follow other butterflies, birds, and lures of a variety of shapes, sizes, and colours specially prepared by the experimenters. Experiments of this kind enabled Tinbergen and his colleagues to discover that the female's colour was not very important; indeed the males were more stimulated by red or black lures than by grey or brown ones, which were in colour closer to the species itself. Shape was important. Circular or rectangular lures were pursued to the same degree as butterfly-shaped ones, but rectangles with sides of a similar length were preferred to long, narrow ones. Size also had an effect, with large lures, being pursued more than normal-sized ones, which in turn were pursued more than small ones. The way in which flight was simulated was also significant: Lures that moved in a skipping or rotary fashion were more successful than were those following a straight path. These experiments show that, in approaching the female, the male responds solely to visual stimuli and that his response can be enhanced experimentally by using *supernormal stimuli*, such as a larger-than-normal female lure or heightened coloration.

When a receptive female is pursued, she will sooner or later drop to the ground. The male then alights beside her and moves round in a semicircle so that he faces her. This is the first phase in a strict behavioural sequence, during which the male first of all vibrates his wings persistently, then moves his antennae in large circles and increases the wing movement to produce a fanning action, and finally rests his antennae on the ground, arches himself, and brings his forewings forward so as to enclose the female's antennae. The purpose of this is to excite the female with a pheromone produced by two narrow bands of densely packed *androconia* on the male's forewings. Next the male moves round in short stages to make contact with the female's abdomen. If she is receptive, she raises her wings slightly and uncovers the tip of her abdomen, and mating takes place at the first attempt. However, she may at any moment show that she is not receptive by rhythmically opening and closing her wings. The female's positive or negative response depends on her physiological state (virgin or fertilized) and the ambient temperature. A refusal interrupts the male's behavioural sequence. He may attempt to start again from the beginning, but if he makes several unsuccessful attempts in a row, the sequence becomes more irregular; the male may then skip one or more phases.

9/2
3/1
Ø 8 cm
2/1
Ø 6 cm
3/2
Ø 4 cm
Normal
Ø 2 cm
1/2
1/4
1/6
40 30 20 10
40 30 20 10
Critical response distance

Plate 66 **Stages in the courtship of the queen (*Danaus gilippus berenice*)**

Courtship in this species is mainly controlled by olfactory stimuli. The male queen has small, black pockets on the hind wings and extrusive, yellow, hair pencils at the end of the abdomen, which produce a specific pheromone that encourages the female to mate (see the text): (1) the male (above) stimulates the female's antennae in flight with his partially evaginated hair pencils. (2) When they have settled, the male continues to stimulate the female with his hair pencils, which are now fully evaginated. (3) The male circles over the female, who remains at rest. (4) The male drops onto the female feeling her with his antennae. (5) Copulation. (6) Postnuptial flight. (Adapted from L. Brower, J. Brower, and Cranston, 1965.)

Mating, which on the average lasts a little over an hour, is followed by a varying amount of time devoted to oviposition.

Courtship procedures of the type described above, although varying in length and complexity, are the rule among butterflies. However, the approach made by the male in response to the female's visual stimuli is often more selective than it is in *Hipparchia semele*. In the silver-washed fritillary (*Argynnis paphia*) the male's approach is triggered by the rapid movement of black spots on an orange background, produced by the beating of the female's wings. The same effect may be enhanced experimentally, as Magnus has done, by rotating at the correct speed a cylinder covered with wings taken from females. The key stimulus for the Japanese swallowtail (*Papilio xuthus*) is the series of transverse yellow bands running across the black background of the wing. In the wild males are readily attracted to artificial models of black cardboard with suitably spaced yellow stripes. Visual stimuli in the ultraviolet part of the spectrum, invisible to our eyes, are important to many pierids. So, for example, the males and females of the cabbage white (*Pieris rapae crucivora*), studied in Japan by Y. Obara and T. Hidaka, appear very different if photographed through a filter that lets through only light that is close to ultraviolet (maximum wavelength 360 mμ) although to our eyes and probably to those of predators they both seem white. The two Japanese authors have observed that in these butterflies the entire courtship procedure is controlled by visual stimuli. The different degree of ultraviolet reflection appears to depend on the differing amounts of pteridine pigments on the wings of the two sexes. However, sexual communication between sulfur butterflies (subfamily Coliadinae) takes place partly by ultraviolet visual stimuli and partly by pheromones produced by glandular areas near the front margins of the male hind wings. These are normally covered, but during one phase of courtship the male deliberately opens his wings and releases the aphrodisiac.

To be effective, pheromones must be highly volatile (a characteristic associated with low molecular weight), and they must also be released at just the right moment since the insect has only a limited supply. However, their molecular weight cannot be too low because a high specificity, with the occurrence of thousands of variants in the different species, can only be the result of a high number of carbon atoms (generally between ten and seventeen with molecular weights of 100 to 300). The chemical nature of male pheromones varies in complexity from simple aliphatic compounds (acids and aldehydes) to aromatic compounds (alcohols, aldehydes, terpenes, and dihydropyrolized alkaloids). Female pheromones are usually mixtures of two or more compounds, and in addition to the substances mentioned above they may include ketones with over twenty carbon atoms. Most known male pheromones are effective only over short distances and act by making the female receptive for mating or by inhibiting her from flying. However, those produced by females act over long distances and attract males. In males the glandular structures are situated on the wings, the legs, the thorax, or the abdomen and are generally derived from hypertrophic trichogen cells whose scales have become transformed into androconia, or brushlike tufts or hairs. The latter structures are usually protected within pockets or cuticular invaginations and emerge only during courtship. In females, on the other hand, the glands are situated on the abdomen and almost always, in moths, in the intersegmental membrane between the eighth and ninth segment.

The use of male pheromones in courtship is clearly illustrated by the example of the queen, *Danaus gilippus berenice* (Plate 66), which was studied by Lincoln Brower and Jane Brower. They released virgin, laboratory-reared females and recorded the behaviour of the males. The

Plate 67 **Stages of courtship in *Ephestia cautella***

The sexual behaviour of this pyralid [15 to 20 mm (0.6 to 0.8 in)], which causes considerable damage to cereals and other food stocks, has been studied under the microscope after having been filmed. The whole sequence is divided into different phases and takes place within the space of a few seconds: (1, 2) Male and female of *E. cautella*. Courtship can be divided into five stages: display (3), acceptance or rejection (4, 5), encounter (6–11), genital contact (12–14), and copulation (15–17). The male displays to the female by raising his abdomen (3). The female may accept (4) or reject him (5) after using her antennae to palpate the tufts of odoriferous scales at the base of the male's wing. During the encounter the male arches his abdomen and stands in front of the female (6, 7, dorsal view; 8, side view), pressing his head frontally beneath his partner's, a process that includes a brief touching of the palps (9, enlarged ventral view), and strongly curving his abdomen back to touch the female's head with his everted genitals (10, 11). The male then presses his abdomen forwards until it touches the female's genitals (12–14). Full copulation takes place when the male has turned round (15, 16) so that the two partners are back to back (17). (Adapted from Barrer and Hill, 1977.)

coloration of the male is similar to that of the female (see plate), but it differs in having a small black pocket on the hind wing and at the tip of the abdomen a pair of conspicuous tufts of hairs that can be evaginated. The complete behavioural sequence comprises five phases prior to mating and four phases from mating to final detachment. The male first approaches the female after having located her visually (phase 1). He then, in the second phase, extends his brushes and repeatedly touches the antennae and the head of the female while still in flight. The female responds by flying more slowly and settling on the grass. The male increases the intensity of his action with his tufts of hair and flies in a semicircle, performing a kind of dance (phase 3). The female responds by fluttering her wings or lowering them. At this point the male retracts the tufts of hairs and drops onto the female (phase 4). This phase varies in length but it is often followed almost immediately by a mating attempt (phase 5). If mating is successful, it takes place in several phases, a postnuptial flight, fertilization, which occurs subsequently, and final detachment. The tufts of hair, which cover the female's antennae with pheromone, clearly play a crucial role in this sequence.

At the time the experiments were carried out (1960 to 1961) the relationship between the glandular pockets and the abdominal brushes was unclear, and similarly the chemical nature of the pheromone was unknown. Subsequent experiments have shed light on various points. First of all, T. E. Pliske and Thomas Eisner found that males whose tufts of hairs had been removed experimentally performed their courtship normally but failed to attract the females. The same authors have established that the brushes release a ketone in the form of a powder, which stimulates special sensilla on the female antennae. A specific amount of the ketone, subsequently named *danaidone*, is necessary to stimulate the female. A group of German researchers have investigated the role of the pockets in a related species, *Danaus chrysippus*. From time to time the brushes are inserted into these pockets and this contact is essential for the production of danaidone in physiologically normal quantities. It appears that an alkaloid, which is assimilated directly from the food plant (*Asclepias*) during the caterpillar stage, accumulates on the brushes. This alkaloid, known as *danaidol*, is a precursor of danaidone and can be converted into the latter by enzymes in the wing pouches. These structures thus act like tiny chemical laboratories, in which the butterfly produces the aphrodisiacs it needs to attract females.

Plate 67 illustrates another example of a behavioural sequence in which the production of male pheromones plays a decisive role. It involves the small moth *Ephestia cautella*, a harmful pest that causes much damage to dried foodstuffs. Courtship is divided into many short phases, which take place within seconds.

In many moths, the courtship sequences are greatly simplified, and mating takes place shortly after the male, which may have been attracted over distances of several miles by pheromones, locates the female. Substances such as bombicolin, produced by the female *Bombyx mori* (the silk moth), or gyplure, the pheromone of the gypsy moth (*Lymantria dispar*), are highly attractive. The male's receptors are situated on the bipectinate antennae and are capable of transmitting a nerve impulse even if they are stimulated by just a single pheromone molecule.

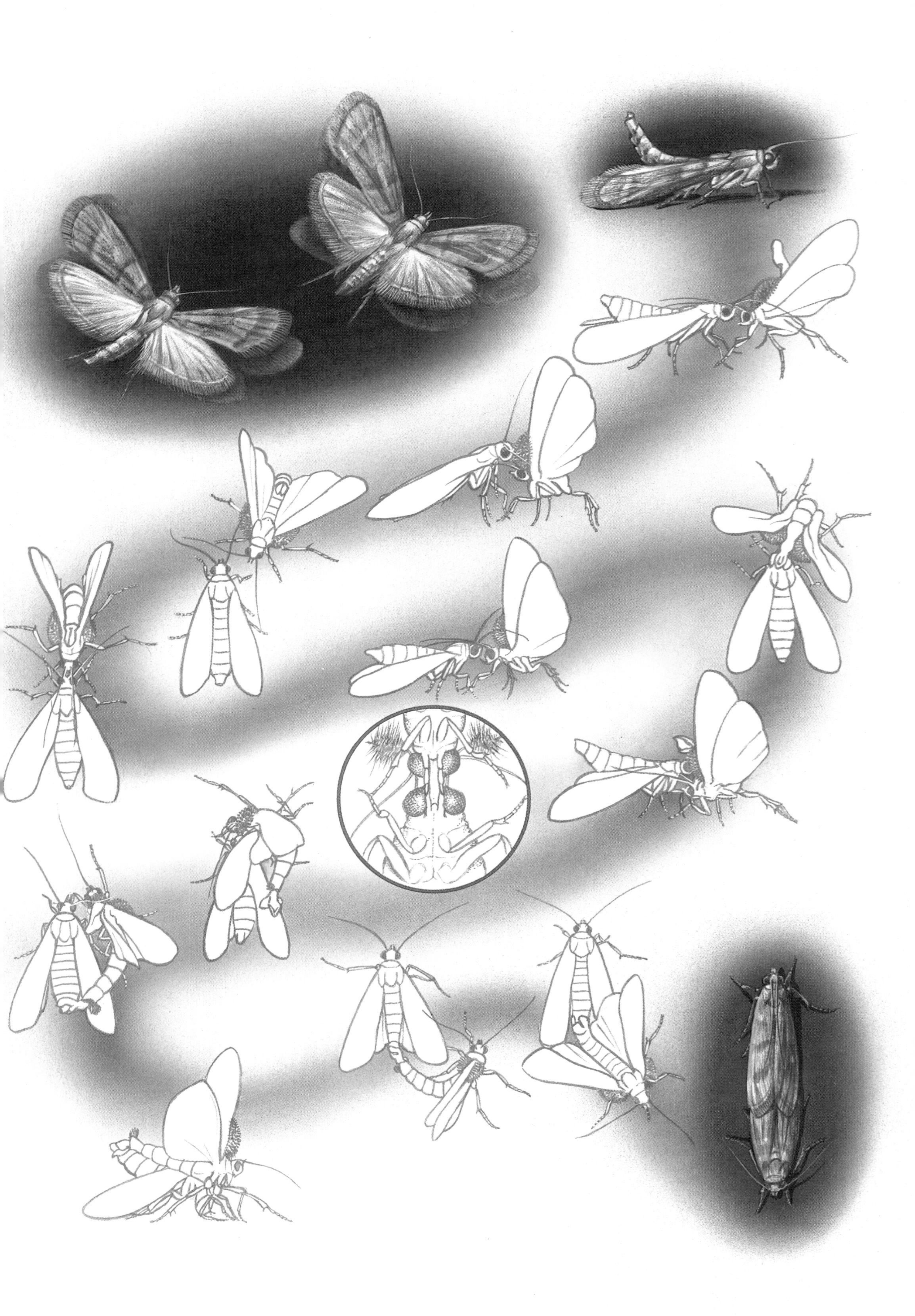

Populations, Demography, and Migrations

Plate 68 **Seasonal variation**

Lepidoptera inhabiting a single area do not necessarily share the same flight periods and reproductive cycle. Species phenology is variable, depending to differing degrees on the phenology of the caterpillar host plant, on the ontogeny of the individual (for example, those which have hibernated as pupae emerge earlier than those which have hibernated as larvae or eggs), and on competition with other species. The season succession of seven butterfly species found near Sapporo (Japan) is illustrated in Figure A. All have only a single annual generation. B shows how the adults of two multiple-brooded *Pieris* from the same community vary through the seasons. The peaks in terms of numbers seen in May, July, and August correspond to the usual three annual generations.

The various species of butterflies and moths are not evenly distributed in time and space. As a result of season, climate, amount of food resources, and type of habitat, as well as genetic characters of species themselves, there are marked differences in the numbers of individuals between one species and another. Collectors are particularly interested in these differences, and the distinction often made between rare — and hence much sought after — butterflies and moths and common ones sometimes reflects real demographic differences. One should not, however, confuse the concept of rarity (meaning a paucity of individuals) with other situations: For example, many species of butterfly are considered rare either because they are localized geographically or because their ecological habits are not well known. However, in the proper habitats or in the areas where they live these species may be much more abundant than many others. Within just one species it is possible to find marked variations in the size of populations from different areas or fluctuations within a population from one season or year to another.

Seasonal Changes

Seasons are one important cause of variations in numbers or the presence or absence of a species or population. In temperate regions the life cycle of butterflies and moths is governed by the alternation of two well-defined seasons, winter and summer, which are characterized by marked differences in temperature and photoperiod, divided by transitional seasons. It is commonly thought that most species spend the winter months, with their low temperatures and short days, in the egg stage, as larvae in diapause, or as pupae, and the summer months in the adult state. In the tropics, in particular close to the Equator, the day length is virtually constant throughout the year, and seasonal temperature differences are reduced to the point where the range of temperatures over a single day is greater than that between seasons. In these regions some species are present as adults in all months. However, seasonal variations of a greater or lesser degree occur, but they usually involve the alternation of rainy seasons with dry seasons. Such variations are more marked in the savannahs and the deciduous tropical forest than in the evergreen tropical rain forest ("jungle").

In many species, both temperate and tropical, the adult does not fly through the entire warm season but generally for a much shorter period. The average lifespan of an adult butterfly is often only a few days or weeks although there are much longer-lived ones. Some heliconines and danaines, for example, may live to the venerable age of six months. In parts of North America the mourning cloak (*Nymphalis antiopa*) may live eleven months. Most of this time, however, is spent in aestivation and hibernation.

The flight period is principally determined by the cycle of the food plant. Thus in Europe the two festoons *Zerynthia polyxena* and *rumina* fly early, in April, when the *Aristolochia* plants, which they use for oviposition and subsequent larval food, are in flower, and they are then absent for the rest of the summer. Another important factor in determining the timing of the flight period in a particular species is the state in which it spends the winter. In general the first species to fly in the summer are those which overwintered as adults. In North America the mourning cloak (*N. antiopa*), several anglewings (*Polygonia*), and some lady butterflies (*Vanessa*) fly as early as the first fine days of February and March. Next come species that overwintered as pupae, followed by those which did so as larvae, and finally those species which passed the winter as eggs.

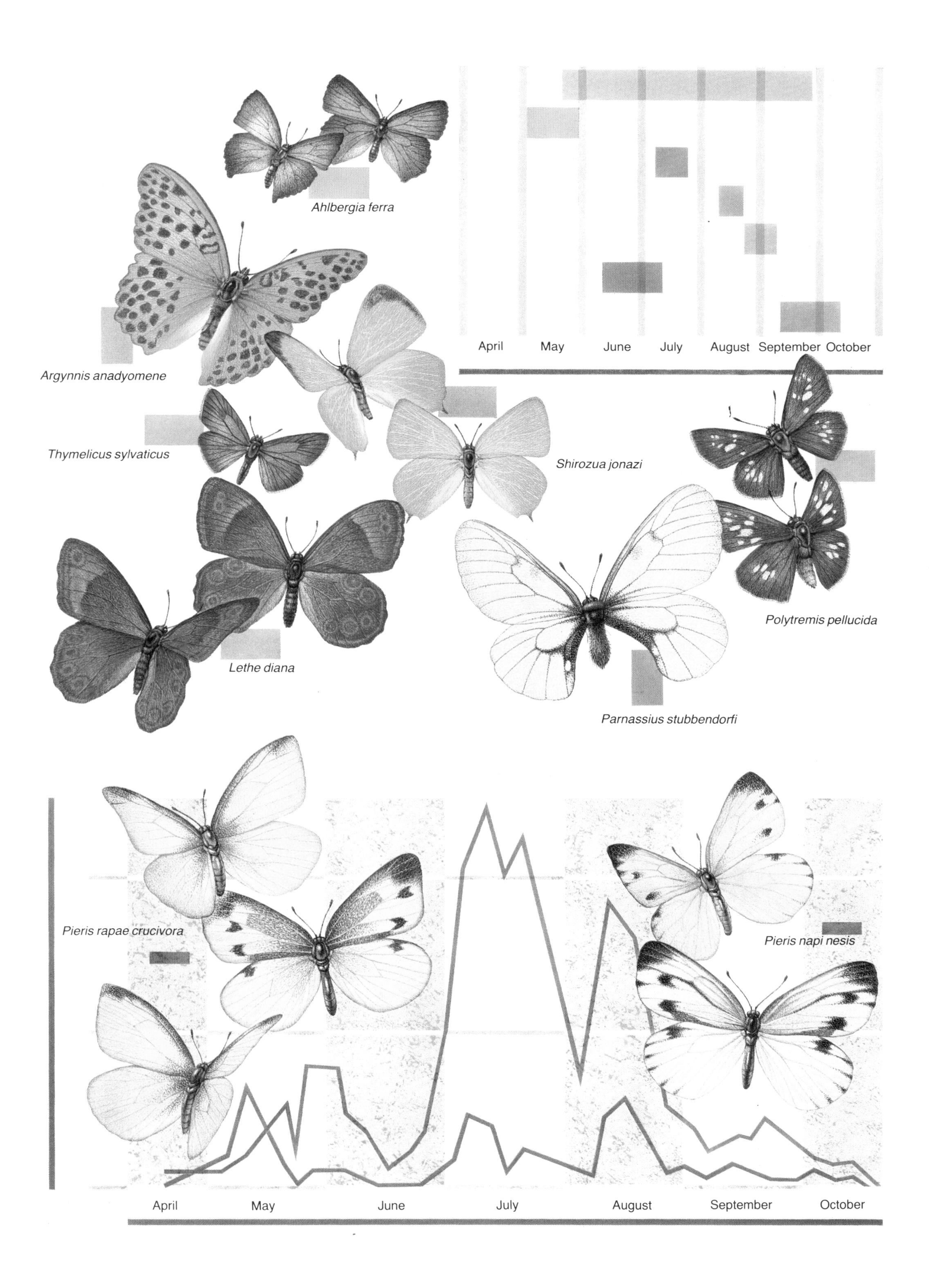
Ahlbergia ferra
Argynnis anadyomene
Thymelicus sylvaticus
Shirozua jonazi
Lethe diana
Polytremis pellucida
Parnassius stubbendorfi
April
May
June
July
August
September
October
Pieris rapae crucivora
Pieris napi nesis
April
May
June
July
August
September
October

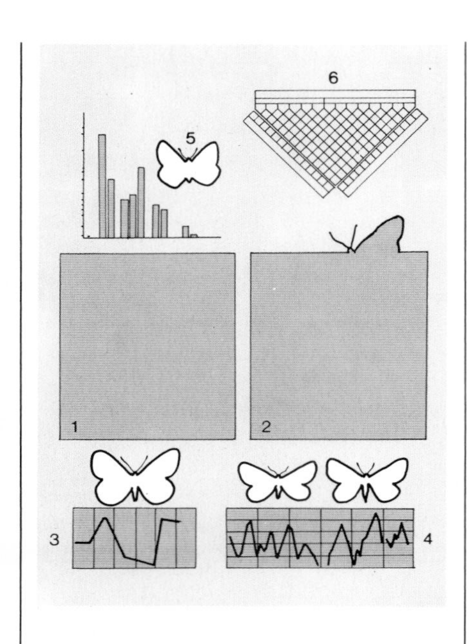

Plate 69 **Population sizes and fluctuations over periods of several years**

Lepidoptera populations can vary in density in both space and time. In certain parts of Mexico and California it is possible to see millions of hibernating monarchs (*Danaus plexippus*, Nymphalidae) in a few acres of woodland (1). In Petaloudes on Rhodes tens of thousands of *Callimorpha (= Euplagia) quadripuntaria* (Arctiidae, 2) flock together. Both species are migratory, and gather in such amazing numbers that they have become tourist attractions. The populations of the marsh fritillary (*Euphydryas aurinia*, Nymphalidae) in Cumberland, England, studied by E. B. Ford, exhibited considerable variations in numbers over the course of fifty years (3). The geometer *Bupalus piniarius* also varied greatly in numbers over the period 1880 to 1940 (4), but in this case there was a cyclical variation (the *y* axis gives the product of the logarithm of the number of pupae and square meters). This type of research naturally requires reliable methods of estimating butterfly populations. Figure 5 shows variations in numbers of an English population of the common blue *Polyommatus icarus*), based on the mark-release-recapture method (see the text). Diagram 6 gives the details of daily counts, with the number of individuals marked and released (*M*), the total caught the next day (*T*), and the proportion of the latter that were marked (*R*). The daily figure for the total number of individuals (*N*) can be calculated from the formula $N = M \times T/R$.

Generations

Plate 68 illustrates some examples, taken from a study by M. Yamamoto, of species with rigorously determined phenologies (that is, the timing of the various stages in the development of a species), which emerge in the course of the summer at a site near Sapporo in Japan. With the exception of *Lethe diana* the species shown fly for a short part of the season and produce just one generation annually (*univoltinism*; Plate 68A).

It is, however, quite common for species to produce more than one generation a year (*multivoltine*), and indeed *Pieris rapae crucivora* and *P. napi nesis*, which are found in the Sapporo area, produce three generations (Plate 68B). In *P. rapae crucivora* the first generation emerges in late May and early June. The eggs produced by these females give rise to rapidly developing larvae, which result in a second generation of butterflies in July. These in turn produce a third generation, which emerges in August. The larvae produced by the third generation pupate before winter's onset, and the first generation of the following year emerges from overwintering pupae. *Pieris napi nesis* has a similar annual cycle.

The number of generations is also affected by latitude and local climatic factors, and so different populations of the same species may produce different numbers of generations a year. Thus, for example, western North American populations of *P. napi* are univoltine; the Finnish populations always produce two generations; while the Italian ones produce three, as do the Japanese ones mentioned above.

The life cycle of a species is often synchronized with environmental conditions by diapause, which in turn is governed by photoperiod. The European arctiid *Euprepia pudica* is well adapted to the Mediterranean climate and passes through a diapause in its sixth larval instar, coinciding with the period of summer drought, when the grasses upon which the caterpillar feeds have dried up. The adults emerge at the end of September or the beginning of October, when they can find suitable grasses for oviposition. This cycle is governed by the photoperiod experienced by the early larval instars. The sixth instar enters diapause only if the earlier stages were subjected to long days (more than twelve hours of light) and short nights. When the Czechoslovak entomologist Spitzer experimentally subjected young larvae to longer nights than days, he found that they did not enter diapause and that adults emerged precociously.

Fluctuations over More Than One Year

In addition to seasonal variations there are annual variations or fluctuations over several years in the numbers of a species in a given area. There are a variety of causes for this, and not all are well understood.

Certain species, which are termed *biennial*, complete their life cycle over two years. This is particularly common among lesser fritillaries (*Boloria*), arctics (*Oeneis*), and alpines (*Erebia*), which are adapted to the extreme conditions of the Arctic or alpine tundra, where the harshness of the climate and the short season during which food plants are available do not allow the caterpillars to develop fully in a single season. In biennial species where the generations do not overlap years in which there are fritillary adults in flight alternate with ones where there are none. In the polar (*B. polaris*), for example, adults are observed only during odd years in some areas of Norway and Alaska and in Manitoba, while in other parts of Alaska, northern Canada, Greenland, and Finland they occur only in even years. In other biennial species, such as the Arran brown (*E. ligea*), adults fly every year but with marked fluctuations from one year to the next.

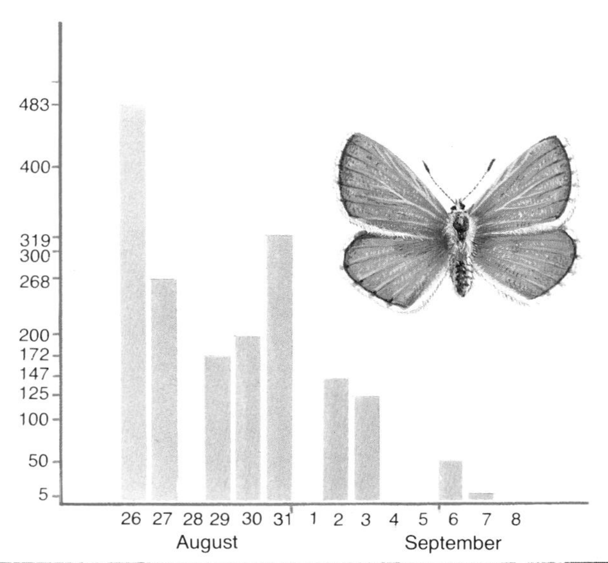

483
400
319
300
268
200
172
147
125
100
50
5
26 27 28 29 30 31 1 2 3 4 5 6 7 8
August
September

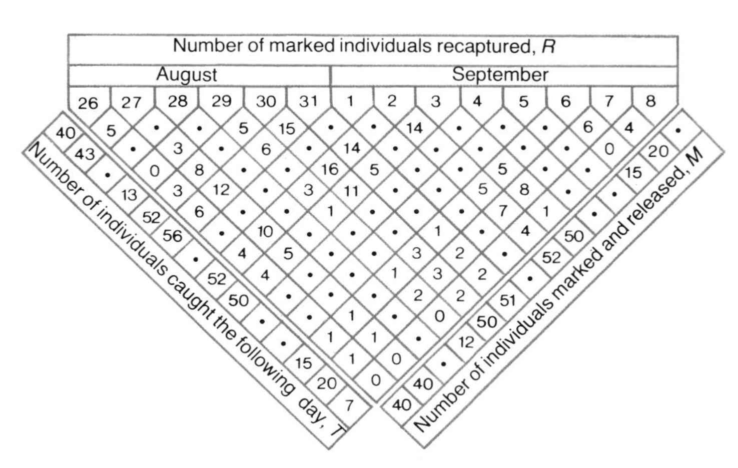

Number of marked individuals recaptured, R
August
September
26 27 28 29 30 31 1 2 3 4 5 6 7 8
Number of individuals caught the following day, T
Number of individuals marked and released, M

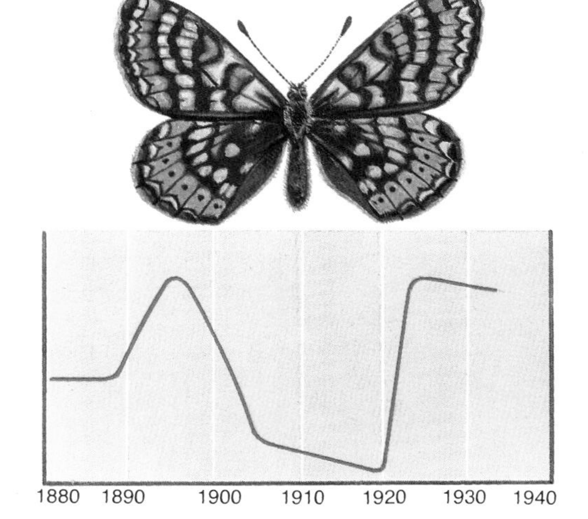

1880 1890 1900 1910 1920 1930 1940

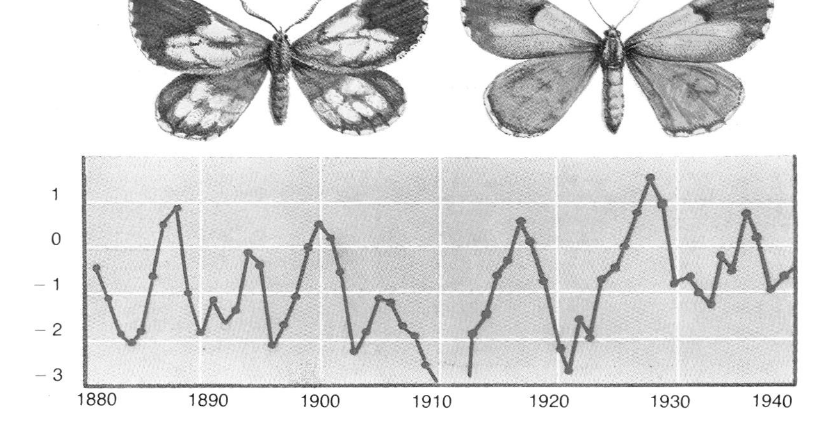

1
0
−1
−2
−3
1880 1890 1900 1910 1920 1930 1940

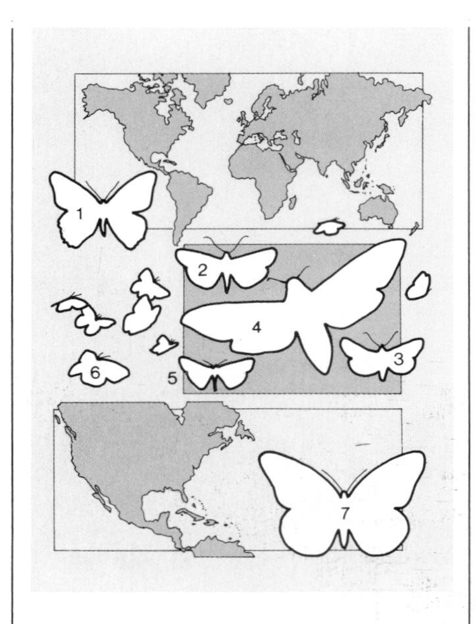

Plate 70 **Migration**

Migration is fairly widespread among Lepidoptera. It is not at all clear how this may have developed although many observations suggest that it was an adaptation designed to combat the difficulties of unstable environments. The painted lady (*Vanessa cardui*, Nymphalidae) is more or less cosmopolitan in distribution and one of the best-known migrants (1). In some years it migrates in both the Old World and North America. In winter it breeds in central Mexico and in warm areas north of the Sahara in Africa, migrating in spring through much of the United States and southern Canada and all over Europe as far as Iceland and northern Scandinavia. There are also migration routes into India over the high mountain passes and into Africa along the west coast.

Other migrants found over huge areas include the noctuids *Agrotis ipsilon* (2) and *Plusia gamma* (3), the death's-head hawkmoth (*Acherontia atropos,* 4) which is often transported by ship, and the arctiid *Utetheisa pulchella* (5). The pierid *Catopsilia pyranthe* (6), which is found over the whole Oriental region, migrates en masse in India and Ceylon, swarming in millions together with individuals of other species, moving with the northeast monsoon.

The most studied migrant butterfly is the monarch (*Danaus plexippus,* 7), fully described in the text. The light brown areas on the map show where the butterflies hibernate, the western populations in California and the eastern in Mexico with a tiny proportion in Florida. Areas with resident populations of monarchs are shown in grey.

It is more common for fluctuations in numbers to take place over a variable number of years. A recent study by Ekholm has shown that in one locality in Finland almost all the species of butterflies have, over a twenty-year period from 1947 to 1968, undergone striking variations in numbers, with periods of abundance or scarcity recurring at three-, five-, six-, seven-, or ten-year intervals. It appears that these variations may be due largely to climatic factors and in particular to the temperature and the relative humidity during the summer.

Only a few cases of longer-term fluctuations have been well studied. The English geneticist E. B. Ford was able to document the rise and fall in numbers of an isolated colony of the marsh fritillary (*Euphydryas aurinia*) over a period of fifty years (Plate 69). From 1881 the population gradually increasd to a peak in 1897 and then went into a steep decline until 1906. Between 1906 and 1920 the population remained very low and bordered on extinction. In 1921, however, there was a population explosion, and it returned to the levels of abundance of thirty years before, staying at the same level for the following four or five years. For almost three decades in California Paul Ehrlich and his students have studied local population fluctuations and shifts in the bay checkerspot (*E. editha*).

More regular fluctuations were recorded in Germany between 1880 and 1940 for the geometrid *Bupalus piniarius* (Plate 69) and three other Lepidoptera that damaged pine forests. The significance of such variations has not been satisfactorily explained, but it is likely that parasites played an important role. In the case of *E. aurinia* it was probably a braconid wasp and in the case of *B. piniarius*, an ichneumonid wasp and a tachinid fly.

Censuses of Butterfly Populations

The periodic censuses that form the basis of these studies are not easy to carry out. First of all it is important to define the limits of the study population and to establish whether it is *closed*, that is, isolated from others, or *open*, that is, subject to frequent exchanges of individuals with other populations. If the population is open, it is difficult to obtain wholly satisfactory estimates, and the level of abundance is often measured in relative terms, giving, for example, figures for the numbers of larvae or adults per unit of area. With closed populations, on the other hand, it is possible to arrive at an estimate of the total number of individuals by means of a variety of methods. The most widely used involve marking and recapture. Such methods are based on the "dilution" of a known number of individuals in the total population once they have been marked and released. Marking can be done simply: All it involves is placing a spot of paint on the wings, after which the individuals are released so that they can mix with the other, unmarked, individuals. Several hours later a sample of butterflies is recaptured (containing both marked and unmarked individuals), and the size of the total population can be obtained by multiplying the number of individuals marked and released (M) by the total number of those recaptured (T) divided by the number of recaptured marked individuals (R). More sophisticated methods take into account the fact that populations change day by day as new individuals emerge and others die. An example is illustrated in Plate 69.

Migrations

The migratory habits of the various species of Lepidoptera fall between two extremes: Populations of Edith's checkerspot (*Euphydryas editha*) are so sedentary that individuals do not move more than 600 m (2,000 ft) in

their entire life, whereas the Monarch (*Danaus plexippus*) or certain *Vanessa* regularly make trips of over 2,500 km (1,500 mi). How should the intermediary cases be classified? At what point is it valid to talk of migrations? Is a swallowtail that flies a few miles a migrant? As with other animals only those species capable of long outward flights and then able to return to the areas where they breed should be regarded as migratory. Under such a definition there are more than 200 species of migratory butterflies and moths. In butterflies the migratory phenomenon is most widespread among the Pieridae, the Danaidae, the Nymphalidae, and to a lesser extent the Lycaenidae and the Hesperiidae. In moths migration is particularly common in the sphingids and the noctuids (night flying) and the uraniids (day flying).

It is difficult to obtain comprehensive information on butterfly and moth migration. Most of the information comes from careful collation of sometimes fragmentary data. It is quite common to observe masses of butterflies flying over mountain passes in both temperate and tropical regions, but for most species little is known of the destination or the significance of these movements. It is known that in the northern parts of Europe and North America many species make seasonal southward movements (in the same way that birds migrate). In particular certain butterflies fly south in the late summer or autumn and north from North Africa or South or Central America in the spring. However, unlike birds, a particular individual is not able to undertake both the outward and return trip — one leg being undertaken by the offspring.

The Monarch, *Danaus plexippus*, the best known butterfly, is an exception to this. It is distributed through almost the whole of the Americas from southern Canada to Paraguay. However, in North America it is present only during the summer, feeding on milkweeds (*Asclepias*); depending upon the climatic conditions, it produces a variable number of generations. In the autumn the butterflies begin to move south, initially in small groups and then in flocks of many thousands of individuals of both sexes. The western populations fly southwest, while the eastern and central populations fly south or south-southwest. The western populations winter in small areas along the Pacific coast of California, between San Francisco and Los Angeles. These sites contain tens of thousands of the butterflies clustered on the trunks and branches of eucalyptus and Monterey pines, where they spend the winter in a state of semihibernation. In the spring they mate and both sexes fly north to reoccupy the vast area that they abandoned the previous autumn. On the return flight they halt to deposit eggs on any milkweeds that they come across.

The populations from the eastern United States hibernate in a different area, but despite extensive research involving the marking of thousands of monarchs to follow their route, their hibernation site remained a mystery until a few years ago. The marking, which consisted of attaching a numbered label to the forewing, nevertheless provided important information. A marked butterfly was recovered 125 km (77 mi) away from the site of its original capture that same day. It was found that the marked butterflies were flying towards Mexico, but the traces disappeared at the Texas state line. In 1976 as a result of persistent research by the Canadian zoologist J. A. Urquhart and cooperators the winter quarters of the monarch were finally discovered in a small 1.5-ha (3.7-acre) valley at an altitude of 3,000 m (10,000 ft) in the Mexican state of Michoacán. Again the butterflies were crowded together on the trunks and branches of trees (various species of conifers), but the numbers of individuals packed together at the site was quite staggering. An estimate made by Lincoln Brower and colleagues in 1977 revealed that there were more than 14 million butterflies in an area of 1.5 ha (less than 4 acres).

Ecological Relationships

Like other organisms butterflies and moths are highly dependent upon the environment in which they live. During the various stages of their life cycle they are subjected to the influence of the climate and other physiochemical factors that make up their *habitat*. In interacting with other living beings, they form part of a *biotic community* — the collection of organisms in a particular biotope. The habitat and community together form an ecological system, or *ecosystem*. Within the ecosystem each living species has its own role (*niche*), which may, as a first step, be identified by its *trophic level*, that is, its position in the *food chain*. On this basis it is possible to distinguish *producers*, which are mainly plants capable of fixing light energy and synthesizing organic substances (food) from simple, inorganic substances, *consumers*, represented by animals and other organisms that obtain their food, either directly or indirectly, from the producers, and *decomposers*, represented principally by fungi and bacteria, which break down the protoplasm of dead organisms, absorbing some of the compounds and releasing simple, inorganic substances, which may be used by the producers. Primary consumers feed directly on the producers (if they are herbivores); secondary or tertiary consumers (if they are carnivores or parasites) by feeding on other producers benefit from primary production only indirectly. Apart from some notable exceptions (haematophagous or carnivorous species) the Lepidoptera are primary consumers.

Groups of species which feed on the same resource in a given community are known as *guilds*. When one analyzes particular communities, it is always important to consider all species in a particular guild rather than using taxonomically related organisms as the objects of analysis.

In addition to its trophic level the role of a species in the biotic community is more precisely defined by its interrelations as a whole with the other organisms and by *competition* in particular. In other words each species occupies an exclusive *ecological niche*, which characterizes not only its spatial position but above all its relationships with the other animals, plants, and microorganisms that make up the community. In this chapter we shall examine the ecological relationships that involve the Lepidoptera. Hence we shall discuss their dietary habits, first in terms of the relationships between larvae and host plants and certain peculiar adaptations, seeing how the caterpillar-host plant relationship has influenced and directed the reciprocal evolution of plants and Lepidoptera. We then examine the ecological and evolutionary consequences of such antagonistic relationships as competition or of mutually beneficial ones, such as myrmecophily. Finally we shall review the principal predators and parasites of Lepidoptera, leaving to the next chapter an analysis of their adaptations for countering predation.

Relations between Caterpillars and Host Plants

The almost strictly herbivorous diet of the Lepidoptera is of some interest. The only other insect group to show such dietary specificity is the order Homoptera, which has far fewer members. The majority of adult butterflies and moths feed on the sugars in nectar, while the larval stages feed on various higher-plant tissues: only 1 percent of caterpillars feed on mosses, liverworts, lichens, ferns, and nonplant material.

The Micropterigidae feed on mosses as larvae and pollen as adults. Only a few Lepidoptera include lichens in their diet, but some that do are lithosiine arctiids, ctenuchids (*Dixauses*), noctuids (*Bryophila*), and some lycaenids. The larvae of certain tineid moths eat fungi, while others feed on liverworts. The aquatic larvae of pyralids of the genus *Paragyractis* eat

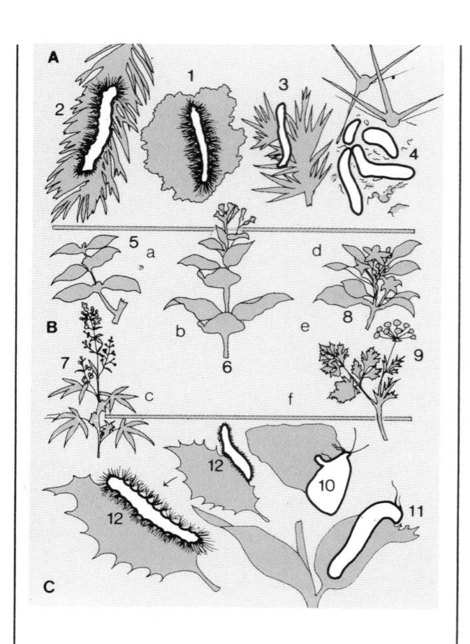

Plate 71 **Lepidoptera and plants**

The evolution of butterflies and moths is closely tied to the radiation of flowering plants although the various species are exploited to differing degrees by the Lepidoptera.

A: Examples of polyphagous species (1, *Lithosia quadra*, Arctiidae, on lichens; 2, *Panthea coenobita*, Noctuidae, on conifers), monophagous (3, *Thera juniperata*, Geometridae, on *Juniperus communis*), and oligophagous (4, *Cactoblastis cactorum*, Pyralidae, on *Opuntia*).

B: Examples of alkaloids and other secondary substances produced by plants as adaptations against phytophagous forms. The illustration shows the formulae of the principal active compounds and the plants that produce them: (5) coffee (*a*, caffeine); (6) tobacco (*b*, nicotine); (7) Indian hemp (*c*, cannabinol); (8) orange; (9) parsley (*d*, methylcavicol, *e*, anethol, *f*, anisic aldehyde). It is interesting to note that the secondary products of the last two plants are identical despite their belonging to two quite unrelated plant families.

C: The secondary substances produced by plants are used by Lepidoptera as specific recognition signals either for oviposition [10, the mustard oils released by cabbage (*Brassica oleracea*) attract females of the large white (*Pieris brassicae*), which lay their eggs on it] or as a means of finding food [11, larvae of the monarch (*Danaus plexippus*) on bloodflower (*Asclepias curassavica*)]. Trichomes and the spiny edges of leaves have in the course of evolution come to limit the ravages of caterpillars. A simple experiment demonstrates the effectiveness of such mechanical protection: (12) A caterpillar is able to eat holly leaves only after the spiny margins of the leaves have been removed. (Partially adapted from P. Ehrlich and P.H. Raven, 1967.)

algae and diatoms on the surface of submerged rocks, and the caterpillars of the American arctiid *Clemensia albata* (Lithosiinae) feed on the alga *Protococcus viridis*, which grows on tree trunks. Finally only a few Lepidoptera eat ferns.

Many of the species that feed on lower plants are not selective and will feed on a vast range of plant families and species. Such a diet is said to be *polyphagous*, whereas species that eat only one plant species are *monophagous*. Those which feed on a few closely related plants — the members of a single family, for example — are *oligophagous*. Monophagous and oligophagous diets have become established in the course of evolution, as a response to the chemical defenses developed by plants (see the examples in Plate 71), by those butterflies and moths that have found ways through enzyme production to neutralize the toxic effects of the defenses. This has resulted in a high degree of specificity between host plant and caterpillar, which makes itself apparent as the so-called botanical instinct. This instinct leads a female to deposit her eggs on a particular plant and governs many other activities. Today we know that this botanical instinct is based on the selective recognition of the glucosides, alkaloids, terpenes, and the like, which are produced as a defense by the plant hosts (allelochemical compounds). This topic will be returned to later.

Species that live on phanerogams (flowering plants) attack herbaceous plants, shrubs, and trees. The caterpillars are able to feed on various plant parts: buds, flowers, fruit, seeds, leaves, and wood. However, although some species display a certain amount of versatility in being able to exploit different substrates, it is much more common for them to specialize in feeding on a particular tissue.

Some species prefer flowers. The females of certain hairstreaks deposit their eggs in flower buds and the larvae emerge and begin to eat the buds before the flowers open. Fruits, which are rich in nutritive substances, are often used by caterpillars, and indeed many of the "maggots" found in fruit are Lepidoptera larvae. Two entomologists, P.J. Chapman and S.E. Lienk, identified no less than forty-seven species of tortricid moths feeding on apples in New York State. Many species attack seeds, even dried ones, and they are a serious pest in stores of grain, maize, beans, and so forth.

Many endophytic species (ones that spend their larval life inside plant tissues) attack bulbs, roots, and rhizomes, whereas others live in the woody tissue of trunks and branches, the norm among the sesiids and cossids; they cause considerable damage to orchards and take two or more years to complete their life cycle, probably because of the lower nutritive value of wood. Larvae of giant yucca skippers (Megathymidae), which live in the deserts of the southern United States and northern Mexico, dig galleries inside the stalks and fleshy leaves of yuccas and agaves. The pupae have considerable mobility and are able to move about in the galleries — which have silk threads running along them — to reach the openings, where the adults emerge. Another group of endophytic species is the miners, which includes the larvae of several microlepidopteran families and, in particular, of the nepticulids, the smallest lepidopterans. The larvae feed on the leaf parenchyma, digging more or less complex galleries in the leaf itself, although they often manage to avoid impairing the functional ability of the leaf and hence cause only moderate damage to the host plant. The adaptations of the nepticulids to this way of life are their small size and the morphological changes that their larvae have undergone: flattening, atrophy of both true and prolegs, and reduction in the number of larval instars. The form of the galleries, which is visible through the leaves, is highly species-specific and is an excellent taxonomic character, helping in the identification of morphologically similar species. The mining habit of nepticulids is ancient, as is shown by American fossils

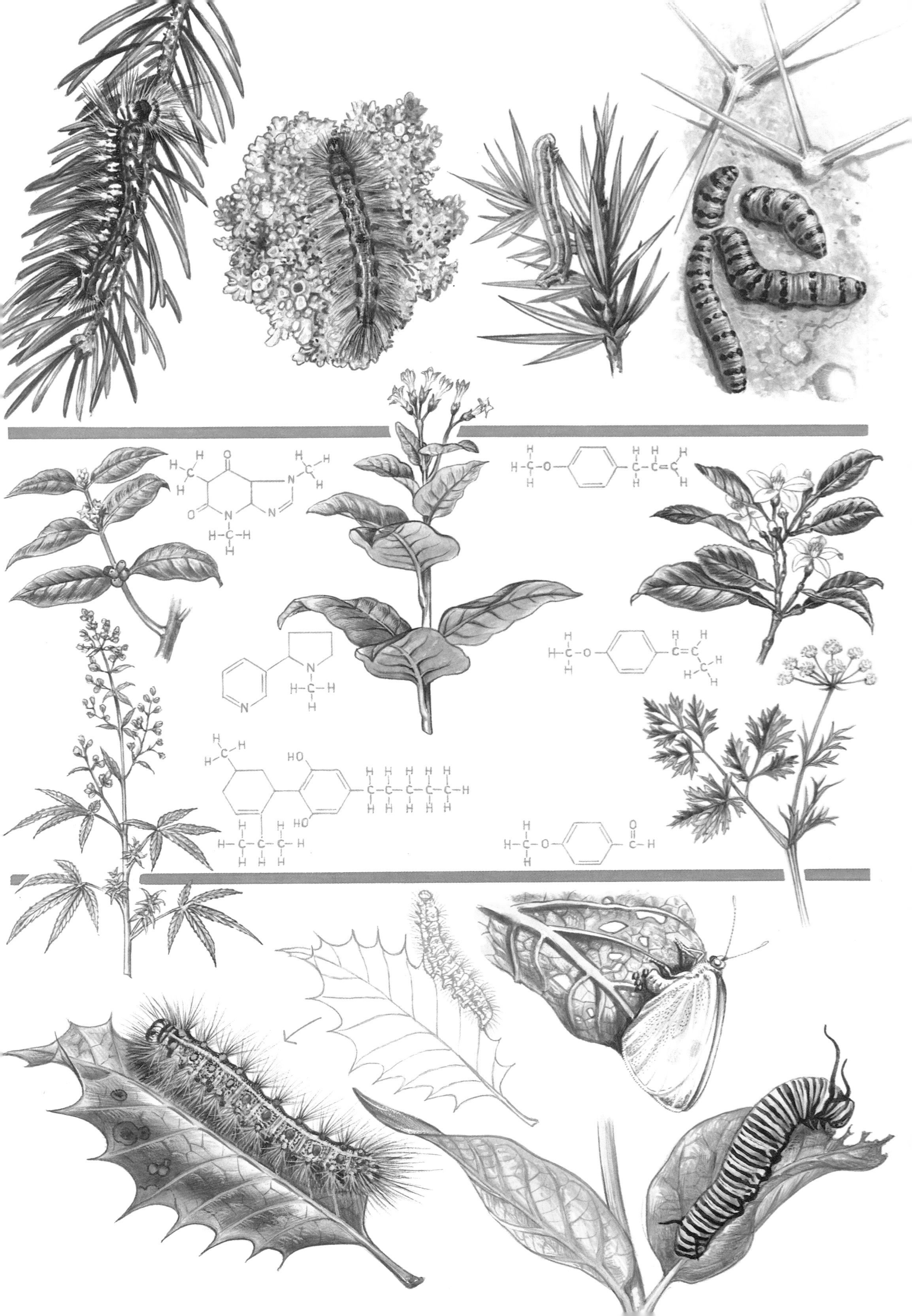

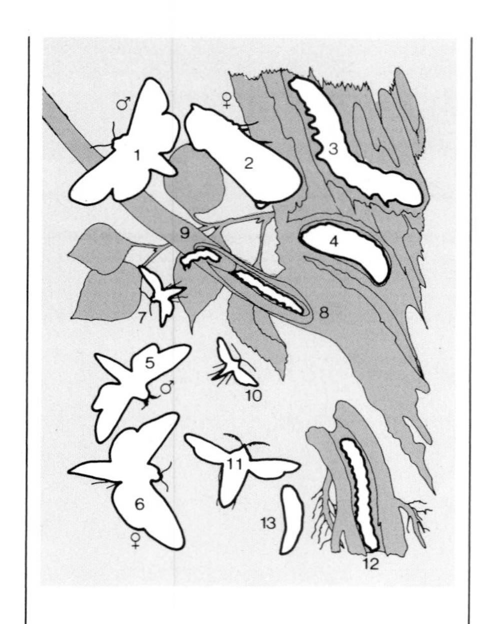

Plate 72 **Wood-eating Lepidoptera**

Caterpillars use various parts of plants and tissues as food. Some families, for example, the Cossidae and Sesiidae, specialize in feeding off woody tissues. The larvae of xylophagous species are generally smooth, robust, and only slightly pigmented. Some species cause considerable damage to fruit trees. (1–4) *Cossus cossus* (Cossidae): male (1), female (2), larva (3), pupa (4). (5, 6) *Zeuzera pyrina* (Cossidae): male (5), female (6). (7–9) *Parathrene tabaniforme* (Sesiidae): moth (7), larva (8), pupa (9). (10) *Synanthedon culiciformis* (Sesiidae). (11–13) Hornet clearwing (*Sesia apiformis*, Sesiidae): moth (11), larva (12), pupa (13).

of oak leaves mined by larvae of *Stigmella* 15 million years ago.

Guilds of leaf miners of given hosts may be analyzed by their trophic guilds. For example, the leaf mining guild on oaks may be broken down in different ways. The miners of coast live oak (*Quercus agrifolia*), studied in California by P. A. Opler, form two guilds, one of which mines newly expanded foliage, while the second mines mature and senescing leaves. The latter is by far the larger guild; within it each species has its own niche, although it may overlap with that of another species. Specialization is accomplished by several means. One mentioned above is by the leaf tissue layer consumed. Some specialize on the central mesoderm, while others feed only on the upper or lower leaf epidermis. The form of the mine differs between species, being a blotch, a flaring trumpet, or linear. Height above the ground or age of the tree are other means by which niche overlap has been prevented through time. One *Phyllonorycter* prefers only very low branches, while a second may be found at any level in the crown. Some of the coast live oak miners are more common in particular ecological formations, such as chaparral.

A certain number of Lepidoptera produce galls, or *cecidia*, on the host plant in the same way as do other, better-known cecidogenous insects, such as the Hymenoptera and Diptera. Galls are tumours produced by the plant as a reaction to the presence of the insect, and they develop in fixed ways. Often oviposition is all that is needed to provoke a reaction in the plant tissues. The shape of the gall is specific and is a useful diagnostic character for identifying the parasite. The majority of known cecidogenous Lepidoptera belong to two families, the Tortricidae and the Gelechiidae. The larva develops inside the gall, which protects it against predators since it is completely cut off from the outside. In some gelechiids, when the larvae reach maturity, they bore a small hole in the wall of the gall and then plug it so that it can be opened only from the inside. This hole is used when the adult emerges. One would have thought that such strategies as this would be highly effective in protecting the larva from predators and parasites, but this is not the case. Certain braconid wasps manage to lay their eggs inside gelechiid eggs before the gall forms and give rise to numerous progeny through the phenomenon of *polyembryony*. These progeny feed on the caterpillar, although allowing it to reach maturity.

However, most Lepidoptera are ectophytic, and the larvae simply eat plant tissues by attacking them from the outside. Most caterpillars which feed externally do so within rolled or folded leaf shelters or do most of their feeding at night — only to return to their hiding places under bark or leaf litter or sewn leaves during the day. Exceptions to this sort of behaviour are aposematic species, those which are distasteful or have stinging spines. These are often brightly coloured and may form clusters to enhance the effect of their warning to potential predators. The softer parts are preferred, particularly by the younger larvae, which tend to attack the leaf parenchyma rather than the epidermis. Caterpillars often attack the leaf margins, positioning themselves astride the leaf. Plants, however, have evolved various defensive mechanisms to avoid or at least to limit the damage caused by caterpillars and other herbivores. Examples of such defenses are the spines of cacti or the serrated and pointed edges of many other plants. The caterpillars of *Lasiocampa quercus* and other species are not normally able to eat the spine-edged leaves of the holly (*Ilex aquifolium*).

The same leaves, however, will rapidly be eaten if the spines are removed as an experiment. The most complex and interesting defensive mechanisms in plants are chemical; and these have had a profound influence on the evolutionary history of butterflies and moths. This subject will be discussed in the chapter on coevolution.

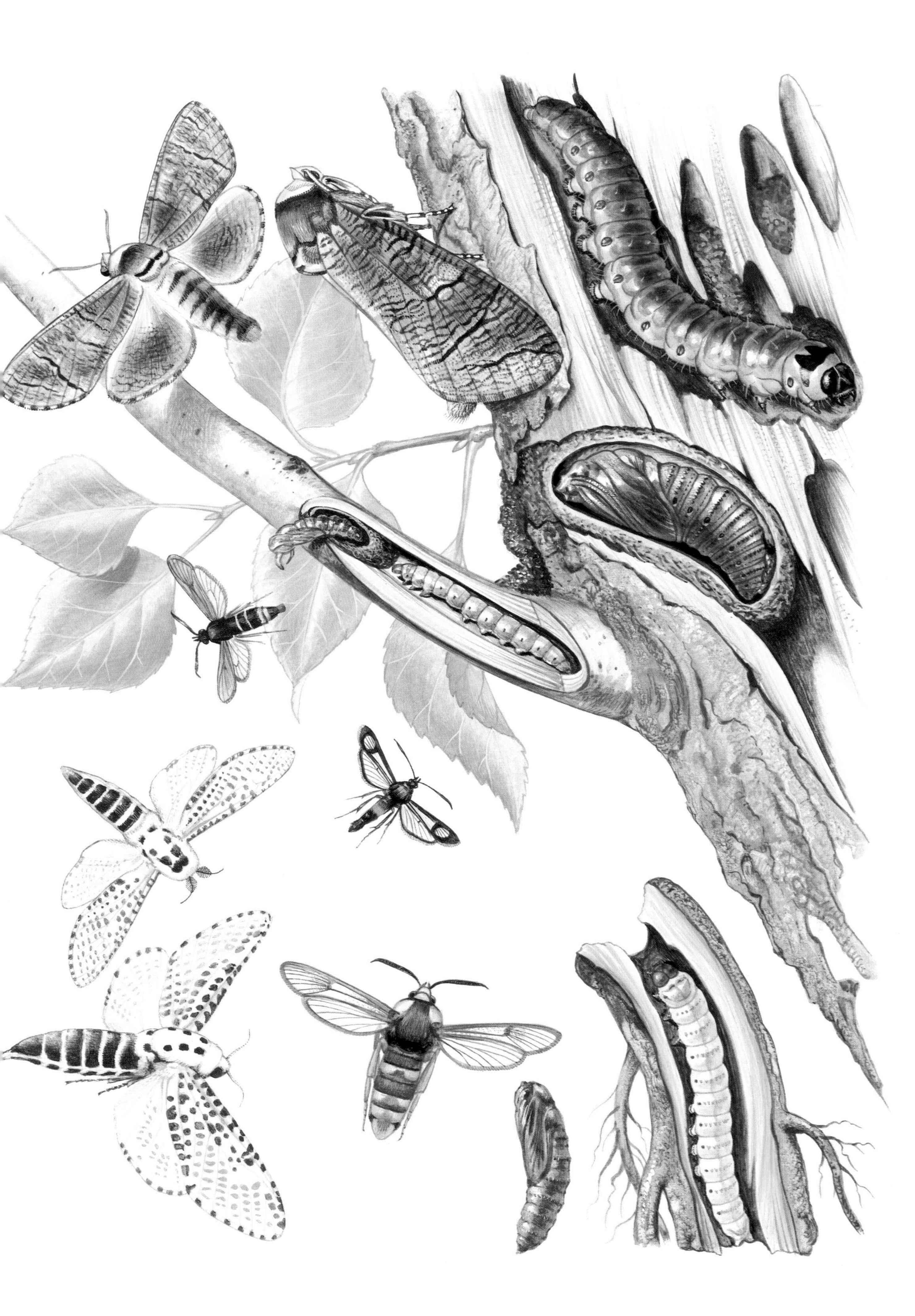

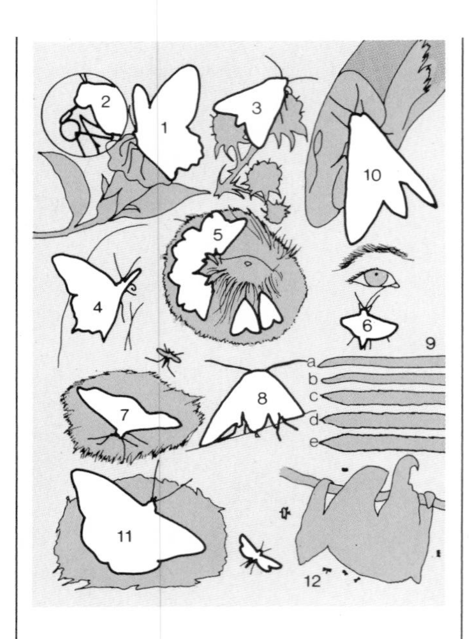

Plate 73 **Unusual dietary habits among butterflies and moths**

The mouthparts of butterflies and moths are designed for sucking, and the great majority feed on nectar. However, a considerable number of species have adapted to very different diets. Many longwings (*Heliconius*) and various *Parides* swallowtails eat pollen as well as nectar, using it as a source of valuable amino acids, and it is not uncommon to see them with a sheath of pollen around their proboscises (1,2, *Parides arcas mylotes*). Many noctuids feed on the juice of fermenting fruits, piercing the epicarp with their strong proboscis [3, the dark sword gloss (*Agrotis ipsilon*) on blackberries]. Various Lepidoptera feed on substances that are animal in origin, such as sweat (*Vindula* sp., Nymphalidae, on man), lachrymal fluid, tears (5, *Lobocraspis griseifusa*, Noctuidae, on *Bos javanicus*; 6, *Filodes fulvidorsalis*, Pyralidae, on man), and the blood of mammals (7, *Nobilia* sp., Geometridae, on *Bubalus bubalis*; 8, *Calpe eustrigata*, Noctuidae, on man). Figure 9 illustrates the distal portion of the proboscis in five species of noctuid, each characterized by a different diet: (*a*) *Plusia gamma*, nectarivorous; (*b*) *Lygephila craccae* sucks up the juices of fruit without piercing the epicarp; (*c*) *Scoliopteryx libatrix* pierces the epicarp of various fruits, as does (*d*) *Calpe thalictri*; (*e*) *Calpe eustrigata* blood feeding, the only lepidopteran capable of piercing the skin of mammals. Finally several species feed on rotting animal matter: (10) *Sphinx luscitiosa*, Sphingidae, on dead fish (*Catastomus* sp.); (11) purple emperor (*Apatura iris*, Nymphalidae) on dung. The small *Bradipodicola* moths, Pyralidae, spend a large part of their time in the fur of sloths (12). (Figures 5 to 9 are drawn from photographs by H. Bänzinger, 1971, 1972.)

Unusual Dietary Habits: Caterpillars

Although the general rule is that caterpillars feed on plant tissues and the adults on nectar, there are exceptions. The caterpillars of a number of species, especially among the microlepidoptera, are adapted to eat other substances, which have been processed to a greater or lesser degree but are derived from plant matter. These include the wood of rotting trunks, corks, dried fungi, grains, flour, biscuits, pasta, chocolate, paper, garments, and so on. The feeding habits of certain tineids, pyralids, momphids, and gelechiids are well known, feeding as they do on substances of animal origin such as wool, wax, various furs, feathers, stuffed animals, collections of insects, and so on. Some tineids feed on the droppings of mammals and birds. The adults of some pyralids (*Bradypodicola* and *Cryptoses*) live in the fur of xenarthrans and of the three-toed sloth in particular (Plate 73); the larvae are specialized for eating the excrement of these animals. There are, in addition, a number of species that are actual carnivores, hunting other insects. These include the larvae of certain Asiatic noctuids (*Eublema*), which prey on scale insects, and those of the Hawaiian geometrids *Eupithecia staurophragma*, *E. orichloris*, and *E. rhodopyra*, which attack various Diptera, using their cryptic coloration to avoid detection by their prey.

Predatory habits are particularly common among the Lycaenidae, the feeding patterns of which range from the occasional cannibalism of the larvae of the Iolas blue (*Iolana iolas*), large blue (*Maculinea arion*), and green hairstreak (*Callophrys rubi*) to the specialized, aphid-based diet of the harvester (*Feniseca tarquinius*), whose larvae feed on the aphid, genera *Schizoneura* and *Pemphigus*. In fact all Miletinae and Liphyrinae display carnivorous habits, and various African species attack jassid and membracid homopterans, mimicking the "caresses" that ants use to obtain the secretions of sugary liquid. Many carnivorous lycaenids have evolved myrmecophilous habits as a result of their ability to produce special secretions that encourage the ants to accept them as guests in their nests, where they prey on the aphids raised by the ants and frequently on the eggs, larvae, and pupae of their hosts as well.

The larvae of the noctuids *Nepenthophilus tigrinus* and *Eublemma radda* display the unusual habit of living in the pitchers of carnivorous plants of the genus *Nepenthes*. Here they are able to withstand the proteolytic action of the juices produced by the plant and thus to deprive it of its victims. Similarly, certain tineids and gelechiids haunt spiders' webs, feeding off the spider's eggs and the remains of its prey. In addition to these indirect forms of parasitism there are instances of true parasitism; for example, the larvae of certain pyralids feed on attacid caterpillars.

Adult Feeding Habits

Adult Lepidoptera, if they feed at all, are either nectar feeders, pollen feeders, or saprophagous. For most species, nectar provides sugars for ready energy, but in some others a diet of pollen and nectar, which contains low concentrations of the protein-building amino acids, may actually contribute to egg development in long-lived species such as *Heliconius* and some hawkmoths. Saprophagous species may feed on fluids in carrion, dung, rotting fruit or fungi, or bird droppings. Butterflies and moths that have longer adult lives tend to be saprophagous. Nectar feeders have much longer proboscises than the others, and their lengths tend to correspond to the lengths of the flowers they visit. Most adults are not specialists on particular species of flowers, but select among an array available in their location during their flight period. For example, adults in the different yearly generations of the same species will have to select different nectar sources, even at the same locality. In some instances, even though suitable caterpillar hosts are present, some local environments may

not be capable of supporting certain butterflies because of a paucity of nectar plants. In other instances, seemingly ideal localities with seas of flowering plants may be devoid of butterflies due to the absence of caterpillar hosts.

Some kinds of specialization can be seen in flower choice. For example, some species choose flowers on plants only above a certain height, while others, which normally fly only within a few inches of the ground, visit only flowers of prostrate plants. Some species select only flowers oriented in a certain way; for example, the European cabbage white (*Pieris rapae*) selects only flowers oriented to the side. Some groups of species seem to select flowers of certain colours. For example, flowers visited by hawkmoths are frequently white or at least pale-coloured, and flowers visited by skippers are most often pink, purple, or white. Some species assist in pollinating the plants their caterpillars eat. Some examples are the Monarch (*Danaus plexippus*), a frequent visitor to the milkweed flowers of its caterpillars only food plant, and the *Euphilotes* blue, which most often takes nectar from the buckwheat flowers that serve as its caterpillar's only food.

Unusual Dietary Habits: Adults

Exceptions to the typical nectar-based diet of adult butterflies and moths occur. Some feed on or supplement their diet with pollen. The Micropterigidae, for example, which have chewing mouthparts, are able to do this; and certain butterflies (heliconiines and papilionids of the genus *Parides*) are often found with their proboscises coated with pollen, which they use as a source of amino acids (Plate 73). However, not all butterflies and moths are harmless visitors to flowers. *Acherontia atropos* (Sphingidae) steals honey from hives, and many species of butterflies and moths feed on the juices of rotten, ripe, or unripe fruit. The noctuids display a whole range of adaptations from nectarivorous species to fruit-eating ones, some of which suck up juices from fruits of which the skin has already been broken by vertebrates, while others have a proboscis that is strong enough to pierce the epicarp. *Agrotis ipsilon* is able to pierce only the thin epicarp of blackberries and raspberries, but *Othreis fruttoria* can penetrate the rind of citrus fruit. Another series of adaptations has developed from the primitive nectarivorous condition towards an animal-based diet. This may have arisen among independent groups of forest-dwelling species because from time to time the adults would come across edible excretions, exudates, and other secretions left by mammals or other vertebrates. This probably led to lepidopterons' actively searching out such substances and becoming adapted to sucking up urine, excreta, decomposing bodies, cutaneous secretions, and so on (Plate 73). In all probability it was such habits that gave rise to the moths that specialize in feeding on the tear fluids of large mammals — human beings included — by inserting their proboscis between the eye and the lid. About thirty species of tropical geometrids, pyralids, noctuids, and notodontids habitually suck the lachrymal fluids of buffaloes, cattle, deer, elephants, tapirs, wild boar, and other ungulates. In Thailand and Malaysia the noctuid *Lobocraspis griseifusa* and the pyralid *Filodes fulvidorsalis* occasionally alight even on people. Leucocytes and epithelial cells (which can be digested because they have a protease) as well as bacilli, cocci, spirochaetes, and other pathogenic microorganisms have been found in their midgut. The most startling adaptation was undoubtedly that discovered to his cost by the Swiss entomologist Hans Bänzinger: The oriental noctuid *Calpe eustrigata* feeds on blood by piercing the skin of various mammals (including — at least experimentally — man) with its robust proboscis. This "vampire" is the only member of its genus to display this habit; but the comparison with the related *Calpe thalictri* shows how, with relatively modest structural and behavioural modifications, the

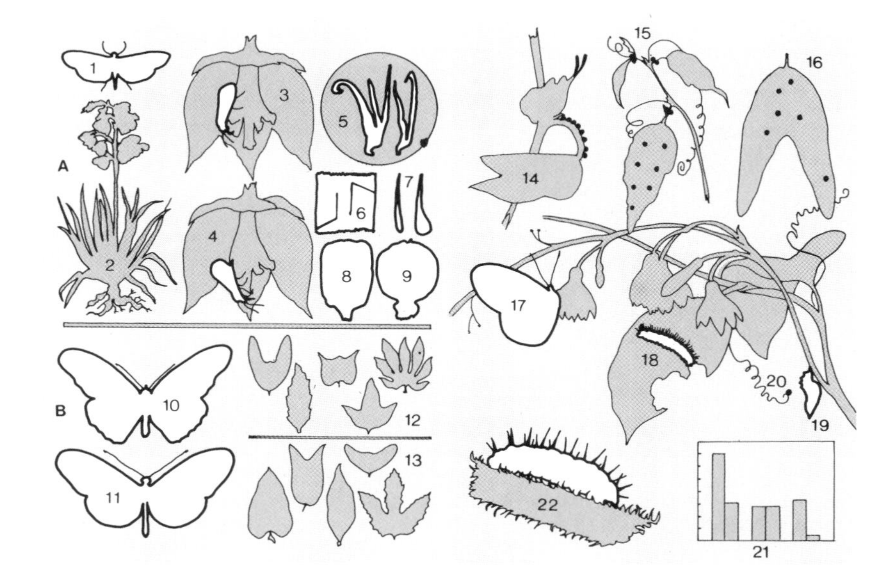

Plate 74 **Coevolution of Lepidoptera and plants**

The extensive interactions among lepidopterans and plants have in several cases resulted in their adapting in response to reciprocal selective pressures. This phenomenon is known as coevolution. The plate illustrates two examples of coevolution discussed in the text, the first being mutually beneficial, the second highly antagonistic.

A: The yucca moth (1, *Tegeticula yuccasella*) is, with a few related species, the only organism capable of ensuring the pollination and reproduction of yuccas, Liliaceae (2; *Yucca filamentosa*). This is made possible by the transformation of the female's maxillae, in particular of the stipe (5, left, female, right, male). The female carries the pollen collected from the anthers of one flower (3) on her maxillae to the stigmas of another flower (4), in whose ovary she will deposit her own characteristic eggs (6, 7). Subsequently the larvae eat only some of the seeds (8, section of the capsule of *Y. whipplei* with many seeds remaining intact; 9, exit holes of *Tegeticula* larvae).

B: Some aspects of the coevolution of Heliconiinae and *Passiflora*: (10) Silver spot (*Dione juno*), Trinidad; (11) *Heliconius pachinus*, Costa Rica; (12) differences in leaf shape in five species of *Passiflora* in Trinidad; (13) differences in leaf shape in five species from Costa Rica. The diversity of leaf shape seems to have been evolved as a means of deceiving the females of *Heliconius*, which choose the site for depositing their eggs by sight. Likewise, *Passiflora* has evolved extrafloral nectaries (14, 15), which attract ants capable of attacking the eggs and larvae of *Heliconius*. They also develop false eggs on the stipules (14) or leaves (15, 16), which again deceive females searching for a free site on which to lay their eggs. (17–20) Various stages in the development of *H. nattereri*, a rare Brazilian species. (21) The results of an experiment conducted by K. Williams and L. Gilbert, which demonstrates the ability of *H. cydno* to discern colours. *A* gives a comparison between the frequency with which eggs were laid on leaves without eggs (approximately 70 percent, left) and on those with eggs (approximately 30 percent, right). The experiment was repeated, but the eggs, which are normally yellow, were painted green; it was found that there was no difference in the frequency of oviposition (*B*). When the butterfly was given the choice of laying its eggs on leaves bearing a green egg (*C*, left-hand column) or a yellow one (right-hand column), the females again showed a clear ability to discern colours. (22) *Passiflora adenopoda* is a particularly common and widespread member of its genus. It bears hooked spines that are capable of ripping the integument of any caterpillar.

evolution from a fruit-eating to a haematophagous diet is possible (Plate 73). It is not pure fantasy to think that in the future bloodsucking moths may spread, specializing and diversifying in order to exploit such a widespread and abundant food source as the blood of mammals and other vertebrates. However, a major obstacle to the dissemination of this hypothetical adaptation is such formidable competitors as mosquitoes and other Diptera, Rhynchota (bed bugs and related forms), Siphonaptera (fleas), Anoplura (lice), and mites.

Coevolution of Lepidoptera and Plants

Returning to the interaction between the Lepidoptera and host plants, we have already stressed that the impact of caterpillars on plants is enormous and is probably greater than that of other herbivorous animals. Now we shall try to quantify it. It has been calculated that by the time it is fully developed a caterpillar has eaten twenty times its own weight in plant matter. When one realizes that the world population of certain species numbers thousands of millions or even millions of millions of individuals in every generation, that many of these species have several generations a year, and that, further, there are more than 150,000 phytophagous species, one can see that butterflies and moths consume thousands of millions of tons of plant matter every year. Given this sort of selective pressure coupled with the pressure exerted by other herbivores, it becomes clear why plants have developed suitable defenses in the course of millions of years of evolution. We have already discussed such obvious mechanical defenses as thorns, but the main method used by plants is the production of toxic and repellent chemicals that, as they play no physiological role in the life of the plant, are clearly aimed against their predators. The chemicals include alkaloids, quinones, essential oils, cardiac and cyanogenic glycosides, terpenes, and raphides (calcium oxalate crystals). Certain substances, such as pyrethrum extracted from chrysanthemums, have been widely used in the past as powerful natural insecticides. Other substances, such as the alkaloids (almost an "invention" of the angiosperms), are not insecticidal but act as drugs that affect insect behaviour. Many substances in common human use — caffeine, nicotine,

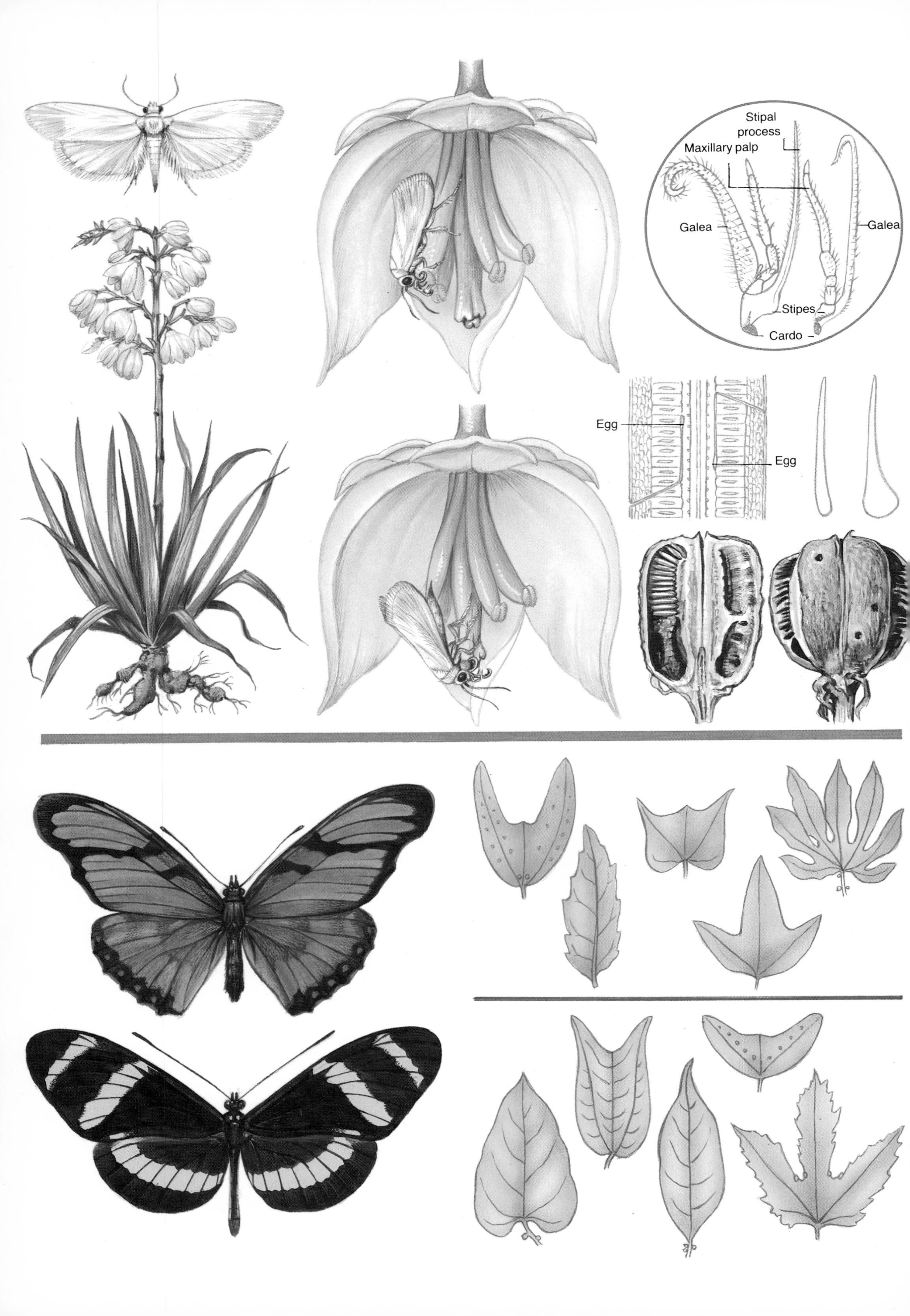

Stipal process
Maxillary palp
Galea
Galea
Stipes
Cardo
Egg
Egg

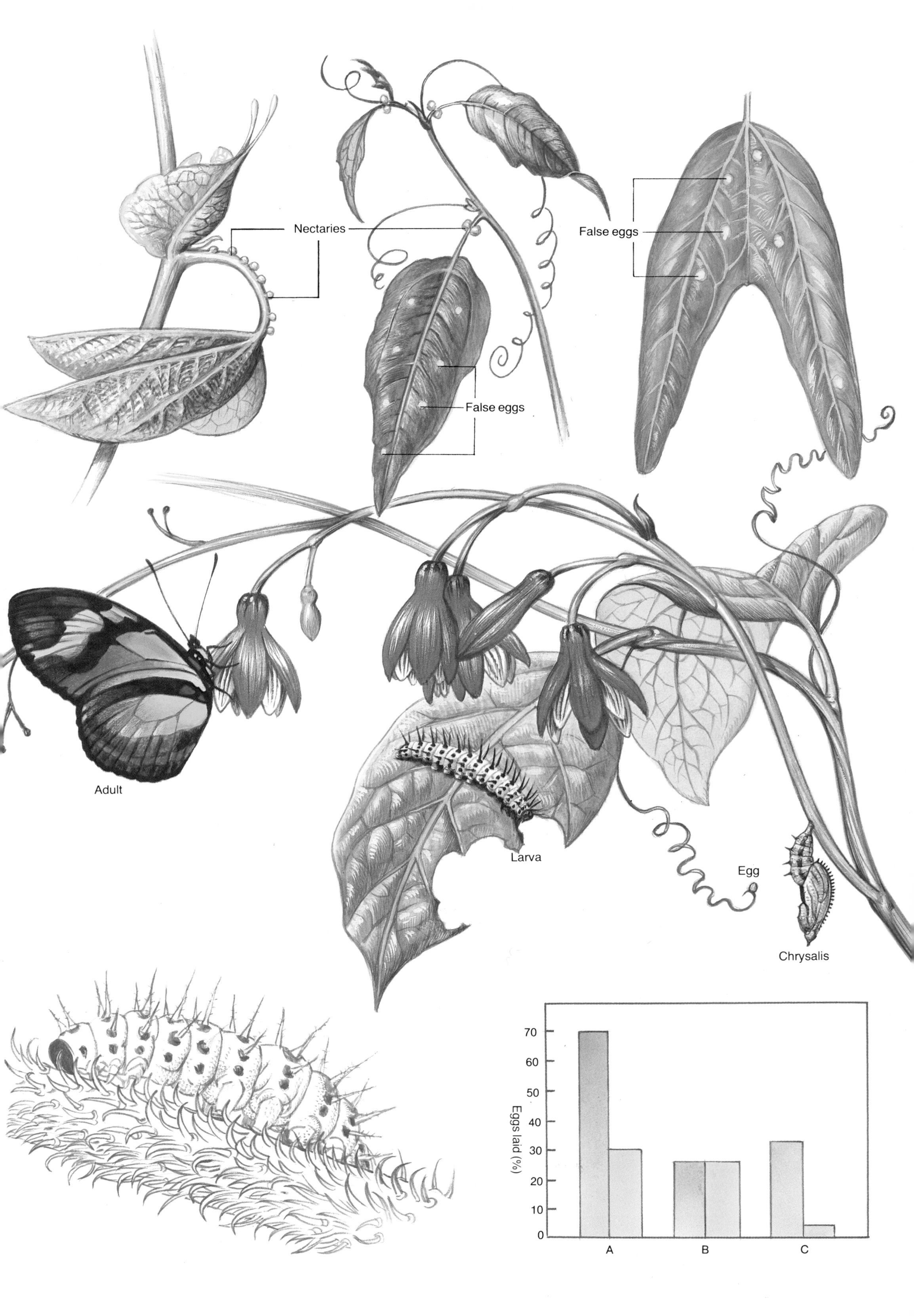

Nectaries
False eggs
False eggs
Adult
Larva
Egg
Chrysalis
70
60
50
40
30
20
10
0
Eggs laid (%)
A
B
C

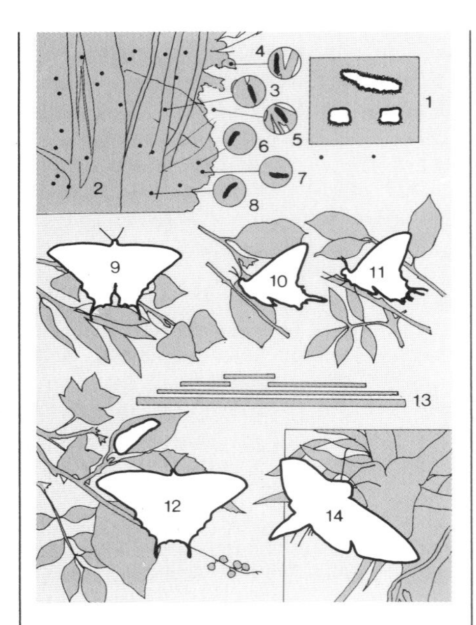

Plate 75 **Competition and ecological niches**

The ability of any species to live and breed in a given ecosystem is determined on the one hand by the presence of a set of suitable conditions and on the other by its capacity to compete with other species. These factors, taken as a whole, define the species' ecological niche. The caterpillars of the two related species *Amata phegea* and *A. ragazzii* differ in that one has a mask on its head and the other does not (1). When they live in the same area, as they do in the Alban Hills (2), they occupy different heights and make use of different food resources. The larvae of *A. ragazzii* occur mainly higher up on trunks and branches, where they eat bark (3), shoots (4), or young leaves (5); the caterpillars of *A. phegea* live on the ground and feed on the leaves of blackberries (6), dead leaves (7), and mosses (8). The western tiger swallowtail (*Papilio rutulus*, 9), the pale swallowtail (*P. eurymedon*, 10), and the two-tailed swallowtail (*P. multicaudatus*, 11) coexist in large areas of the western United States, and in the larval stage they feed on different plant species: willow (*Salix*), poplar and aspen (*Populus*), alder (*Alnus*), and the like (*P. rutulus*); wild plum (*Prunus*), coffeeberry (*Rhamnus*) (*P. eurymedon*); and ash (*Fraxinus*), hop tree (*Ptelea*), and wild plum (*P. multicaudatus*). The related *P. glaucus* (12), which replaces these three species to the east, feeds on most of the food plants of the three western species. The latter also fly at different seasons (13), another factor that differentiates their niches and strengthens their reproductive isolation.

The ecological niche of a species is also determined by reciprocal relations, such as those between flowers and pollinating insects. The Madagascar sphingid *Xanthopan morgani predicta* (14) is, thanks to its long proboscis, the only insect capable of pollinating the orchid *Angraecum sesquipedale*.

marijuana, opium, and quinine (Plate 71) — are alkaloids.

The principal result of this chemical battle has been to drive many plant parasites off and reduce the amount of harm done. However, it has also led to the increased specialization of those Lepidoptera that have been able to respond biochemically to neutralize the toxic effect of various substances.

The evolutionary success and tremendous diversification of the angiosperms, which from the Cretaceous period to the present have been the dominant form of vascular plant, seem to be the result of the development of effective chemical defenses. On the other hand the great adaptive radiation of butterflies, moths, and many other insects is probably the evolutionary result of the chemical struggle between plants and phytophagous animals. As the phytophagous forms gradually evolve ways of reacting biochemically to plant defenses (for example, by synthesizing particular enzymes capable of countering the toxic effects), they start to use the substances produced by plants for their own benefit, assimilating them in some way into the natural history of their own species. The molecule produced by the plant not only may become the specific recognition signal used by the caterpillar when feeding or by the adults when depositing their eggs ("botanical instinct") but also may be used as a means of mutual recognition by adults when courting. Indeed, as has already been mentioned, certain pheromones — danaidone, for example — are synthesized by butterflies (*Danaus*) from alkaloids directly assimilated from milkweed food plants.

There are numerous examples of botanical instinct. The Capparidaceae and Cruciferae, families that contain many familiar plants, such as the caper in the former case and cabbages, cauliflowers, and radishes in the latter, have evolved special biochemical defenses by using thioglucosides (better known as mustard oils). These are repellent to many insects but not to most whites, a subfamily of pierid butterflies with many oligophagous species, which specialize in feeding on the Capparidaceae and Cruciferae. It has been shown experimentally that it is precisely these substances that provide the specific signal by which the caterpillar or the adult when about to deposit her eggs recognizes the proper host plant (Plate 71). The caterpillars of *Pieris* will not normally accept other plants, but they will voraciously devour plants they would normally not touch or even pieces of paper if they are impregnated with sinigrin or sinalbin, two of the commonest mustard oils. In the Papilionidae an entire tribe, the Troidini, feeds on plants belonging to the Aristolochiaceae, a family whose members contain a powerful alkaloid; and a group of species in the genus *Papilio* (tribe Papilionini) is linked to the Rutaceae, the genus *Citrus* in particular. It is interesting to note that some of these *Papilio* species, for example, the common North American black swallowtail (*P. polyxenes*), also feed on the Apiaceae plants that at first sight seem to have little in common with the Rutaceae. However, the members of the two plant families in fact produce the same essential oils [methylcavicol, anethol, and anisic aldehyde (Plate 71)]. Today the biochemical similarity between the Rutaceae and the Apiaceae, something that butterflies discovered long ago, is regarded with interest by botanists, who see a possible evolutionary link between these families.

One other aspect of the coevolution of plants and butterflies deserves attention: The substances produced by plants for defense can be used by butterflies to ward off predators. Thus those species which have evolved the ability to metabolize allelochemical compounds have simultaneously acquired immunity from many predators, which find them toxic, repellent, or otherwise unappetizing. This condition has probably in turn led to the appearance of striking patterns of coloration that act as signals to warn predators (aposematic coloration). It is easy to see how instances of

Amata ragazzii
A. phegea
Papilio eurymedon
P. rutulus
P. multicandatus
P. glaucus
May
June
July

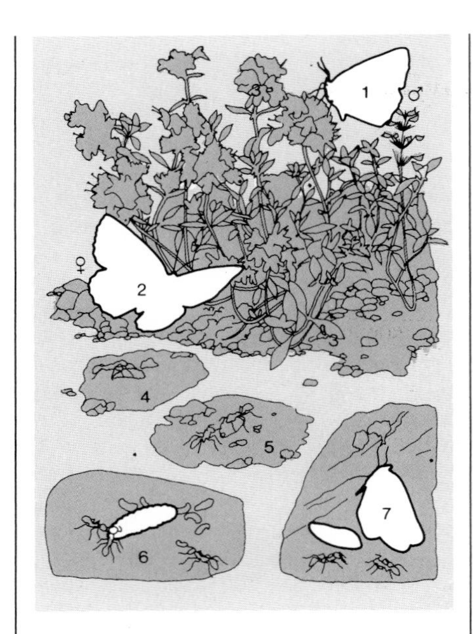

Plate 76 **Mutually beneficial relationships: myrmecophily**

This plate illustrates some aspects of the relationship between the large blue (*Maculinea arion*) and ants (see the text). (1–3) *M. arion* (Lycaenidae) on wild thyme (*Thymus serpyllum*): male (1), adult female (2), first-instar larva (3). (4) Ant stroking the myrmecophilous glands of a third-instar larva with its antennae. (5) Larva of *Maculinea* expanding its thorax and thus stimulating the ants to carry it into their nest. (6) The mature larva in an ant nest, occupied with feeding on ant larvae while it is tended by the workers. (7) The emergence of *M. arion*.

mimicry might arise on the basis of these bright colours.

The term coevolution should, according to some biologists, be restricted to those instances where the evolutionary changes in two interacting species (for example a plant and a herbivore) are the direct result of selective pressures that the two species exert on each other. It is not easy to identify such examples in the wild because of the multiple selective pressures resulting from complex interactions with various competitors, predators, parasites, commensals, and so on, as well as with the physical environment. Nevertheless, several cases have been studied in detail in which the reciprocal influence of the butterfly or moth and the host plant has clearly influenced the evolution of certain characters. Two examples are illustrated in Plate 74. The first involves the interaction between the yucca moth, *Tegeticula yuccasella*, and some members of the genus *Yucca* in the southern United States and Mexico, where the moth and its relatives are the only pollinators. A female of *T. yuccasella* will deposit her eggs by piercing the ovary of the yucca flower with her ovipositor. The same individual then pollinates the plant by depositing on the stigma packets of pollen that it has collected earlier with its modified maxillae. The *Yucca* is hence strictly dependent on the moth for its reproduction; on the other hand *Tegeticula* can survive only thanks to the plant, whose reproductive success it ensures not only by pollinating it but also because its larvae eat only a limited proportion of the seeds in each capsule, leaving the rest to develop. In this instance coevolution takes the form of symbiosis.

Highly specific, mutually advantageous relationships are quite common between orchids and their respective pollinating insects, many of which are sphingids. Charles Darwin observed a close correspondence between the depth of the flower's nectary and the length of the proboscis of the pollinating moth; this correspondence even led him to postulate the existence of a sphingid with a particularly long proboscis of at least 30 to 35 cm (12 to 13.75 in), which would be able to reach the bottom of the deep nectaries of flowers of the Madagascan orchid genus *Angraecum*. Such a moth was in fact discovered some years later by Lord Rothschild and C. Jordan, who gave it the fitting name of *Xanthopan morgani predicta* (Plate 75). The interaction between the Heliconiinae and species of *Passiflora* in neotropical forests, which has recently been studied by L. Gilbert, is markedly antagonistic. *Passiflora* species have few enemies apart from these heliconiine, and many of the adaptations of these plants seem to have been evolved in response to the selective pressure exerted by these butterflies. In the first place the Heliconiinae have overcome the chemical test, having found a way to protect themselves against the alkaloids produced by *Passiflora* and indeed to exploit their toxicity to ward off predators. The members of *Passiflora* for their part have developed various ways of avoiding attack: They display a tremendous variety of leaf shapes, produce false eggs, which deceive the female butterflies searching for a site on which to lay their eggs, put forth caducous stipules in the form of tendrils, which cause the failure of any eggs laid there, and even produce nectaries outside the flowers, which attract species of ants. These act as guards, attacking any eggs or caterpillars of *Heliconius*.

Competition and the Ecological Niche

Competition is one of the major forms of interaction that governs the structure and dynamics of biotic communities. There are essentially two forms of competition: *intraspecific*, between members of the same species and occurring when searching for food or a mate and so on, and *interspecific*, between different, sympatric species. Some aspects of

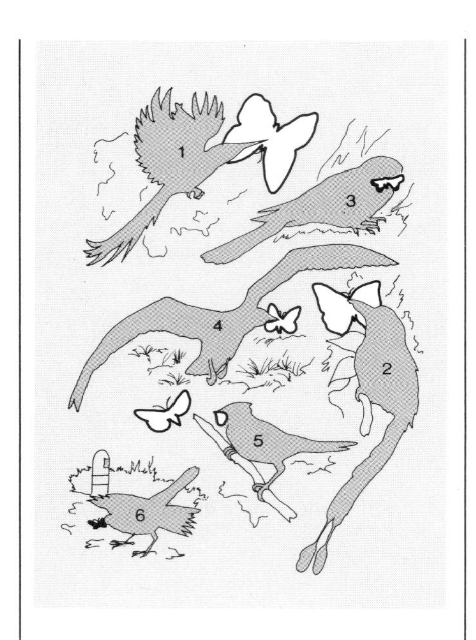

Plate 77 **Predators of Lepidoptera: birds**

Birds are among the major predators of butterflies and moths. The plate illustrates a number of examples of predation taken from literature or observed by the authors. Not all the butterflies and moths are shown in natural postures.

(1) Jacamar (*Galbula ruficauda*, Galbulidae) catching *Morpho achilleana* (Nymphalidae) in Brazil; (2) motmot (*Baryphthengus ruficapillus*, Momotidae) preying on a ghost brimstone (*Anteos clorinde*, Pieridae), Mexico; (3) The Australian nightjar (*Aegotheles cristata*, Aegothelidae) with *Anthela connexa* (Anthelidae) in its beak; (4) Black-headed gull (*Larus ridibundus*, Laridae) preying on *Lasiocampa quercus* (Lasiocampidae). Although gulls are not specialized insect eaters, some populations are at least partly adapted to such a diet. This is the case with the black-headed gull, which, according to H. B. D. Kettlewell's observations, is the principal predator of *Lasiocampa* in northern Scotland. (5) The blue jay (*Cyanocitta cristata*, Corvidae) attempting to catch an underwing (*Catocala*, Noctuidae). (6) The pied wagtail (*Motacilla alba*, Motacillidae) pecking and then releasing a *Zygaena ephilates* (Zygaenidae), an inedible species.

intraspecific competition, such as territorial behaviour, have already been examined. This section will show how interspecific competition for space and food resources causes increased specialization in both caterpillars and adults or at least a separation of the ecological niches of the species involved. It has been shown by laboratory experiments and further supported by mathematical models that two or more similar species cannot permanently coexist in the same environment, using the same space in the same way at the same time and also the same food source. This is the *principle of competitive exclusion*, according to which two species cannot occupy the same ecological niche. The result of competition is therefore either the departure or the extinction of the species that succumbs or the evolution of different morphological, physiological, or behavioural characters, which enables the species to use the habitat and the food sources differently. We have already seen this with the *Erebia tyndarus* group, whose members live at similar altitudes when allopatric but tend to stratify vertically when they coexist sympatrically on the same mountain (Plate 23). We have observed another example of spatial separation and of divergent food sources in central Italy; this is seen in the larvae of two *Amata* species (Ctenuchidae). In areas where they coexist the caterpillars of *A. phegea* and *A. ragazzii* occupy different levels in oak and chestnut woods. The first species tends to occupy the forest floor, feeding on mosses and dead leaves, whereas *A. ragazzii* occurs mainly higher up on the trunks and branches, where it feeds on bark and buds. Preliminary research carried out in areas where only *A. phegea* is found indicates that in the absence of *A. ragazzii* it tends to spread up into the zone usually favoured by *A. ragazzii* (Plate 75).

Subdivision of a niche has been observed by C. L. Remington in the North American *Papilio glaucus* species group. The western tiger swallowtail (*P. rutulus*), pale swallowtail (*P. eurymedon*), and two-tailed swallowtail (*P. multicaudatus*) coexist over large parts of the western United States from the West Coast to the Rocky Mountains, and both as larvae and adults they diverge in their food plants and in the season that they fly, a period strictly limited for each species. On the other hand the widely distributed eastern tiger swallowtail (*P. glaucus*), which has no competitors, has a widely based diet and with several generations a year flies from spring to autumn (Plate 75).

Butterflies, Moths, and Ants

Unusual interspecific relationships exist between butterflies and ants. These relationships mainly involve lycaenids, but there are a few examples from other families (Pieridae, Riodinidae, Ethmiidae, Pyralidae). Such relationships cover a whole range of situations from simple predation to symbiosis. The caterpillars of the Indo-Australian lycaenid *Liphyra brassolis* live in nests of the arboreal ant *Oecophylla smaragdina* and devour its larvae. They are protected from host attack by the odour of the nest, with which they are impregnated, and by their extremely smooth integument. At emergence the adult is covered with a down of scales, which are pulled off by the jaws of the pursuing ants.

A number of more specialized lycaenids have larvae with *myrmecophilous glands* on the seventh or eighth abdominal segment; these secrete sugary substances. Various members of the genera *Miletus*, *Aslauga*, and *Triclema* are accepted in this way by ants and are able to feed on the aphids that the ants raise. The larvae of the Asiatic species *Miletus boisduvalii* eat up to 300 larvae each before completing their development, but the ants leave them in peace, appeased by the sugary secretions of one or other of their guests.

The large blue (*Maculinea arion*) is the best known myrmecophilous form (Plate 76). A Palaearctic lycaenid, it has been much studied in England, where in recent years it has declined steadily, finally becoming extinct in 1979. The female deposits her eggs on *Thymus serpyllum*, and the larvae initially feed on the flowers of the plant (with a certain tendency towards cannibalism of younger larvae). After the second moult myrmecophilous glands develop, and the larvae excite the attention of various species of ants, in particular *Myrmica scabrinodis* and *M. laevinodis*, which begin to tend them. After the third moult the larvae move around on the ground, abandoning their food plant, and it is at this stage that the ants, stimulated by the glands and by the behaviour of the caterpillars — they arch their swollen backs and thoracic segments — take them in their mandibles and carry them into their nest. In the ants' nest the caterpillars continue to grow at the expense of the mature larvae of *Myrmica*, with which they live and overwinter. In May the caterpillars pupate without further moults. At emergence the adult of *M. arion* bears downy scales, which come loose in the jaws of the pursuing ants and enable the butterfly to escape. The myrmecophily of *M. arion* is interesting because it is an instance of a reciprocal relationship (the ants "milk" the caterpillars for a sugary liquid produced by the myrmecophilous glands) associated with a sort of parasitism or predation, from which the caterpillars gain an advantage and the ants are penalized.

Parasites and Predators

As we saw at the beginning of the chapter, the structure of a community is largely based on the food chain — the trophic relationships of each species with the other members of the community. It is therefore logical to suppose that the Lepidoptera, being important and numerous primary consumers, might themselves be eaten by a large number of species. This is in fact so. There is no stage in a butterfly's or moth's development when it is not attacked by more or less specialized enemies. First, however, one needs to distinguish parasites from predators even though the division between the two is not always sharp: Unlike predators parasites do not kill their host rapidly; the female deposits eggs inside the host, which remains alive, so ensuring that food is available throughout the parasite's development.

Leaving aside microorganisms, such as bacteria and viruses, which cause a significant degree of mortality among caterpillars but which are still little known, most lepidopteran parasites are tachinid or dexiid flies, or ichneumonid, braconid, or chalcid wasps. The best known braconid examples are *Apanteles*, which has numerous members more or less adapted to various hosts; these tiny wasps deposit their eggs by piercing the thin skin of young caterpillars with their fine ovipositors. *Apanteles zygaenarum* may parasitize burnet moths despite the fact that the caterpillars can release high levels of toxic hydrocyanic acid. The parasite is able to do this because its tissues contain the enzyme rodanase, which catalyzes the transformation of the hydrocyanic acid into thiocyanate, removing its toxicity. The chalcids are probably the most specialized hymenopteran parasites, whose tendency towards parasitism is so marked that they frequently lay their eggs in the eggs or larvae of other parasitic Hymenoptera. Such instances of *secondary parasitism*, or *hyperparasitism*, can lead to a chain in which the caterpillar may be acting as host not only to its own parasite but to third- or fourth-degree *hyperparasites*. Less specialized, but no less harmful to the Lepidoptera, are the tachinid (Diptera) parasites. These do not necessarily deposit their eggs in or on the caterpillar; if they do not do so, the larvae themselves find the host.

Plate 78 **Other predators of Lepidoptera**

Examples of vertebrate and invertebrate predators, including mammals (1, 2), reptiles (3–7), amphibians (8, 9), insects (10, 11), and spiders (12, 13): (1) *Plecotus auritus* (Chiroptera, Vespertilionidae) pursuing *Thyatira batis* (Thyatiridae), Europe; (2) the prosimian *Arctocebus calabarensis* (Lorisidae) holding a *Sphingomorpha* (Noctuidae), West Africa. (3) An anole (*Anolis carolinensis* Iguanidae) with a great spangled fritillary (*Speyeria*, Nymphalidae), United States; (4) the American whiptail lizard *Cnemidophorus lemniscatus* (Teiidae) with the checker spot (*Chlosyne eumeda*); (5) another North American reptile, an alligator lizard (*Gerrhonotus multicarinatus*, Anguidae), with an *Acontia* (Noctuidae); (6) the green lizard, *Lacerta viridis* (Lacertidae), catching the southern gatekeeper (*Pyronia cecilia*, Nymphalidae, Satyrinae), Italy; (7) the Asiatic gecko (*Gekko gecko*, Gekkonidae) with its prey *Spodoptera exigua* (Noctuidae). (8) A tree toad, *Hyla arborea* (Hylidae) lying in wait for *Euclidia glyphica* (Noctuidae); the moth is enlarged; (9) a bullfrog (*Rana catesbeiana*, Ranidae) attacking an eastern tiger swallowtail (*Papilio glaucus*, Papilionidae), United States. (10) Among flies the robberflies (Asilidae) are the most active predators of butterflies and moths: *Asilus crabroniformis* with its victim the heath fritillary (*Mellicta athalia*, Nymphalidae); (11) the mantid *Pseudocreabotra wahlbergi* grasping a white. (12) A Mexican lynx spider (Oxyopidae) devouring the swallowtail *Parides montezuma*, a species that birds and other predators find unpalatable; (13) a member of the genus *Araneus* (Araneidae) with its victim, a six-spot burnet (*Zygaena filipendula*, Zygaenidae), Italy.

The commonest situation is for the parasite to pupate within the larval or pupal cuticle of the butterfly or moth — much to the disappointment of entomologists who discover emerging a small hymenopteran or black fly instead of the expected lepidopteran. Before ending this short and far-from-complete survey of butterfly and moth parasites, we should mention the habits of the small mite *Myrmonyssus phalaenodectes*, which gathers in large numbers on the tympanal organs of certain noctuids, eating the tympanal membrane and probably also the haemolymph of their host. In most cases the parasites have been found only on the tympanal organ of one side, a habit that has possibly been selected since it increases the chances of moth survival and hence, ultimately, of the parasite.

The predators of the Lepidoptera, of both caterpillars and adults, are too numerous to survey even briefly. They are present to a greater or lesser extent in every class of vertebrates from fish to mammals. Among the latter bats, insectivores, and monkeys take the lion's share, but the major vertebrate predators are certain birds and lizards. It is curious that at the turn of the century, when great debates over mimicry raged, many naturalists denied the importance of birds as predators. To convince the sceptics, it was necessary to produce volumes of experimental evidence, backed up by copious photographic and cinematographic documentation. However, there are still more predators among the invertebrates, arthropods in particular. Spiders take a heavy toll, with thomisids lying in wait on flowers, aranaeids spinning webs, large oxiopids hiding in bushes, and small, mobile salticids. These arachnids also have the advantage that for the most part they are immune to the chemical defenses of the butterflies that they devour by sucking up the predigested tissues, after having ensnared their prey and wrapped it up in the web. In addition to spiders other principal predators of butterflies and moths are to be found amongst the mantids, robber flies, and the reduviid and pimatid Hemiptera. Moreover, caterpillars and adults are often attacked and eaten by tiger beetles, ground beetles, wasps, and numerous species of ants and other Hymenoptera as well as by myriapods and scorpions.

Strategies against Predation and Mimicry

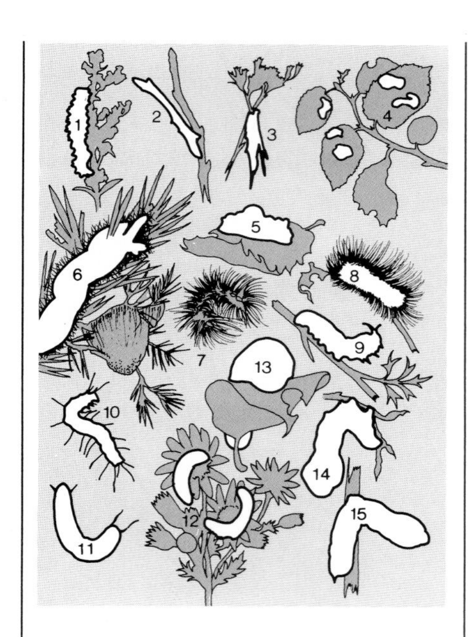

Plate 79 **Strategies against predation in caterpillars**

Adaptations and cryptic disguises achieved by imitating branches, leaves, bird droppings, and so on: (1) *Cucullia artemisiae* (Noctuidae); (2) *Selenia bilunaria* (Geometridae); (3) *Pachytelia villosella* (Psychidae); (4) *Acronicta alni* (Noctuidae), first instar; (5) *Phobetron pithecium* (Limacodidae).

Stinging hairs and repellent glands: (6) nest and caterpillars of the pine processionary caterpillar moth (*Thaumetopoea pityocampa*, (Notodontidae); (7) *A. aceris* (Noctuidae); (8) megalopygid caterpillar (Megalopygidae); (9) swallowtail (*Papilio machaon*, Papilionidae), with osmeteria visible.

Warning or aposematic coloration: (10) *A. alni* (Noctuidae), mature larva. Note that the coloration is very different from the earlier instars (4). (11) Monarch (*Danaus plexippus*, Nymphalidae, Danainae); (12) cinnabar moth (*Tyria jacobaeae*, Arctiidae).

False eyes and mimicry of a snake's head: (13) Spicebrush swallowtail (*P. troilus*, Papilionidae); (14) *Leucorampha* sp. (Sphingidae); (15) *Rhyncholaba* sp. (Sphingidae).

Because Lepidoptera have been subject to the attacks of a host of predators ever since they came into existence, at various stages in their development they have had to evolve an incredible variety of defensive adaptations, some extremely sophisticated. Every possible protective strategy has been adopted by one life-cycle stage or another: an armoured exoskeleton, stinging spines and hairs, the ability to flee from a predator or hide in fixed or movable shelters, cryptic coloration, and so forth. These phenomena provide some of the clearest evidence in support of the role played by natural selection in the development of adaptive morphological, physiological, and behavioural characters; and their study has contributed greatly to the understanding of the mechanisms of evolution.

Butterflies, Moths, and Bats

This chapter will deal principally with adaptations to counter animals that hunt by sight, that is, those which rely upon their eyes to find their prey; but butterflies and moths also display a series of adaptations to counter predators that hunt by hearing or by scent. One example of this is the interaction between certain moths and bats, their main predators. The latter can detect their prey, even in complete darkness, by emitting a series of ultrasonic pulses; the echo that they receive establishes the direction and the distance of the insect. This sonar is sufficiently sophisticated to enable them to locate and capture insects smaller than a mosquito. However, many moths (in particular numerous arctiids, noctuids, and geometrids) reduce their chances of being caught thanks to their sensitive tympanal organs, which can detect the ultrasonic emissions of a bat more than 30 m (100 ft) away — which is too far for the bat to intercept the moth. When they detect the signal, geometrids and noctuids escape by dropping down or changing their course suddenly. Many arctiids, on the other hand, respond by emitting a rapid series of sharp, metallic sounds at high frequency. Such a response is not in fact suicidal, as one might assume, since bats interpret these ultrasonic emissions as warning signals and alter direction before making contact, as has been shown by the American researcher Dorothy Dunning. The reason for the signals is that arctiids have effective chemical defenses and are generally distasteful to vertebrates; the signal enables a bat that "knows" this to recognize inedible prey.

The Art of Concealment

Turning now to predators that hunt by sight, we find that the first defensive strategy is avoidance of being seen at all. Many caterpillars keep well out of sight and feed only at night, while many butterflies and moths conceal themselves if they are pursued. Various satyrines disappear into the middle of bushes, some riodinids alight on the undersurface of leaves, and considerable numbers of nocturnal moths spend the day hidden away in cracks. The caterpillars of *Limenitis* (Nymphalidae) build winter shelters by rolling up a leaf and strengthening its attachment to the stem with silk, whereas the caterpillars of various families (Coleophoridae, Tineidae, Psychidae) live in mobile cases made of seeds or of silk and assorted materials (pieces of dried leaves, sand, and so on).

However, the commonest way in which butterflies and moths avoid the attention of their predators is cryptic mimicry, whereby structural (shape and colour) and behavioural features blend in with their surroundings or parts of them. This may be achieved by cryptic coloration, that is, having the same coloration as the background and using its shape and coloration

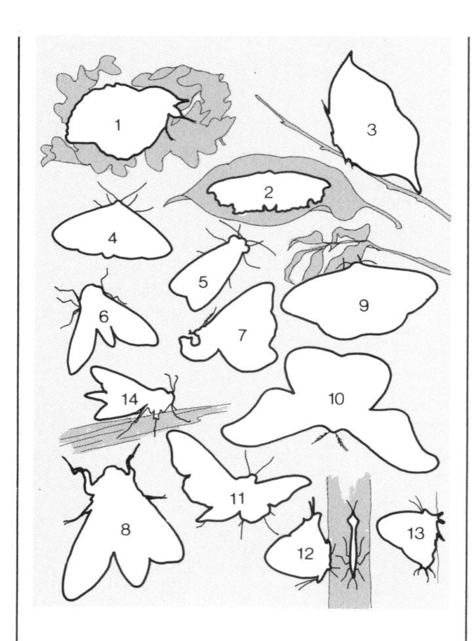

Plate 80 **Strategies against predation in adult Lepidoptera**

Cryptic adaptations, mimicry of leaves: (1) *Gastropacha quercifolia* (Lasiocampidae); (2) *Draconia rusina* (Thyrididae); (3) Dead-leaf butterfly (*Kallima inachus*, Nymphalidae).
Aposematic coloration and the exuding of repellent substances: (4) Virgin tiger moth (*Apantesis virgo*, Arctiidae); (5) *Utetheisa pulchella* (Arctiidae); (6) *Rhodogastria bubo* (Arctiidae).
Warning behaviour: (7) warning posture of a saturniid, displaying the coloration of its abdomen.
Sound production: (8) death's-head hawkmoth (*Acherontia atropos*, Sphingidae).
Display of eyespots: (9) Io moth (*Autometris io*, Saturniidae); (10) *Anthela eucalypti* (Anthelidae); (11) eyed hawkmoth (*Smerinthus ocellata*, Sphingidae).
False head and antennae on the hind wings: (12) *Arawacus aetolus* (Lycaenidae); (13) *Hypolycaena erylus* (Lycaenidae).
Batesian mimicry: (14) *Sesia vespiformis* (Sesiidae).

to reduce the effect of shade, or by disruptive coloration, which breaks up the animal's outline and makes it less apparent. Many caterpillars reduce the effect of shade by being paler on the underside. It is clear that behaviour plays a crucial role in an animal's ability to remain unnoticed. The ability to remain still during daylight, the ability to select a substrate that matches the animal's coloration, and the tendency to take up a position that allows maximum concealment are all modes of behaviour that are selected for, together with coloration and other structural features.

Geometrids are very good at concealing themselves in this way, and many of them eliminate shading completely by pressing the edge of their wings against rocks or bark covered with lichens. They often rest with the axis of their bodies running at right angles to that of a branch or trunk so as to maximize the effect of their coloration — which is partly disruptive and partly cryptic — against the rough, cracked surface of the bark. Experiments have shown that moths can actively select a suitable background to alight on. In England, H. B. D. Kettlewell has shown that the pale (*typicus*) and dark (*carbonarius*, *insularis*) forms of *Biston betularius* differ in their choice of habitat, each as a rule preferring the background (pale or dark) that suits their own coloration. The stimulus by means of which the moth recognizes the substrate is not necessarily visual.

Theodore Sargent has carried out experiments with the American *Catocala ultronica* (Noctuidae) and *Melanolophia canadaria* (Geometridae) and has shown that these species are guided by tactile stimuli in the choice of a suitable position on the substrate. The moths were placed in a tube and given the choice of two white surfaces with black stripes. In one case the stripes, which consisted of adhesive tape, were vertical, and in the other horizontal. The researcher then recorded the proportion of moths that were able to position themselves correctly, that is, so as to conceal themselves most effectively against the striped background. In *Melanolophia* the correct position is with the axis of the body at right angles to the stripes since the black stripes on the wings are predominantly horizontal. Sargent noted that a very high proportion (23:5) of individuls adopted the correct position. If, however, the substrate was covered with a thin, transparent film of cellulose acetate, the moths positioned themselves at random, showing that they were unable to discern the background pattern.

Cryptic Disguises

One form of camouflage is a series of adaptations that allows the moth, caterpillar, or pupa to mimic in colour and shape objects that a predator would ignore in its habitat. These might include bits of wood, thorns, green or dried-up leaves, droppings of birds or other animals, and feathers. Twigs are often mimicked by caterpillars, notably by geometrid larvae, which, when resting, grip the substrate with only their anal prolegs and hold their bodies rigid, away from the substrate (Plate 79). Some American notodontids (*Schizura*) rest during the day with their heads and their wings pointing downwards close to their bodies, which stick out at right angles from the surface they are resting on, usually a branch or a trunk. The result is a convincing resemblance to a branch. Another notodontid, the European *Phalera bucephala*, mimics a cut twig in the way in which it wraps its wings around its body and in the coloration of the tip of the thorax and the wings (Plate 56). The mimicking of thorns is the speciality of chrysalids of orange tips and marbles (Pieridae).

Leaves are also models for both caterpillars (Plate 79) and butterflies (Plate 80). Three genera of tropical nymphalids, *Anaea* from Central and

Plate 81 **Batesian mimicry: *Papilio dardanus***

The African mocker swallowtail (*P. dardanus*) is one of the most extensively studied examples of Batesian mimicry. It is widely distributed in tropical Africa and Madagascar; its various populations and races display marked differences. In Ethiopia and Madagascar there are ancestral populations that are not mimetic and have but a single form (Monomorphic). In these the female (2) has tails and yellow and black coloration very similar to those of the male (1). Elsewhere the females differ from the males (whose coloration remains constant) and mimic one or more inedible species depending on the population. Hence there exist polymorphic populations in which there are various female forms characterized by widely differing mimetic coloration. The genetic basis of this polymorphism is rather complex and is determined by the variation of a series of associated genes (supergenes) and by modifying genes, which increase the degree to which the mimic resembles its model. The function of polymorphism in Batesian mimics such as *P. dardanus* is to reduce the number of mimics to each model and thus to increase the selective advantage. The plate shows two of the many mimetic female forms (3, *trophonius* form, 4, *hippocoon* form) and their respective models [5, the plain tiger (*Danaus chrysippus*), 6, the friar (*Amauris niavius*)]. The mimic (*Hypolimnas misippus* is a nymphalid with nonmimetic males (8) and polymorphic females (7), which mimic *D. chrysippus* to a greater or lesser degree; it often takes part in the mimetic complexes in which *P. dardanus* is involved.

South America, *Kallima* from the Oriental and Ethiopian regions, and *Doleschallia* in the Oriental and Australian regions, provide considerable insight into natural selection and adaptation. In all three cases the striking coloration of the upper wing surfaces contrasts with their cryptically coloured undersides — brown, grey, or black — which have delicate, variegated patterns hiding the venation. In *Kallima* and *Doleschallia* as well as in some *Anaea* a dark central band runs from the tip of the forewing to the end of the tail of the hind wing and resembles the midrib of a leaf. When these butterflies are at rest their tails resemble pedicels, and they adopt poses on a branch that perfect the appearance of a dried-up leaf. Further touches, such as false spots of mildew or "bites" taken out by unknown herbivores, complete the illusion. A. R. Wallace noted that *Kallima* is clearly visible in flight but suddenly disappears when it lands; it is thus very difficult to locate exactly. Members of the Pieridae family also frequently mimic leaves. The morphology not only of the wings but also of the body and the antennae in the well-known genus *Gonepteryx* provides an excellent example. Among the moths the mimicking of leaves occurs in a greater variety of ways and crops up in a number of families. There are notable instances in the Lasiocampidae (Plate 51), Noctuidae (Plate 60), and Thyrididae (Plate 80) and also among some geometrids, sphingids, and saturnids.

One rather bizarre disguise is the imitation of bird droppings. Such a curious form of mimicry might simply be set down as one of Nature's many caprices if it were not so common not only among butterflies and moths of various families but also among many species of caterpillar. The white and black droppings of birds are a striking feature of many environments: They are clearly visible and stand out on leaves, but they are certainly of no interest to birds or other predators. Like leaves and twigs they are thus a component of a habitat that is completely overlooked by even the most keen-eyed predator. It is therefore not surprising that on many separate occasions natural selection has favoured the development of coloration, forms, or behaviour in various species and at different stages that mimic droppings. Prime examples among adults include the European geometrid *Eupithecia centaureata*, the drepanid *Cilix glaucata*, and Neotropical microlepidoptera of the genera *Stenoma* (Stenomidae). Additional examples are the chrysalids of the black hairstreak (*Strymonidia pruni*), the caterpillars and chrysalids of *Limenitis* (*L. archippus*, *L. arthemis*), the caterpillars of various *Papilio* (*P. demodocus*, *P. memnon*), those of the African bombycids (*Trilocha*), and those of the European noctuid *Acronita alni*.

Flash Coloration and Eyespots

Another defensive mechanism frequently used by butterflies and moths is the sudden, momentary display of brightly coloured wing areas. The effect may surprise and disorient predators. Many noctuids, particularly *Catocala* and some *Noctua*, display striking red, blue, or yellow colour only on their hind wings. The forewings are cryptically coloured and at rest completely conceal the hind pair. If these moths are disturbed, they suddenly display their brightly coloured hind wings, causing the predator to hesitate for an instant, and thus allowing the moths to escape. This is known as *flash coloration* and may be used for other purposes besides defense. Thus the striking, shimmering blues of male *Morphos*, which are solely on the upper surface of the wings, may be used as a sociosexual signal to members of the same species as well as a means of disorienting predators. *Morpho* species fly in zigzags, beating their wings slowly and covering a considerable distance with each beat. The undersurface of the

Plate 82 **Batesian mimicry: *Papilio memnon***

Another good example of Batesian mimicry is provided by the Oriental swallowtail (*P. memnon*). Again only the females are mimics, and they display a high degree of polymorphism and geographical variation. The genetic control of the various forms has been studied principally by the English geneticists C. A. Clarke and P. M. Sheppard, who also researched the polymorphism of *P. dardanus*. The various female forms are controlled, as in Plate 81, by a close association of a series of gene loci, each of which has several alleles, which act as a single unit of inheritance (a supergene). The models for the various forms of *P. memnon* are members of *Atrophaneura*, a genus of Oriental swallowtails that eat pipevines (Aristolochiaceae) and are protected from predators. The range of *P. memnon* extends from India and Sri Lanka to the Philippines and the Moluccas and from southern Japan southwards to Java. In Japan the females are neither polymorphic nor mimetic. The plate shows three of the numerous female mimetic forms, together with their respective models, from three separate regions: (1) *A. varuna* female, Malaysia; (2) *P. memnon, butlerianus* form, Malaysia. (3) *A. coon* male, Java; (4) *P. memnon, achates* form, Java. (5) *A. sycorax* male, Sumatra; (6) *P. memnon, anceus* form, Sumatra.

wing is dark and cryptically coloured, and so the overall effect when the butterfly is flying in the darkness of the forest is a regular disappearance from sight as the cryptic underside is raised and then a reappearance as a flash of blue in unpredictable positions.

Eyespots are another form of signal for disorienting predators. They resemble large eyes and either intimidate the predator or draw its attention to parts of the body that are less vulnerable than the head. This kind of highly conspicuous signal is found on the hind wings of many butterflies and moths and on the thorax or abdomen of caterpillars of several families. The false eyes are often large and lens-shaped, with a series of paler concentric rings around a dark, pupillike center; the result is that they always resemble the eye of a vertebrate. Eyespots are most effective when exposed suddenly and directly to the source of danger. A predator then receives the impression of being stared at and threatened by some kind of vertebrate. The sphingid *Smerinthus ocellatus* exposes the false eyes on its hind wings only if it is disturbed; otherwise it remains motionless with its eyespots covered by the cryptically coloured forewings (Plate 80). A large number of *Automeris*, many of which have leaf-shaped forewings, do the same. The caterpillars of several species are able to increase the flow of hemolymph to the thorax, expanding the thorax and hence enlarging the eyespots. In quite a number of forms, particularly sphingids, the expanded thorax with its eyespots resembles a snake's head (Plate 79). When disturbed, these larvae lash the air with their "heads" and pretend to strike. Eyespots are not always used to intimidate predators. In many cases they divert the attack by misleading the predator as to the direction of flight. Large numbers of butterflies and moths (lycaenids, nymphalids, satyrines, and the like) have small eyespots at the tips of their hind wings. Birds often direct their attack at these spots; when the butterfly is thus threatened, it flies off in an unexpected direction, escaping unharmed or, at worst, losing a piece of wing in its predator's beak.

False Heads

Many hairstreaks (*Thecla*, *Arawacus*, *Talicada*, *Atlides*, *Oxylides*, and so forth) draw the attention of predators to the tips of their hind wings, where their false eyes are merely one part of an extensive appendage formed by the tails that simulates a complete head with antennae. The true head is in fact partially concealed by the forewings, which at rest are held closed together above the back. Various morphological features and details of behaviour draw the predator's attention to the false head. In the Neotropical hairstreak (*A. aetolus*; see Plate 80) a series of conspicuous dark lines runs down the underside of the wings, converging on the false eyes and the tail. As soon as members of this and other species land, they rub their hind wings together, lowering and raising the false antennae. They also rapidly turn around so that the false head points away from the direction they came from or points upwards if they are resting on a vertical support. An instant before it lands, the African species *O. faunus* actually turns through 180° in flight. These measures are in fact effective in confusing predators, as can be seen from the large numbers of butterflies and moths that are caught bearing the marks of attacks aimed at their false heads.

The American entomologist R. K. Robbins found that out of about a thousand theclines caught in a Colombian forest the hind wings of more than 8 percent showed signs of having been attacked. This proportion rose to about 22 percent if only species such as *A. aetolus*, which have very highly developed false heads, were taken into account.

Plate 83 **Batesian mimicry in the Indo-Australian area**

Both Batesian and Müllerian mimicry are well developed in tropical ecosystems, where a great variety of predators are present in large numbers. Three examples of Batesian mimicry are illustrated, in which the mimics are swallowtails (*Papilio*) whose coloration is very different from that typical of their genus. Their models belong to different genera and families: (1) *Danaus sita* (Nymphalidae, Danainae), Malaysia; (2) *P.* (= *Chilasa*) *agestor* (Papilionidae), Malaysia. (3) *Alcides agatnyrsus* (Uraniidae), New Guinea; (4) *P. laglaizei* (Papilionidae), New Guinea. (5) *Atrophaneura priapus* male (Papilionidae), Java; (6) *P. lampsacus* (Papilionidae), Java.

Predators

By now it should be apparent that predators are an important selective factor in the evolution of such numerous, sophisticated defense mechanisms. However, one should not ignore the corresponding changes in predators themselves. Just as much as the Lepidoptera they are actors in a system of coevolution, and their recognition mechanisms certainly evolve in step with the adaptations of their prey. False eyes, false heads, and the mimicry of leaves are probably highly effective against unprepared predators not resident in an area or against polyphagous predators with a broadly based diet. They do not, however, provide complete protection against specialists.

It is unlikely that generations of specialized, resident insectivores could afford to ignore such a potential resource as camouflaged or cryptically coloured insects. It is known that insectivorous birds have a precise image of their prey, and they will concentrate on particular forms that — even if they are difficult to find — are more abundant than others. Consequently one can see why insects — and many Lepidoptera in particular — have evolved other "primary" defensive mechanisms, based on mechanical or chemical properties, which make them dangerous, distasteful, or otherwise inedible to their predators.

Toxic and Repellent Substances

Poisonous prickles, stinging hairs, and repugnant secretions are feared even by human beings because of the painful irritations, stings, and rashes that they can cause on the skin. Some Lepidoptera have hollow glandular hairs through which they inject toxic substances. This feature is most highly developed in the Megalopygidae and the Eucleidae, but examples occur in other families. The caterpillars of certain notodontids have large glands capable of spraying formic acid over a distance of several inches. The osmeteria of *Papilio* have already been mentioned. In the processionary caterpillars the stinging substances are found mainly in the excreta smeared on the spines and hairs. To protect the eggs, many lymantriids have tufts of stinging hairs on the tip of the abdomen of adult females. The eggs may also be protected by toxic substances that they contain: Those of certain *Zygaena* have high levels of linamarin and lotaustralin, two precursors of hydrocyanic acid.

The diversity of toxic and repellent substances found in butterflies and moths appears to be even greater than it is in other insects although possibly this is simply because they are better known. As shown in the previous chapter, the toxins are frequently assimilated from food plants and then either metabolized or stored in the tissues as they are. The former case is illustrated by certain sphingids, such as *Daphnis nerii*, which lives on oleander (oleandrin), or *Manduca sexta*, which lives on tobacco (nicotine). Here the defensive mechanism is associated with the secretion of the toxin. In the second case, which is much commoner, a wide variety of substances has been found in various families: glucosides and mustard oils (Pieridae), cardenolides and aristolochic acids (Papilionidae), calactene, calotropine, and other cardenolides (Danainae), pyrrolizidine alkaloids (Arctiidae), and oleandrin and other cardenolides (Ctenuchidae). Sometimes the production of these substances is restricted to certain glands; in other instances the toxins or their precursors are carried by the hemolymph, as in *Zygaena*, Acraeinae, and Heliconiinae (Nymphalidae). All these Lepidoptera are capable of releasing substantial amounts of hydrocyanic acid. Either it may ooze out as a result of lacerations of the cuticle caused by a predator or droplets may be

Plate 84 **Mimetic complexes in a South American forest**

The plate shows species from a variety of families that are involved in four mimetic complexes typical of the stratification that is found in various South American forests. The group of butterflies with transparent wings that live in the lower layers of vegetation is both widespread and highly characteristic. It is interesting that the transparency is achieved by different structural means in the various families: In some the scales are translucent, in others they are missing, and in others they are arranged at right angles to the wing's surface.

(1) *Heliconius clytia* (Nymphalidae, Heliconiinae), (2) *H. doris* (Nymphalidae, Heliconiinae), (3) *Eurytides pausanias* (Papilionidae); (4) *Archonias bellona* (Pieridae), (5) *H. xanthocles* (Nymphalidae, Heliconiinae), (6) *Pericopis phyleis* (Hypsidae); (7) *Mechanitis isthmia* (Nymphalidae, Ithomiinae), (8; *Tithorea harmonia* (Nymphalidae, Ithomiinae), (9) *M. egaensis* (Nymphalidae, Ithomiinae), (10) *H. hecale* (Nymphalidae, Heliconiinae), (11) *Chetone angulosa* (Hypsidae); (12) *Thyridia confusa* (Nymphalidae, Ithomiinae), (13) *Gazera linus* (Castniidae), (14) *Ituna phenarete* (Nymphalidae, Danainae), (15) *Dismorphia orise* (Pieridae).

deliberately secreted at the thoracic joints and the base of the antennae. Glucosidase, an enzyme, is responsible for this reaction, hydrolizing the cyanogenic glucosides (linamarin, lotaustralin) in the hemolymph. These glucosides, at least in the case of the Heliconiinae and the Acraeinae, are not found in the food plants and must consequently be synthesized by the insect. In many chemically protected forms the action of the toxins described above is compounded by irritation caused by such pharmacologically active amines as histamine and acetylcholine.

When a toxic or repugnant species is caught or pecked at by a predator, the prey may be permanently damaged and its chances of survival thus reduced even if it is not eaten. Consequently chemical defenses are much more effective if the predator, before it strikes, is able to recognize, as a result of past experience, the toxic or repellent qualities of its potential prey. Protected species, therefore, almost always display clear identifying signals, which "warn" the predator. Most common is the evolution of striking patterns and coloration, which are known as warning, or aposematic, coloration.

Sounds and odours are also used to warn predators that hunt predominantly by hearing or smell. Warning coloration often consists in combinations of colours such as white, black, red, and yellow, which provide maximum visibility against the greens and browns of the natural habitat. Instead of hiding, aposematic species flaunt themselves, are indolent in habit, and are often gregarious. Unlike cryptic forms they are usually diurnal; this is true even of many diurnal moths.

Predators quickly learn to avoid aposematic species. Numerous experiments have shown that only a few attacks on insects with warning coloration are needed to ensure that they are recognized as inedible and avoided on all subsequent encounters. It is therefore advantageous for aposematic forms to display their coloration by moving their wings slowly. Equally beneficial is the flocking together of members of the same species, whether adults or caterpillars, since this maximizes the warning signal. The protracted mating, lasting several hours, that occurs in certain aposematic species, such as *Zygaena* and *Amata* (Ctenuchidae), may have a similar significance.

True Mimicry

An understanding of aposematic defense mechanisms provides the basis for a discussion of mimicry. The adaptations involved in mimicry allow at least one species, the *mimic*, to closely resemble a species protected by warning coloration, the *model*, as faithfully as possible to deceive a predator.

There are two types of mimicry: In *Batesian mimicry* the mimic is an edible species that resembles and behaves like a different, inedible, or otherwise protected species (the model), which may be completely unrelated. This type of mimicry clearly benefits the mimic, which almost parasitically exploits the protection gained by the model. To be protected, the Batesian mimic must usually fly at the same time and in the same place as the model, and usually it must be less common than the latter. *Müllerian mimicry*, on the other hand, consists in the reciprocal resemblance of two or more unrelated and inedible species. The distinction between mimic and model is only historical, the mimic being the species that evolved its coloration later than the model, although all the species involved are generally referred to as *comimics*. Müllerian mimicry benefits all the species involved: Predators have to learn to avoid only one warning signal instead of several, and the toll taken while they learn is more or less evenly distributed among all the species of comimics.

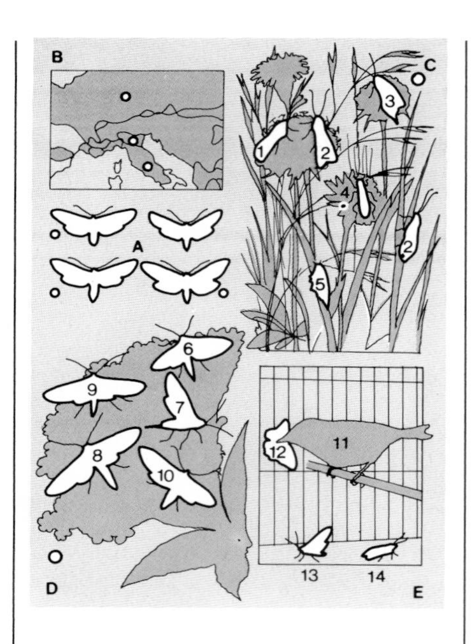

Plate 85 **Mimicry in *Zygaena ephialtes***

The phenomenon of mimicry is not so common among temperate butterflies and moths. *Zygaena ephialtes*, a species already known for its polymorphism (Plate 14), is an instance that has been extensively studied in Italy by the authors of this book and by L. Bullini and P. Ragazzini. Its four principal forms are shown in Figure A, and their ranges are outlined in Figure B. As with other members of *Zygaena*, *Z. ephialtes* can release large doses of cyanide gas, acetylcholine, and histamine, so making itself unappetizing to birds and other predators. In central Europe, therefore, the red peucedanoid form belongs to a Müllerian mimetic complex as do other red and black *Zygaena* as well as beetles and treehoppers (Cercopidae) with similar coloration (C): (1) *Z. punctum*, (2) *Z. lonicerae*, (3) *Z. filipendulae*, (4) *Trichodes apiarius* (clerid coleopteran), (5) *Cercopis sanguinea* (Homoptera). In central Italy, however, *Z. ephialtes* does not belong to the same mimetic group even though there are numerous members of *Zygaena* (see Plate 32) and other red and black insects. Instead it occurs as the yellow ephialtoid form (*coronillae*), mimicking *Amata phegea* (Ctenuchidae) and other *Amata* species (D) characterized by black, white, and,yellow coloration: (6–8) *A. phegea*, (9) *A. ragazzii*, (10) *Z. ephialtes*. *Amata* species are undoubtedly good Müllerian models since they are unpalatable, and insectivorous birds quickly learn to avoid them. In doing so, they leave the yellow ephialtoid form of *Z. ephialtes* alone, confusing it with *Amata*, and instead attack (E) other edible insects: (11) *Liothsix lutea*, being kept in a cage for the purpose of an experiment, (12) the meadow brown (*Maniola jurtina*), the control, (13) *A. phegea*, (14) *Z. ephialtes*, *coronillae* form.

The study of mimicry began in the second half of the last century, shortly after the publication of Darwin's *The Origin of Species*. Its founding fathers were two great travelling naturalists, Henry Walter Bates (1862) and Fritz Müller (1879). Since then scientific research into, and debate about, mimicry has played a major role in the study of evolution. According to the great English geneticist R. A. Fischer, mimicry represents the most important and significant application of the Darwinian theory of evolution and the most convincing proof of the operation of natural selection. The study of mimicry in butterflies and moths has passed through several phases: Initially it was mainly concerned with the identification and description of instances of mimicry, but subsequently, in response to the objections of numerous sceptics, a considerable amount of experimental work was done to prove that mimicry arises in response to, and evolves through, the pressure of predators. At the same time research has been undertaken into the genetic basis of mimetic signals, the results of which have been essential in elucidating the evolutionary dynamics involved. Recently experimental research has concentrated on the behaviour and ecology of both mimetic species and their predators. The understanding of the biochemical mechanisms on which mimicry is based represents one of the most lively fields of current research and provides a broader view of the coevolution of plants, butterflies, and predators. Mimetic relationships have also been studied theoretically by means of mathematical models. Plates 81 to 85 are a short survey of mimicry examples.

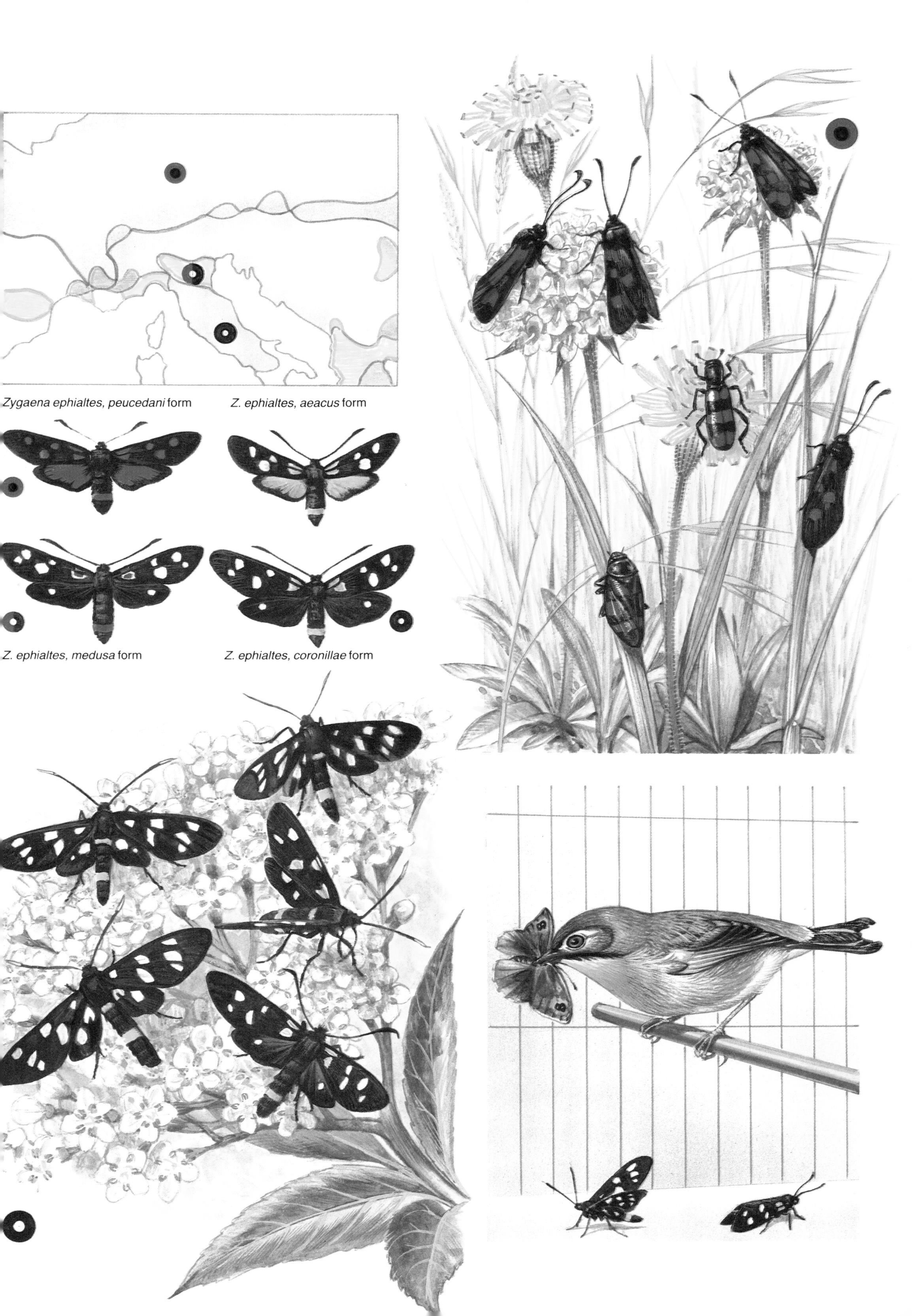

Zygaena ephialtes, peucedani form
Z. ephialtes, aeacus form
Z. ephialtes, medusa form
Z. ephialtes, coronillae form

Ecological Distribution

Plate 86 **Species diversity in a temperate wood**

The accompanying histograms and the butterflies illustrated in both this plate and Plate 87 refer to two random samples of butterflies taken during three hours in a deciduous, temperate forest on Mount Ceriti, Lazio, Italy, and in a tropical rain forest in the state of Veracruz, Mexico. The graphs show the number of individuals of each species caught, in groups of decreasing abundance. The numbers beside the dots refer to the species illustrated in Plates 86 and 87. The species richness in the temperate woodland is relatively low. Though 185 butterflies were caught in all, they belonged to only 24 species as against the 171 butterflies belonging to 60 species in the tropical sample. In the temperate-forest sample, again, certain species like (1) the meadow brown (*Maniola jurtina,* Nymphalidae), (2) the little skipper (*Thymelicus flavus*, Hesperiidae), the heath fritillary (*Mellicta athalia*), and (3) the ilex hairstreak (*Nordmannia ilicis*, Lycaenidae) were very common and accounted for almost half of the specimens caught. Nearly all the species from the tropical sample, by contrast, were rare, and each one was represented by a very small number of specimens. (4) Woodland grayling (*Hipparchia fagi*, Nymphalidae); (5) holly blue (*Celastrina argiolus*, Lycaenidae); (6) dark green fritillary (*Mesoacidalia aglaia*, Nymphalidae); (7) Berger's clouded yellow (*Colias australis*, Pieridae); (8) large tortoiseshell (*Nymphalis polychloros*, Nymphalidae); (9) comma (*Polygonia c-album*, Nymphalidae); (10) scarce swallowtail (*Iphiclides podalirius*, Papilioniadae); (11) nettle-tree butterfly (*Libythea celtis*, Libytheidae); (12) mourning cloak (*N. antiopa*, Nymphalidae).

In analyzing the distribution of butterflies and moths in various environments, one should keep in mind the concept of the *biotic community*, defined earlier as the complete array of living organisms in a particular biotope. The word *biotope* refers to a definite geographical area, of whatever size, characterized by unique ecological conditions. An organism that belongs to a given community must interact, more or less directly, with the other members of the community and be subject to the same constraints and stresses imposed by the physical environment, such as temperature, humidity, soil type, availability of water, seasons, and so on. The community has already been discussed from a functional point of view as have certain types of interaction involving the Lepidoptera. The overall effect of the various interspecific interactions, both anatagonistic and mutually beneficial, is a dynamic equilibrium, as a result of which after a certain point the community tends to maintain a fairly constant mix of species. This chapter will examine the structure of biotic communities in general and how they vary in space and time, dealing subsequently with fluctuations in the numbers of butterfly and moth species making up such communities and the factors that govern these changes in relation to the availability of ecological niches in different environments. Finally, some of the principal biomes will be described: the major zones of the earth, which may be characterized by their predominant plant communities; and we shall examine their populations of butterflies and moths.

Spatial Structure of Communities

A community is an open, hierarchically organized structure: Larger communities contain smaller communities consisting in secondary combinations of species. The larger communities comprise an enormous number of individuals and species; they attain a high level of organization and are fairly independent of neighbouring communities. Smaller communities, by contrast, are dependent to a greater or lesser degree on the larger communities. Of course even the largest communities are dependent on outside energy sources — in most cases the sun. As an example of a lesser community we may cite the collection of organisms living on a rotting trunk in a beach forest. In this instance the greater community is the forest itself with all the organisms that inhabit it. Not all the organisms in a given community have the same degree of importance in shaping its character, and we almost always find so-called *dominant organisms*, which exercise a determining role on the other organisms. The pressure exercised by the dominant members on the others is at the trophic level. Generally the dominant elements at ground level are those plants which are most abundant and which constitute the chief source of food for the organisms that feed on them. Temperate forest communities are characterized by a single or a small number of tree species that lend their name to the community: oak forest, cork grove, beech wood, and so on. In the same way certain species of grass are dominant in meadows. This dominating capacity of plants can also be seen in their ability to produce radical modifications in a number of physical features of the environment, such as light, humidity, movement of air, and surrounding space, and in other features, to which the other organisms are forced to submit. In aquatic habitats animals often play the dominant role.

It should be noted that the concept of dominance can be applied at each trophic level of a given community. Thus there will be one or more dominant species among both primary and secondary consumers and again among the decomposers. In each of these groups the dominant species will be those through which the greatest part of the ecosystem's energy passes and is transformed. The number of dominant species varies depending

Species diversity in a temperate wood

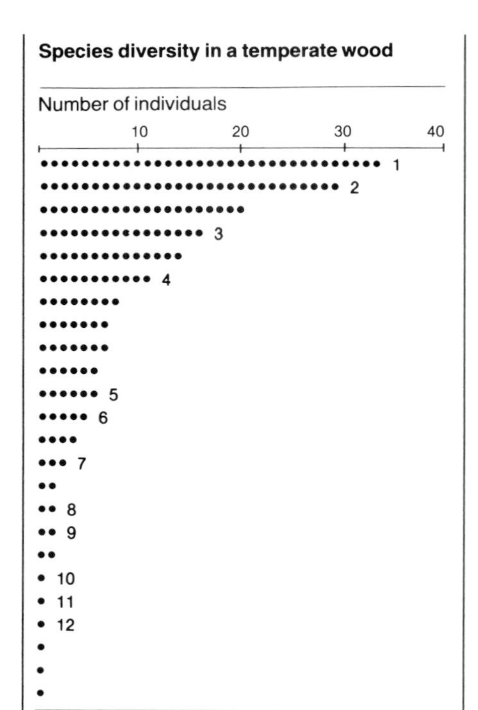

Plate 87 **Species diversity in a tropical rain forest**

The histogram shows the diversity of species in a sample of butterflies collected in a Mexican tropical rain forest (see caption to Plate 86 for explanation). This great species richness is general in the tropics not only among Lepidoptera but also among almost all animal and plant groups. As the histogram shows, the high number of species contrasts with the small number of individuals in each. (1) Gulf frittillary (*Agraulis vanillae*, Nymphalidae); (2) dogface (*Colias cesonia*, Pieridae); (3) white peacock (*Anartia jatrophae*, Nymphalidae); (4) sweet oil (*Mechanitis menapis*, Nymphalidae); (5) *Eurytides epidaus*, Papilionidae; (6) many-banded dagger wing (*Marpesia chiron*, Nymphalidae); (7) orange-barred sulphur (*Phoebis philea*, Pieridae); (8) *Dryadula phaetusa*, Nymphalidae; (9) *Parides montezuma*, Papilionidae; (10) blue wing (*Myscelia ethusa*, Nymphalidae); (11) Orion (*Historis odius*, Nymphalidae); (12) *E. branchus*, Papilionidae; (13) giant hairstreak (*Pseudolycaena marsyas*, Lycaenidae); (14) Grecian shoemaker (*Catonephele numilia*, Nymphalidae); (15) *Pyrrhogyra otolais*, Nymphalidae; (16) *Euptychia hesione*, Nymphalidae; (17) tropical checkered skipper (*Pyrgus oileus*, Hesperiidae); (18) Guave skipper (*Phocides polybius*, Hesperiidae); (19) rare blue doctor (*Rhetus arcius*, Riodinidae); (20) Mexican yellow (*Eurema mexicana*, Pieridae); (21) *Pteronymia cotytto*, Nymphalidae; (22) *Thecla* sp., Lycaenidae; (23) king shoemaker (*Prepona demophon*, Nymphalidae).

upon the type of community. In agricultural communities pioneer communities settling in a previously unoccupied habitat contain a few dominant species. Tropical communities, by contrast, often contain a great many species, each represented by a small number of individuals, none of which clearly dominates the others. The physical characteristics of the environment mostly change gradually on a geographical scale, abrupt environmental boundaries, such as exist between land and water or such as those between the hypogeal and epigeal environments at the mouth of a cave, being much less common. The band of varying width that marks the boundary between two communities is known as the *ecotone*. This is an area of tension where the species of the adjoining communities are in competition. In an ecotone the physical conditions of the environment are usually intermediate, differing from those of both contiguous communities. Frequently an ecotone will contain more potential sources of food, a greater variety of shelters, and in general more ecological riches. The result is that the variety and density of organisms in an ecotone are often greater than they are in the original communities. This phenomenon is known as the "boundary effect." One of the best-known examples of an ecotone is the boundary between forest and grassland. Human beings tend to create and maintain this type of community around the areas where they live. Thus when they settle in forest zones, they reduce the forest to groups of trees interspersed with meadowland or cultivated fields; and when they settle on the open plains, they tend to plant trees, creating the same type of ecotone. Gardens represent a particular type of ecotone that gives shelter to many species of butterflies and moths, particularly in the tropics.

Most communities are recognizably structured in terms of the space occupied by their component organisms. One very common structure is known as *stratification*, with variation occurring vertically. Thus in a broad-leaved forest five layers can usually be distinguished: hypogeal, ground, field or herbaceous, shrubby, and tree. In tropical forests, where the tree crowns (canopy) are concentrated at varying heights, a greater number of strata is found. The distribution of other organisms and especially of the Lepidoptera is closely linked with plant stratification. In the two lower strata, where heterotrophic organisms predominate, there takes place the larger part of the process of decomposition and the reduction of organic material not utilized by other members of the community. In the higher strata producers and primary and secondary consumers predominate. Most animal species are to a large extent tied to a certain stratum, but some may move vertically from one to another, frequently with the alternation of night and day. Stratification increases the number of niches available in a given surface area, and hence the more stratified a community, the greater its species richness.

The Community in Time: Ecological Succession

The composition of a community may vary in time as well as space. Over a given time the environment may be modified as a result of climatic or physiographic change or through the activity of the organisms that live there. These modifications in habitat may affect the number of individuals of dominant species, thereby allowing a new community to develop, which gradually replaces the previous one. The process by which one community replaces another in a given area as a result of a more or less orderly sequence of events is known as *ecological succession*. Usually a series of temporary communities, or *seral stages*, in the succession leads up to a more stable community, known as the *climax*, at equilibrium with local conditions. In such climax communities new species no longer settle and modify the shape of the environment.

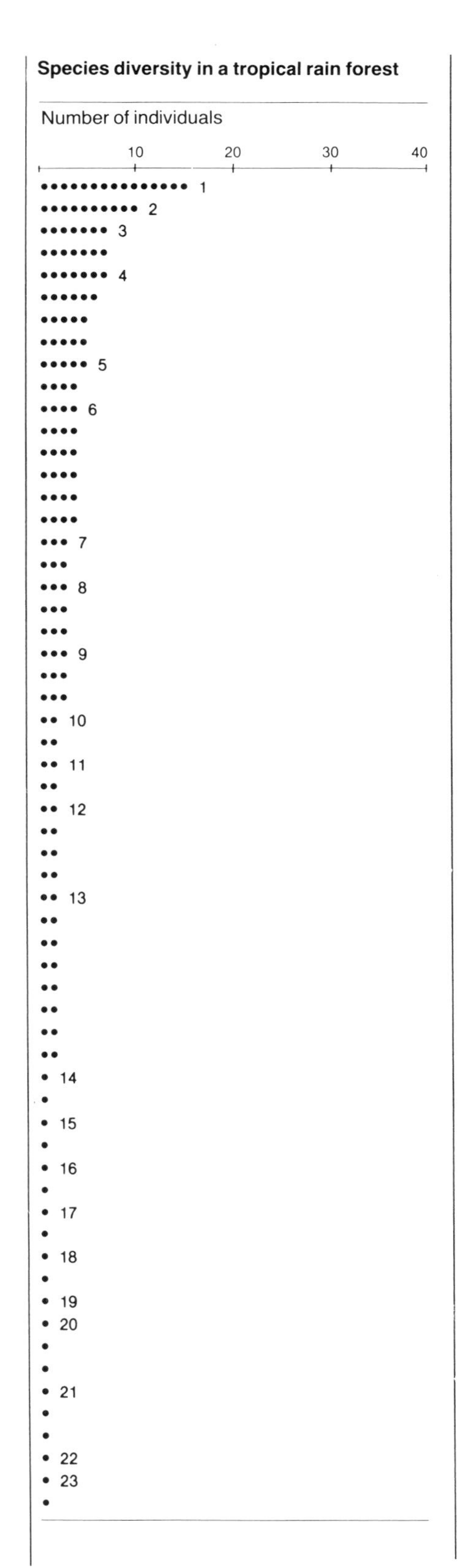

Implicit in the succession is a regular progression of vegetable forms from grasses to shrubs and finally trees. The plants of the first and final stages use different means of reproduction and growth. The former can reproduce and disperse quickly and are thus able to colonize environments that have not been recently exploited or disturbed. The latter disperse and reproduce more slowly, growing and prospering only in particular environmental conditions. As the succession proceeds, the community becomes more and more complex, and stratification and the number of species increase. In both the animal and vegetable kingdoms the number of specialized, "stenoecious" organisms with specific environmental requirements also increases at the expense of the euryoecious, more opportunistic organisms with a wide ecological tolerance, which predominated in the first seral stages. There is thus a certain logic in the composition of the species of a community, which depends (at least in part) on the type of seral succession that takes place in an area with particular soil and climatic conditions. In some communities, particularly arid ones or ones on steep slopes, continual physical disturbance prevents attainment of a climax; these are referred to as *disclimax*. Such disclimax communities, in areas such as western North America, have rich butterfly communities.

Fire is a periodic phenomenon to which many communities have adapted. Fire disclimax communities include prairies, savannahs, chaparral, and some dry-adapted woodlands: The North American prairies are well adapted to fire — especially tall-grass prairies. Were it not for fire, these habitats would be overtaken by woody shrubs and small trees. The grass-feeding Dakota skipper (*Hesperia dacotae*), the byssus skipper (*Problema byssus*), the Powesheik skipper (*Oarisma powesheik*), and the violet-feeding regal fritillary (*Speyeria idalia*) are all dependent on periodic fire to keep their habitats open and to provide for suitable hosts and nectar plants. Under present conditions, where wildfires are often suppressed, such other disturbances as mowing and grazing may substitute — at least in part — for the effects of fire.

Plate 88 **Ecological equivalents**

Unrelated animals and plants that inhabit similar environments in different zoological regions not uncommonly occupy the same kind of ecological niche and display a remarkable similarity in appearance and behaviour. These are known as cases of *ecological equivalence*. The plate illustrates a few examples of ecological equivalents among tropical-forest butterflies of Central and South America, Southeast Asia, and Africa. All the species illustrated are nymphalids but are not closely related and belong to different subfamilies. Thus some Neotropical Brassolinae (1, 2) are typified by the presence of conspicuous odoriferous glands and eyespots and by their crepuscular behaviour; they have perfect equivalents in the Amathusinae of the Oriental region (3, 4), with which they also share the habit of eating rotten fruit on the ground. Another remarkable case of convergent adaptation is seen in the Neotropical Heliconiinae (5, 6) and some African forest Acraeinae (7, 8). Similarities between the two subfamilies include wing shape, the choice of Passifloraceae as their food plants, chemical protection from predators, a slow, lazy flight, a tendency to congregate in groups, and the formation of complex mimicry chains with other butterflies. (1) Mort bleu (*Eryphanis aesacus*), Mexico; (2) bark (*Opsiphanes batea*), underside, Brazil; (3) *Thaumantis klugius*, Borneo; (4) *Thauria eliris*, Borneo; (5) Julia (*Dryas iulia*), Central and South America; (6) *Heliconius cydno*, Colombia; (7) elegant acraea (*Acraea egina*), Equatorial Africa; (8) *Bematistes poggei*, Equatorial Africa.

Species Diversity

From the preceding material the reader will already have inferred the importance of the diversity of species as a distinguishing characteristic of biotic communities and as a criterion by which they may be compared. This is true not only of animal and plant species as a whole but also of individual animal and plant groups. Thus the Lepidoptera provide a textbook example of the general pattern. In particular moths that can be caught with light traps in a given biotope furnish highly useful information about the complexity and state of equilibrium of the entire community and are frequently used by ecologists for this purpose. Butterflies are an equally useful source of information although in their case it is harder to obtain an unselective, random sample that does not favour one species or another and prejudice assessments of their relative abundance.

Diversity can be measured either by simply registering the number of species (usually called *species richness*) or by other indicators of varying complexity that, taken together, can provide information both on the number of species and on the relative numerical importance of each. One widely used index of diversity, derived from information theory, is that devised by Claude Shannon and Norbert Wiener. The diversity, H, of a community or sample of it is calculated from the formula $H = - \Sigma \hat{pi} \log \hat{pi}$, where $\hat{pi}$ represents the relative frequency of occurrence of a species i in the sample; the value of i may vary from I to S, where S is the total number of species in the sample. Of two communities with an equal number of species one may give a greater value for H than the other does if the total number of individuals belonging to it is more evenly divided among the different species. The index of diversity will be lower if the majority of individuals belongs to a single dominant species.

One rather interesting fact is that species richness exhibits geographical variation, increasing towards the equator. This general pattern is quite independent of the trophic level or the group under consideration; tropical communities are thus typified by a proportionately higher index of diversity than are temperate ones. The data compiled by D. F. Owen in his book *Tropical Butterflies* illustrate this tendency. On the basis of accurate counts made in various years he compared the number of species of butterflies in a garden near Freetown in Sierra Leone with the number in an English garden in Berkshire and in one in Michigan. In the African garden 115 species were recorded, compared with 23 in Michigan and 16 in Berkshire (the count covered only the Papilionidae, Pieridae, and Nymphalidae). Clearer still is the comparison of total butterfly species found in two large areas of comparable size, one, Liberia, tropical and the other in Michigan, temperate. Whereas 720 species were recorded in Liberia, Michigan had only 136. Many similar examples could be cited. It is interesting to compare Plates 86 and 87, which deal with two forest biotopes that have not been subjected to much alteration, one temperate, in central Italy, and the other tropical, in Mexico; the number of individuals has been taken into account. Many hypotheses have been advanced to explain local and latitudinal variations in species richness. Causal factors suggested include time, that is, age of the community, climate, degree of spatial variation, competition, predators and their effect in rarefying the species, and stability of environmental weather conditions and their predictability. None of these mechanisms is sufficient by itself to explain the phenomenon, and indeed it is likely that they are all interdependent and that each plays a certain part in determining the degree of species diversity. Without great detail it is worth while to emphasize the part played by two factors that are certainly important in contributing to the increase of ecological niches. The first of these is time: Many scholars believe that the time needed for a temperate community

Plate 89 **Arctic tundra**

The arctic tundra is one of the harshest biomes for animal life. Nevertheless, several Lepidoptera are permanent residents and display special adaptations (see the text). The arctic tundra merges southward into the northern forests (taiga). The plate illustrates some representative Lepidoptera of the Eurasian and North American tundra together with a few others (6, 9, 11, 18) that are tied more to the cold forests: (1) Eversmann's parnassian (*Parnassius eversmanni*, Papilionidae); (2) *Byrdia rossii*, Lymantriidae; (3) *Sympistis lapponica*, Noctuidae; (4) dusky-winged fritillary (*Boloria improba*, Nymphalidae); (5) arctic fritillary (*B. chariclea*, Nymphalidae); (6) Frejya's fritillary (*B. freija*, Nymphalidae); (7) Labrador sulphur (*Colias nastes*, Pieridae); (8) Greenland sulphur (*C. hecla*, Pieridae); (9) great northern sulphur (*C. gigantea*, Pieridae); (10) Booth's sulphur (*C. boothii*, Pieridae); (11) blueberry sulphur (*C. pelidne*, Pieridae); (12) high-mountain blue (*Agriades aquilo*, Lycaenidae); (13) arctic alpine (*Erebia rossii*, Nymphalidae, Satyrinae); (14) Melissa arctic (*Oeneis melissa*, Nymphalidae, Satyrinae); (15) Melissa arctic larva (*O. melissa*, Nymphalidae, Satyrinae); (16) Jutta arctic (*O. jutta*, Nymphalidae, Satyrinae); (17) arctic grayling (*O. bore*, Nymphalidae, Satyrinae); (18) Macoun's arctic (*O. macounii*, Nymphalidae, Satyrinae).

with a high number of species to become saturated is greater than that which has elapsed since the last glaciation. Conversely tropical communities are more mature because the absence of disturbances comparable to glaciation has increased the rates of speciation and natural processes of specialization, and hence the creation and occupation of a larger number of niches. The other factor is the stability of the climate and food resources, which is typical of many tropical environments in contrast to temperate ones. When the climate is stable and there is no winter, there are often some plant species that produce leaves, flowers, fruits, and seeds during any season. Phytophagous organisms can thus specialize so that they depend upon a single resource because it is more predictable. In this way the number of niches can increase.

Biomes

The biosphere is usually subdivided by ecologists into a small number of biomes. A *biome* is a collection of communities characterized by a particular type of climax community that remains relatively stable in the climatic conditions of a given region. Biomes consist of large regions of the earth characterized on the basis of ecological factors: tundra, taiga, or coniferous forest, temperate rain forest and deciduous forest, grassland, desert, and tropical rain forest. Biomes always comprise a complex of communities, including of course some still in the first stages of their succession towards the climax typical of the biome. Biomes are the outcome of an equilibrium established between the organisms of a region and its climatic conditions, and so they vary as the climate changes with latitude and height above sea level. We shall now examine the characteristics of a few biomes and their populations of Lepidoptera.

Tundra

The tundra is the biome that lies between the tree line and the permanent ice. In the northern hemisphere the tundra is circumpolar and stretches without a break from the north of Scandinavia as far as the Siberian Chukchi Peninsula and from Alaska to Labrador in North America, taking in the southern coastal strip of Greenland. In the southern hemisphere the climatic conditions favourable to the development of tundra are found in a band around the Antarctic circle, where there is very little land above sea level so that the southern tundra is found only in Tierra del Fuego, Graham Land in the Antarctic, and on one or two subantarctic islands. Similar conditions and comparable vegetation are found in mountain areas above tree line, and this is known as alpine — as distinct from arctic and antarctic — tundra; unlike the former the physiographic character of the latter two is predominantly flat or gently undulating, with depressions filled with lakes or ponds of varying size. The limiting factors are determined chiefly by the sun's annual cycle. Depending on the latitude, the polar night may last weeks or months; in winter the soil is completely frozen, the climate is dry, and what little precipitation there is falls in the form of snow. Summer is short, and the topsoil is unfrozen for less than three months. Beneath this the soil never thaws, forming an impermeable layer known as *permafrost*. The summer climate is quite pleasant because of the length of the polar day, with temperatures rising to 7 to 10°C, not usually exceeding a maximum of 15°C; it is humid and often windy. Average precipitation is below 250 mm (10 in). The antarctic tundra is less continental in climate with warmer winters and cooler summers, but it is more frequently battered by violent winds.

Of all the biomes on earth the tundra is the antithesis of the tropical rain forest in its near absence of vertical stratification. The typical vegetation of

Plate 90 **Mountain biomes: alpine butterflies**

Mountain vegetation is typically divided into different zones as the altitude increases, culminating in the alpine tundra. This diversity of environments is reflected in a great wealth of species, which in Europe reaches its greatest extent along the chain of the Alps. Illustrated here are some alpine butterfly species found at the mountain and alpine levels: (1) Grison's fritillary (*Mellicta varia*, Nymphalidae); (2) arctic blue, Glandon blue (*Agriades glandon*, Lycaenidae); (3) mountain fritillary (*Boloria napaea*, Nymphalidae); (4) mountain clouded yellow (*Colias phicomene*, Pieridae); (5) de Prunner's ringlet (*Erebia triaria*, Nymphalidae, Satyrinae); (6) alpine grayling (*Oeneis glacialis*, Nymphalidae, Satyrinae); (7) Phoebus parnassian, small Apollo (*Parnassius phoebus*, Papilionidae), (7*a*) larva; (8) Eros blue (*Polyommatus eros*, Lycaenidae); (9) Cynthia's fritillary (*Euphydryas cynthia*, Nymphalidae), (9*a*) larva; (10) sooty ringlet (*Erebia pluto*, Nymphalidae); (11) large grizzled skipper (*Pyrgus alveus*, Hesperiidae); (12) scarce copper (*Heodes virgaureae*, Lycaenidae); (13) Piedmont ringlet (*Erebia meolans*, Nymphalidae, Satyrinae).

the tundra consists of perennial herbaceous plants, mosses, lichens, and low myrtles, heathers, and rhododendrons. Annual plants are not able to develop fully and reproduce in so short a season; so they are reduced in biomass and richness of species. The commonest life forms are pulvinates and cushion plants, shapes also common in deserts and semiarid zones and well adapted to windy conditions. One characteristic feature is the mosaic pattern of vegetation, which is extremely varied over even quite small distances, adapting to apparently insignificant rises and declivities and hence to the water content of the soil and variations in exposure to the sun, which is always low on the horizon. With the advent of summer everything bursts into life, the brief flowering period of the plants taking place almost simultaneously and lending sudden gaiety to the landscape. Animal life too changes greatly, and gradually there appear vast numbers of insects that have spent the winter in hibernation as eggs, larvae, nymphs, or pupae in ponds, lakes, and icy bogs or in the soil and vegetation of the more protected ravines. Quantities of Collembola, Trichoptera, and Odonata emerge — and above all mosquitoes, which swarm in tens of thousands. Enormous as this biomass is, the species richness is quite low. Tens of millions of migratory birds — ducks, geese, swans, many shore birds, and a few passerines — flock to the tundra to nest, while the vast vegetable biomasses, which are partly due to the fertility of the coastal strip, attract tens of thousands of reindeer (or caribou on the American continent) from the taiga, followed by their faithful predators, the wolves. The few permanent vertebrates include water voles, squirrels, lemmings, arctic hares and foxes, willow ptarmigans, and the snowy owl, many of which fluctuate in numbers over a period of years.

As for butterflies and moths there are few in the arctic tundra and even fewer — and less-known species — that have adapted to the antarctic tundra. Some have a circumpolar distribution and are found in both Eurasia and North America. Others are found in isolated zones, with arctic populations and with populations in the high alpine habitats of the great Eurasian massifs, the Himalayas, the Pamirs, the Altai Mountains, the Caucasus, the Alps, and the Pyrenees, and in the Rocky Mountains in America. Certain relict species are also found in the Alps and in the high mountains of Africa. The flight season in the arctic tundra is very short and in general the numbers reach their peak during the second half of June, coinciding with the longest day. However, this may change from year to year, as may actual numbers, neither of which can be predicted with any certainty. The result is that collectors who have devoted one or two weeks to working in the Arctic often return empty handed.

Arctic and alpine butterflies and moths are usually unassuming in appearance. They are small and dark or drab in colour, and the scale covering of the wings is reduced; all these features contrast sharply with the luxuriant shades and aposematic colorations so typical of many tropical Lepidoptera. The basic reason for this difference lies in the means of natural selection. The coloration and other adaptations observable in tropical butterflies and moths are first and foremost a response to predation and competition, that is, to biotic factors, whereas at high latitudes and altitudes natural selection occurs through the rigours of the physical environment. In the Lepidoptera of the tundra coloration is a function of heat regulation, as is the body's being clad in a thick "coat" of long scales. The behaviour of arctic and alpine butterflies and moths is rigorously determined by the sun. As soon as the sky clouds over, they stop flying and search for shelter. Few species fly in windy weather although *Oeneis* flutters rapidly from one shelter to another. Most make short, low flights only when the air is still. Many features of their behavioural adaptation are directed to obtaining the greatest possible benefit from the warmth of the sun. The Lepidoptera sun themselves to

Plate 91 **Deciduous temperate forest**

The butterflies illustrated are representative samples of those to be seen in temperate European forests and in the adjoining ecotones (such as clearings and paths; see also Plate 87): (1) white letter hairstreak (*Strymonidia w-album*, Lycaenidae); (2) purple hairstreak (*Quercusia quercus*, Lycaenidae); (3) heath fritillary (*Mellicta athalia*, Nymphalidae); (4) pearl-bordered fritillary (*Clossiana euphrosyne*, Nymphalidae); (5) speckled wood (*Pararge aegeria*, Nymphalidae, Satyrinae); (6) southern white admiral (*Limenitis reducta*, Nymphalidae); (7) Duke of Burgundy (*Hamearis lucina*, Riodinidae).

the greatest possible extent, making the most both of the direct rays and of the heat trapped in stones or other warm surfaces; many species press their wings against them while raising their abdomens in the air. Another adaptation typical of both arctic and alpine lepidopterans is a tendency to fly by day. Some typically nocturnal genera of geometrids and noctuids have species adapted to the tundra that are diurnal and have correspondingly modified eye structures. The preimaginal instars of the Lepidoptera show quite remarkable resistance to the rigours of the climate. The pupae of the lymantriid *Byrdia groenlandia* can be frozen and thawed out several times at long intervals and can also withstand increases in temperature to over 30°C. The females of several moths emerge with their eggs already ripe, and many are apterous or brachypterous or, even if they have wings, show little propensity to fly. This prevents their being scattered by the wind. In the subantarctic islands, which are constantly battered by violent winds, the males of some species have also become brachypterous. Another notable adaptation of tundra Lepidoptera is the long duration of the larval instar. In the chapter Populations, Demography, and Migrations, we have already mentioned the biennial species that are particularly common among the butterflies. In several moths, however, the larval stage may last as much as three, four, or even more years, with short periods of activity interspersed with long diapauses. The few annual species have only a single generation even though at lower altitudes their populations produce two, three, or more annual generations, subject to photoperiodicity. The American zoologist A. M. Shapiro has found that it is experimentally possible to restore several generations to various subarctic species or populations of whites by keeping them in appropriate photoperiodic conditions. This is especially true of species that have adapted only relatively recently to the extreme conditions of the tundra and so still retain some "physiological memory" of earlier adaptations to different environments.

Species richness is reduced to a minimum in the tundra. The number of lepidopteran species diminishes with increasing latitude, as is true of the great majority of animal species. Only five species of butterflies are found in Greenland (*Colias hecla*, *Boloria polaris*, *B. chariclea*, *Agriades aquilo*, and *Lycaena phlaeas feildeni*), the same species that inhabit the Queen Elizabeth Islands, which lie between 75 and 83°N, only 750 km (450 mi) from the Pole. In this whole area the number of Lepidoptera amounts to eighteen. Most of these are specialized forms, only a few, such as *L. phlaeas*, being ubiquitous, euryoecious species. These forms do not increase much even at lower latitudes with somewhat less restrictive conditions. In these zones the genera adapted to tundra life consist of *Colias* among the pierids (*C. nastes*, *C. hecla*, *C. boothii*), *Boloria* among the nymphalids (*B. frigga*, *B. improba*, *B. polaris*, *B. chariclea*), *Oeneis* (*O. norna*, *O. bore*, *O. melissa*, *O. polixenes*, *O. taygete*) and *Erebia* (*E. polaris*, *E. fasciata*, *E. rossii*, *E. disa*) among the satyrine nymphalids, and *A. aquilo* and *L. phlaeas feildeni* among the lycaenids. The moths best adapted to the arctic tundra include some noctuids (*Anarta richardsoni*, *A. melanopa*, *A. zemblica*), the geometer *Carsia paludata* and other species of the same genus, a few pyralids (*Crambus*), and lymantriids (*Byrdia*).

Butterflies and Moths of Mountain Biomes

The alpine tundra presents a considerably more complex and varied picture both because the lepidopteran populations of the great mountain massifs of the biosphere vary from one region to another and because, even on a single mountain, there is a rapid succession of different species as the zones of vegetation change with altitude. All the tundra genera

Plate 92 **Lepidoptera from aquatic and marshy environments**

Marshes and bogs are often inhabited by a wealth of Lepidoptera: (1) *Chamaesphecia palustris* (Sesiidae), food plant *Euphorbia palustris*; (2) *Paraponyx stratiotata* (Pyralidae, Nymphulinae), food plant *Stratiotes aloides*, (*2a,b*) larva; (3) *Nymphula nympheata* (Pyralidae, Nymphulinae), food plant *Potamogeton natans*, (*3a,b*) larva; (4) *N. stagnata*, Pyralidae, Nymphulinae; (5) *Cataclysta lemnata* (Pyralidae, Nymphulinae), on *Lemma* sp.; (6)*Schoenobius gigantellus* (Pyralidae), food plant *Arundo phragmites*; (7) *Phragmataecia castaneae* (Cossidae), food plant *Phragmites communis*; (8) *Nonagria typhae* (Noctuidae), food plant *Typha* sp.; (9) *Archanara sparganii* (Noctuidae), food plants *Sparganium, Typha*, and the like; (10) *Plusia festucae* (Noctuidae), polyphagous; (11) *Rhizedra lutosa* (Noctuidae), food plant *Phragmites communis*; (12) *Leucoma salicis* (Lymantriidae), food plants *Populus, Salix*; (13) large copper *(Lycaena dispar*, Lycaenidae), oligophagous on *Rumex* and *Polygonum*; (14) cranberry fritillary (*Boloria aquilonaris*, Nymphalidae), food plant *Vaccinium* sp.; (15) false heath fritillary (*Melitaea diamina*, Nymphalidae), food plants *Plantago, Veronica*, and the like; (16) lesser marbled fritillary (*Brenthis ino*, Nymphalidae), food plants *Filipendula, Sanguisorba*; (17) dusky large blue (*Maculinea nausithous*, Lycaenidae), food plant *Sanguisorba officinalis*; (18) violet copper (*L. helle*, Lycaenidae), food plant *Polygonum*; (19) large checkered skipper (*Heteropterus morpheus*, Hesperiidae), food plants *Molinia, Brachypodium, Bromus.*

mentioned above (*Erebia*, *Oeneis*, *Boloria*, *Colias*, *Agriades*, and *Anarta*) have relict populations of the same arctic species or of related species in the high mountain areas of the northern hemisphere (Plates 99 and 100). They are joined notably by the papilionids of the genus *Parnassius*, which contains the best-known and most sought-after alpine butterflies. Some Himalayan species, such as *P. delphius*, *P. simo*, and *P. charltonius*, live at altitudes above 4,500 m (14,750 ft); and *P. acco* has been caught on the slopes of Mount Everest at a height of 5,640 m (18,500 ft). Unlike other alpine butterflies only a few species of *Parnassius* (*P. eversmanni*, *P. apollo*) are also found at lower altitudes or in arctic latitudes; the rest are found typically on the great mountain massifs of the Palaearctic and Nearctic regions. This is also true of the satyrine genus *Karanasa*, which has adapted to the high, arid steppes of the Asian mountains, where it is found at heights of more than 5,000 m (16,400 ft), and of the small Himalayan pierid genus *Baltia*.

The Lepidoptera inhabiting the high Andes are rather unusual. The *paramo*, or alpine grasslands, of this region are populated by many species not closely related to the arctic butterflies and moths; they are probably indigenous to the region, having evolved from tropical ancestors. This is true particularly of such genera of Pronopholini (Nymphalidae) as *Cosmosatyrus*, *Lymanopoda*, and *Pedaliodes*, which recall the alpine *Erebia* species in general appearance and behaviour (Plate 24), in accordance with the principle known as *ecological equivalence*. Other strange Andean Lepidoptera include *Argyrophorus argenteus* and *Panargentus lamma*, metallic-coloured satyrines, and the peculiar lycaenid *Styx infernalis*, the sole representative of the subfamily Styginae. The Andean pierids also contain typical genera such as *Tatochila*, which are perfect ecological equivalents of *Pontia callidice*, *P. sisymbrii*, and other similar Holarctic alpine pierids. A point of contact between the fauna of the Andes and that of the northern Holarctic may, however, be seen in the genus *Colias*, in that a number of its species are also found in the Andes.

Taiga and Deciduous Temperate Forest

In the northern hemisphere the tundra gradually gives way to the taiga, or coniferous forests. This biome forms a band some 1,500 km (950 mi) wide, which, together with the tundra, covers the earth's surface almost continuously from Norway to Kamchatka and across the whole of Canada from Newfoundland to British Columbia. The climate is typically continental, with short summers with long days and the opposite in winter. Summer lasts from three to five months and is hot, whereas winter is cold, dropping at the coldest point in Siberia to less than −70°C. There are only slight differences between the Nearctic and Palaearctic taiga. Vegetation consists predominantly of conifers (*Abies*, *Pinus*, *Larix*, *Picea*) and some broadleaved plants (*Betula*, *Popylus*, *Salix*). The tendency is for a single species to predominate, depending on the zone. The productivity of this biome is not high. Towards its southern edge the taiga gives way to ecotonal vegetation of mixed temperate forest, which merges into the deciduous temperate forest. The latter is a more limited and irregular biome than is the taiga, occupying in the northern hemisphere only the western (central Europe) and eastern (Manchuria, Korea, north and central China, and Japan) portions of Eurasia and the eastern part of North America. This irregular distribution may be explained by the effects of the Pleistocene glaciation, which destroyed the primitive deciduous forests of central Eurasia and the western United States, from which the present temperate forests seem to be descended. Separately derived temperate deciduous forest is also distributed in parts of the southern hemisphere: southern Chile, southeastern Australia, Tasmania, and New Zealand. Winters are milder and summers cooler than they are in the

Plate 93 **Mediterranean maquis**

The typical vegetation of the Mediterranean coastal strip is sclerophyllous and evergreen, the so-called Mediterranean maquis, dominated by holly. Several different types are found, depending on the type of soil and the degree of disturbance caused by human settlement, ranging from mature holly groves to high maquis or the degraded low maquis, or *garrigue*. Among European butterflies the precise indicator of the maquis is the presence of the two-tailed pasha, *Charaxes jasius* (1, male, 2, female, 3, chrysalis, 4, caterpillar). This is a species of African origin, closely tied to the strawberry tree (*Arbutus unedo*). Another characteristic butterfly of this vegetation is the Cleopatra, *Gonepteryx cleopatra* (5, male, 6, female), which is found all around the Mediterranean, with distinct subspecies in Madeira and the Canaries. The southern gatekeeper, *Pyronia cecilia* (7, male) has a very similar distribution. *G. cleopatra* feeds on buckthorn and other *Rhamnus*, while *P. cecilia* feeds on *Deschampsia* grasses. The Spanish festoon, *Zerynthia rumina* (8, female) is found along the coastal strip of Northwest Africa (subspecies *africana*) and in Iberia and Mediterranean France, where it feeds on *Aristolochia pistolochia*. Other butterflies to be seen in the maquis but less closely tied to this biome are the dappled white, *Euchloe simplonia* (9, the Mediterranean equivalent of the related *E. ausonia*), (10) Lang's short-tailed blue (*Syntarucus pirithous*), which is found in many types of environment, and (11) the pigmy skipper (*Gegenes pumilio*), a widespread butterfly, which in Europe prefers coastal dunes.

taiga, especially in zones affected by coastal climate. Rainfall is high and seasonal, annually varying from 500 to over 2,500 mm (20 to 100 in), depending on locality. A degree of stratification is found in deciduous forests, which vary in type with kind of soil, slope and rainfall. In Europe such forests commonly consist of beech, various oaks, chestnut, elm, and maple, but North America is much richer in species. In the north it is dominated by beech (*Fagus grandifolia*) and sugar maple (*Acer saccharum*), sometimes mixed with hemlock (*Tsuga canadensis*). In the southeastern United States uplands the dominant trees are a number of oaks (*Quercus montana*, *Q. coccinea*, *Q. alba*, and so on) and hickories (*Carya*). North America also has some genera in common with the forests of the Far East — gum (*Liquidambar*) and tulip poplar (*Liriodendron*) — that are not found in native European woods. The characteristic genus of south-temperate deciduous forests is the *Nothofagus* genus of beech.

As a whole this biome is fairly productive and highly structured. The great variety of microhabitats favours the development of a high level of animal diversity. Thus we find at the different trophic levels species of mammals, birds, amphibians, reptiles, arthropods, molluscs, and earthworms. The latter play an important part in breaking down vegetable litter. It is interesting that a few species of the lepidopteran *Amata* (Ctenuchidae) also play a considerable part in breaking down the litter of Old World deciduous forests (Plate 75). Both taiga and deciduous forest are home to species of butterflies and moths that have a determining role as primary consumers. *Choristoneura fumiferana* (Tortricidae) attacks the genus *Picea* in Europe and several conifers in North America; the processionary moth (*Thaumetopoea pityocampa*) and the geometer *Bupalus piniarius* both attack pines; all four are veritable scourges of the taiga. Many broadleaved trees of the deciduous forests are subjected periodically to violent attack by the gypsy moth (*Lymantria dispar*; Lymantriidae; see Plate 112); in North America it is mainly limited to the northeastern states. The temperate woodlands contain a rich and abundant lepidopteran fauna. Only some of the species found there can, however, be considered to have specialized for this biome.

As in many other environments only a small fraction of woodland species are *stenoecious*, that is, closely tied to certain plants or ambient conditions; most are adapted to much broader ecological conditions, woodlands being one of several possible habitats. Nor should it be forgotten that the character of the temperate forests today is the product of 4,000 to 5,000 years of ever-increasing interference by man, who has gradually made small and large clearings by felling trees and using wood for charcoal, building houses and roads, and so on. The result has been the creation of a kind of extensive ecotone. Specialized forest butterflies and moths are found in many families and may thus be considered under a range of headings. Only a few species of butterflies are tied to characteristic forest trees. Thus of the roughly twenty-four woodland species found in the British Isles, only three, *Nymphalis polychloros*, *Apatura iris* (Nymphalidae), and *Strymonidia w-album* (Lycaenidae), are tropically dependent on trees. As regards other species such as *Pararge aegeria*, the link with the forest is "structural," being essentially connected with the territorial behaviour of the adult (Plate 64). Thick forest is not a favourable environment for butterflies; few species with few individuals are generally to be found. Moths too are not very diverse, particularly when the forest is monophytic or dominated to a large extent by a single species of tree. Much greater variety is found in clearings, where it is not uncommon to find several dozen Lepidoptera resting on a single flowering bush. Some species that are adapted to this environment tend to spend a large amount of time in the canopy, where they exhibit territorial behaviour, coming down to the ground only at certain times of day. This is

Plate 94 **Arid zones**

The typical vegetation of the semidesert zones of the Near and Middle East consists of sparse cushion plants and small spiny shrubs such as *Astragalus* (Leguminosae), *Zizyphus*, *Paliurus spina-christi* (Rhamnaceae), *Capparis* (Capparidaceae), *Artemisia herba-alta* (Compositae), *Salsola vermiculata* (Chenopodiaceae), and other plants eaten by the Lepidoptera that inhabit such environments. The plate shows a semidesert biotope not unlike those to be found in the Zagros Mountains of Iran, in Lebanon, and in the Arabian peninsula, with some typical Lepidoptera. Highly specialized forms are found in the semidesert together with species of wide ecological tolerance (3) and migratory species (6, 12). (1) Small orange tip (*Colotis evagore*, Pieridae), on *Capparis*; (2) *C. fausta*, Pieridae; (3) Bath white (*Pontia daplidice*, Pieridae); (4) *Anaphaeis aurota*, Pieridae, on *Capparis*; (5) greenish-black tip (*Elphinstonia charlonia*, Pieridae); (6) plain tiger (*Danaus chrysippus*, Nymphalidae, Danainae); (7) *Maniola telmessia*, Nymphalidae, Satyrinae; (8) *Pseudotergumia pisidice*, Nymphalidae, Satyrinae; (9) African ringlet (*Yphthima asterope*, Nymphalidae, Satyrinae); (10) little tiger blue (*Tarucus balkanicus*, Lycaenidae), on *Paliurus spina-christi*; Lang's short-tailed blue (*Syntarucus pirithous*, Lycaenidae), (11) male, (12) female; (13) *Azanus jesous*, Lycaenidae; (14) *Chilades galba*, Lycaenidae; (15) Mediterranean skipper (*Gegenes nostrodamus*, Hesperiidae); (16) *Borbo borbonica*, Hesperiidae; (17) the Middle East hermit (*Chazara persephone*, Nymphalidae, Satyrinae); (18) *Zygaena saadii*, Zygaenidae; (19) *Zygaena olivieri*, Zygaenidae.

the case with *Thecla betulae* (Lycaenidae), the species of *Apatura* (Nymphalidae), and some European and American admirals (*Limenitis*). Many genera and species are common to the temperate forests of both Europe and North America. *Pieris napi* shows a distinct predilection for forests in both continents, and the woodland butterfly *Nymphalis vau-album* also has similar populations in both areas. Others belong to more or less rigidly woodland genera (*Polygonia* and the like) and have separate species on each continent.

A different type of relationship is exhibited in other butterflies, which, while resembling one another in appearance, behaviour, and habitat in both zones, are only distantly related and are classified in different genera. This is true of many species of Eurasian (*Argynnis*, *Fabriciana*, *Mesoacidalia*) and some American (*Speyeria*) nymphalids, which are forest dwelling and live on violets, and of those satyrinae which have many eyespots on their wings, the Eurasian *Aphantopus hyperantus* and the American *Megisto cymela*, which are not in fact related but are similar in appearance, behaviour, and food plants (Poaceae). All these examples illustrate how the faunae inhabiting corresponding unconnected biomes in regions distant from one another develop along parallel lines in response to parallel selective pressures. Similar niches may be occupied by the same species in each region or by closely related species or by species that are more distantly related but have evolved parallel or convergent adaptive characters and act as ecological equivalents.

Other Temperate Biomes

Southwards through the continental masses of the northern hemisphere the taiga and deciduous forests give way to the steppe biome. Steppes and prairies are found chiefly in Eurasia and the central United States east of the Rocky Mountains, a range bounded by 30 and 50°N latitude. Other enormous areas of steppe are to be found in central Australia, South Africa, and Argentina and Uruguay, where the pampas have various peculiar features. This biome presents a wide variation in temperature between day and night and between seasons, with hot summers and cold winters and, excepting the pampas, low rainfall [400 to 500 mm (15 to 20 in) annually]. Other than various grasses the dominant vegetation of the steppes and prairies are various herbaceous plants of the Leguminosae and Compositae. The extensive fauna is dominated by primary consumers, mainly ungulates and rodents, which vary from region to region, and by a few predators. Of the butterflies and moths of the Eurasian steppe the most abundant are various Satyrinae (*Hipparchia*, *Chazara*, *Pseudochazara*, *Melanargia*), various *Colias* (Pieridae), *Melitaea* (Nymphalidae), and several blues and skippers. The North American prairie has more abundant grasses, and as a result hesperiine skippers (*Atrytone*, *Hesperia*, *Oarisma*, *Polites*) are its dominant butterflies. Among the true butterflies the regal fritillary (*Speyeria idalia*) and gorgone checkerspot (*Charidryas gorgone*) are typical prairie species there.

In the western coastal areas of the middle latitudes we find a biome that extends over a small total area and is characterized by Mediterranean, evergreen, broad-leaved forest vegetation. This biome is found chiefly around the coast of the Mediterranean but also in the coastal zones of California, central Chile, and southwest South Africa and Australia. Annual rainfall is relatively low [250 to 750 mm (10 to 30 in)] in these areas, summers are hot and rainless, and winters are wet and mild. The diversity and types of plants composing Mediterranean vegetation change considerably from one region to another, but one feature common to them all is the presence of sclerophyllous trees with hard, evergreen leaves that dominate species of grasses, shrubs, and tuberous or bulbous plants capable of conserving moisture in their root systems. Mediterranean

Plate 95 **Tropical rain forest**

The tropical rain forests of South America contain the greatest number of Lepidoptera in the world. Here the diversity of shapes and colours is at its greatest. The plate shows a selection from a Peruvian forest, with representatives of some of the best known Neotropical Lepidoptera found at various forest strata: (1) *Cithaerias aurorina*, Nymphalidae, Satyrinae; (2) *Agrias claudia godmani*, Nymphalidae; (3) eighty-eight butterfly (*Diaethria clymena*, Nymphalidae); (4) *Morpho aurora aureola*, Nymphalidae; (5) postman (*Heliconius melpomene amaryllis*, Nymphalidae, Heliconiinae); (6) blue doctor (*Rhetus periander*, Riodinidae); (7) *Amarynthis meneria*, Riodinidae; (8) *Theritas tagyra*, Lycaenidae; (9) spear-winged cattle heart (*Parides neophilus olivencius*, Papilionidae); (10) *Callithea davisi tirapatensis*, Nymphalidae; (11) *Hypoleria oreas*, Nymphalidae, Ithomiinae; (12) malachite (*Metamorpha stelenes*, Nymphalidae).

ecosystems are variously named: *macchia* in Italy, *maquis* in France, *mattoral* in Spain, and *chaparral* in California. The original nature of these sclerophyllous forests has been highly modified by human settlement, and this is particularly evident around the Mediterranean, where the primitive forest has given way to agriculture or has degenerated into low scrub or *garrigue*. Species diversity has decreased considerably, especially among the vertebrates. Some typical butterflies of the maquis are illustrated in Plate 93.

Although the chaparral formation of California includes several groups of plants with many species not found in adjoining ares, for example, *Ceanothus* (Rhamnaceae), *Arctostaphylos* (Ericaceae), and *Adenostoma* (Rosaceae), it is generally not considered to have a distinct butterfly fauna. Many butterflies and other Lepidoptera found in chaparral and adjoining areas are also found in similar shrubby habitats or even in quite dissimilar habitats in other western areas possessing more continental climates. Among the butterflies only a few are endemic to chaparral and a few adjacent areas: grey marble (*Anthocharis lanceolata*), Wright's metalmark (*Calephelis wrighti*), Sonora blue (*Philotes sonorensis*), veined blue (*Plebejus neurona*), Hermes copper (*Lycaena hermes*), and the unsilvered fritillary (*Speyeria adiaste*). More typical of chaparral areas in California are several groups of microlepidopterans. Diurnal moths of the genus *Adela* are quite abundant in California although most are found associated with herbaceous plants in annual grassland adjacent to chaparral. A number of leaf-mining moths have undergone radiations on woody plants common in, or adjacent to, chaparral; these include *Tischeria* (especially on *Ceanothus*), *Coptodisca* (Heliozelidae), and *Cameraria* (especially on oaks). Although many chaparral plants have close relatives in Chile, none of the northern Lepidoptera seems to have affinities with those of the southern hemisphere. Many species in the chaparral must be adapted to having their larvae feed during the winter and early spring (the only times when fresh foliage is usually available). In addition the are usually limited to only a single annual generation because of the restricted season of food availability. Another adaptation is an ability to bypass several unfavourable seasons in larval or pupal diapause. J. A. Powell has conducted experiments on prolonged diapause in several of the prodoxine yucca moths of the southwestern United States chaparral.

Desert and Semidesert Biomes

Desert and semidesert biomes are a series of areas scattered across all the continents where rainfall is deficient (not normally reaching as much as 200 mm (8 in) annually) or where rain may occasionally fall heavily but irregularly. The controlling features of desert biomes are abiotic: the erosion of the soil and the climate. It is not easy to distinguish deserts from semideserts, but usually an outer band of semidesert gives way gradually to more arid and less vegetated central desert zones. The characteristic features of a desert are eleven to twelve months of aridity a year, sporadic and irregular precipitation, large variations in temperature between day and night — which in extreme cases may be more than 80°C from the daytime maximum to the nighttime minimum — a high rate of evaporation caused by strong winds, and the extremely degraded condition of the soil. Deserts may be rocky, stony, sandy, clayey, or salty, depending on the subsoil. In the Americas the deserts are concentrated chiefly in western areas of the United States and Mexico (Sonora, Mohave, Death Valley, Baja California), in the western part of South America in Peru and Chile, and in northwest Argentina. In the Old World the largest deserts are to be found in the Sahara, in the Arabian peninsula, in various parts of Asia from Iran to Mongolia, and in southwest Africa between 18 and 28°S latitude. Finally a large part of Australia is semidesert. One of the most

Plate 96 **Tropical mountain forest**

Tropical rain forest is found in the tropics from sea level to a height of some 1,500 m (5,000 ft), gradually giving way, as the altitude increases, to the somewhat different vegetation known as tropical mountain forest and, in some areas, as "cloud forest." In Malaysia, for example, the tropical mountain forest is found above 1,000 m (3,250 ft); a typical feature is the carnivorous plants of the genus *Nepenthes*, and orchids are very common. This is the specific habitat for many butterflies, including the pierid genus *Delias*. Here too species richness is very high. The plate shows a small sample of butterflies caught in the Cameron Highlands in the Pahang district in Malaysia: (1) *Troides cuneifer*, Papilionidae; (2) *Ancema ctesia*, Lycaenidae; (3) *Delias ninus*, Pieridae; (4) *Chilasa agestor*, Papilionidae; (5) *Papilio iswaroides*, Papilionidae; (6) *Parantica (=Danaus) sita*, Nymphalidae, Danainae; (7) *Kaniska canace*, Nymphalidae; (8) *Rapala abnormis*, Lycaenidae; (9) *Polytremis eltolta*, Hesperiidae; (10) *R. iarbus*, Lycaenidae; (11) *Dodona egeon*, Riodinidae; (12) *Ragadia crisilda critolina*, Nymphalidae, Satyrinae; (13) *Vindula erota*, Nymphalidae; (14) Australian fritillary (*Argyreus hyperbius*, Nymphalidae); (15) *Neptis soma*, Nymphalidae.

hostile deserts is the Atacama in Chile, where rain may not fall for periods of more than fifty years. The biotic diversity of desert and semidesert regions is generally very low in the animal and vegetable kingdoms, both of which exhibit special adaptations to these extreme ecological conditions. Individual perennial plants are widely scattered because of the competition for scarce resources of water and food. Forms of adaptation to desert conditions by the vegetation include succulent plants that retain water in their tissues (Cactaceae and Euphorbiaceae), perennial plants with root systems reaching down more than 30 m (100 ft) into the water table (tamarisks, mesquite, palo verde), spiny sclerophyllous plants with reduced leaf-surfaces (ocotillo), such poikilohydrous plants as *Selaginella* and *Cheilanthes*, whose withered leaves immediately become green again with the tiniest quantity of water, xerophytic plants that have adapted to the most extreme conditions of the sandy desert (*Aristida pungens*), and plants that can flower immediately after rain (*Mesembryanthemum*, *Mollugo*, and so on).

The butterflies and moths inhabiting the desert include some forms that have become highly specialized and others that are migratory and found in a wide range of ecological conditions. The extremes of summer drought are usually spent in larval or pupal diapause. The caterpillars are active in spring and autumn. The eggs hatch at the time of greatest plant growth, when the vegetable fibers consumed by the larvae are most tender and full of water. Water loss is reduced to the minimum in the larvae by nocturnal habit (seeking shelter by day) or by special structures like the oral and anal "plugs" found in the diapause (*Euphydryas*). The caterpillars of some desert-dwelling species of *Euchloe* and *Anthocharis* grow very rapidly, becoming fully developed and pupating in less than two weeks. The larvae of the North American giant yucca skippers (*Megathymus* and *Agathymus*) develop inside agave and yucca plants, while those of the yucca moth (*Tegeticula*) find protection by living inside their seeds. Adults too exhibit behavioural and structural adaptation to life in the desert. Some species are crepuscular (Megathymidae), and others fly early in the morning (*Euphilotes*, Lycaenidae). Most Satyrinae have a reduced proboscis, rapidly sucking drops of liquid from fruit, excrement, and soil. Usually desert-dwelling species have a single generation annually, but some produce one in spring and a second in autumn. In many species the signal for emergence is rain so that egg laying is synchronized with the period of greatest plant growth. If no rain falls, the pupa can live several years waiting for favourable conditions. Adult life is usually quite short. Plate 94 illustrates the lepidopteran life of a semidesert zone of Asia Minor.

Tropical Biomes

The characteristic biomes of the tropical and equatorial zones are the dry thorn forest, savanna, deciduous tropical forest, and tropical rain forest ("jungle"). The factors determining which of these biomes is found in a given locality are the nature of the soil and the geographical proximity of oceans and mountains, altitudes, and the prevailing climatic conditions. The amounts of, and the seasonal variations in, rainfall are fundamental and can determine whether a given community will be largely seasonal in type (savanna, deciduous tropical forest) or very stable (rain forest). The tropical rain forest is the biome that gives the greatest delight to naturalists. It is the opposite of the tundra or desert in terms of its intense productivity, its seasonal stability, and the extraordinary richness of plants and animals. Some of the great naturalists of the last century, such as Alfred Russel Wallace, Charles Darwin, Henry Walter Bates, and Fritz Müller, who made such great contributions to the understanding of evolution, gained decisive insights from their firsthand research into the life of the tropical rain forests. This extensive biome is represented above

Plate 97 **Mangrove populations**

The coastal zones in the tropics give rise to special vegetation dominated by mangroves, which form an intermediate area between the sea and the mainland. The floor of the mangrove swamps is composed of vast quantities of organic detritus and mud of varying degrees of solidity, which provides food for a wealth of fauna dominated by the detritus-eating crabs. A variety of butterflies inhabits these swamps in the different continents, and some depend on the mangroves for their larval food (10). Some species common in the mangroves of the Indo-Australian area are illustrated. Butterflies most closely tied to this community are indicated by an asterisk. In Malaysia: (1) black-and-white tiger (*Danaus affinis malayana*,* Nymphalidae, Danainae); (2) *Idea leuconoe chersonesia*,* Nymphalidae, Danainae; (3) *Euploea phenareta phoebus*,* Nymphalidae, Danainae; (4) *Taenaris horsfieldii birchi*, Nymphalidae, Amathusiinae; (5) *Nacaduba pavana singapura*, Lycaenidae; (6) tiny grass-blue (*Zizula hylax pygmaea*, Lycaenidae); *Hypolycaena eurylus teatus*, Lycaenidae, (7) male, (7*a*) female; (8) *Rapala cowani*,* Lycaenidae. In Australia: common eggfly (*Hypolimnas bolina nerina*, Lycaenidae), (9) male, (9*a*) female, (9*b*) chrysalis; dull jewel (*Hypochrysops epicurus*,* Lycaenidae), (10) male, (10*a*) female; this species feeds on black mangrove (*Avicennia officinalis*). In the Americas: (11) mangrove skipper (*Phocides pigmalion*,* Hesperiidae); the larvae of this species feed on red mangrove (*Rhizophora*).

all in the Amazon basin, the Congo basin, western Africa, and the Malaysian-Indonesian region. Smaller expanses of tropical rain forest are also found in other areas within the tropics. The chief climatic conditions are as follows: Day and night temperatures vary between 25 and 30°C — a greater variation than is exhibited seasonally — temperatures remaining almost constant throughout the year with a complete absence of frost; annual rainfall is very high, usually more than 2,000 mm (80 in), and occurs throughout the year although with slight variations in intensity. The constant high temperature and humidity provide the best possible conditions for plant growth. It has been calculated that over 100,000 plant species inhabit the rain forests of the world. This forest may have many strata and is dominated by trees 20 to 30 m (65 to 100 ft) high, many surpassing 50 m (165 ft). The dense canopy, characterized by species with shiny, coriaceous leaves, greatly reduces the amount of light that reaches ground level (sometimes to as little as 1 percent), with the result that the forest floor is rather poor in vegetation. The high temperatures mean that decomposition is extremely rapid, and the soil is rich in carbon dioxide between the flattened, tabular roots that are the supports and buttresses of the great trees. Both the roots and tops of these trees are covered with *epiphytic* plants — mosses, lichens, ferns, bromeliads, orchids, and so on — while *lianas*, including aroids, climbing palms (*Calamus*), bignons, and various strangler figs, as well as many species of twining and tendrillous plants are very common. Most of the plants are relatively seasonal in flower, fruit, and leaf production; but some species may bear flowers, fruits, or leaves at any time. Both trunk growth and leaf cover are continuous. Tropical rain forest proper is found from sea level to about 1,500 m (5,000 ft). Higher up, especially on the steep mountains that hold back the humid air, we find a different type of forest, in which the trees are shorter but which contains as many species as the rain forest. Tree ferns and epiphytic plants are found in greater quantities in these forests, which occur in Central and South America (Guatemala, Venezuela), New Guinea, Borneo, and Malaysia and are known as *cloud forests*.

A greater species richness is found here than in any other biome. Animal life is concentrated chiefly in the upper strata. Tree-dwelling and flying forms are especially abundant not only among insects and birds but also among mammals and some reptiles. The animals are concealed and hard to spot, giving the impression that the forest is uninhabited. In fact most of the animals are incredibly specialized and well protected against predators, and a closer look will begin to reveal myriads of small creatures, especially insects and other arthropods, with cryptic forms, coloration, and behaviour. In primary forests specialized forms far outweigh more general types. The butterflies and moths of the tropical forests are certainly some of the best examples of this. Together with birds and a number of amphibians, they make the greatest use of colour as a means of intraspecific and interspecific communication, many being extremely eye catching with their flash and aposematic coloration. Mimicry, whether cryptic or phaneric, reaches its ultimate expression in this biome. Many species are closely tied to a particular stratum of vegetation; others, closely related among themselves, share out the tropic niches very carefully by using different food plants. The spatial distribution of species is heterogeneous despite the apparent monotony of the environmental conditions, and many Lepidoptera exhibit a scattered distribution. This is not haphazard, however; for the different populations show a certain predilection for their own territories. Single individuals also have their own home range, tending to move along well-defined routes that vary only slightly from one day to another. This type of behaviour has been observed in some morphos and in *Heliconius* and is certainly characteristic of many other forest butterflies and moths. Another important feature is the

Plate 98 **Cave moths**

Caves and underground passages form a highly specialized habitat, characterized by absence of light and hence of green plants, constant temperature, high humidity, and scarcity of food, and are thus less than ideal environments for Lepidoptera. Nevertheless, caves throughout the world are inhabited on a regular or an occasional basis by one or more species of Lepidoptera. The great majority of these troglodytic species live underground only in the adult phase, and only during certain seasons do they spend the diapause there, whether hibernation or aestivation. In Anatolia and other semiarid zones of the Mediterranean basin swarms of Satyrinae can frequently be seen at the entrance to caves during the summer. Caves are also sometimes used for hibernation by other butterflies, such as the large tortoiseshell (*Nymphalis polychloros*), but this practice is more typical of such moths as the noctuid *Scoliopteryx libatrix* (1) and the geometers *Triphosa dubitata* (2) and *T. sabaudiata* (3). Other species may be found on cave walls even in complete darkness, especially during summer: for example, noctuids *Pyrois effusa* (4), *Hypena obsitalis* (5) — see also the graph showing the seasonal variation in a cave in central Italy — and *Apopestes spectrum* (6). A more occasional visitor is the noctuid *Mormo maura* (7), which like the other species above mentioned may also be found in such artificial caverns as catacombs, cellars, and disused mines. Species of the tineid genus *Monopis* (8) are more specialized, spending their entire biological cycle in caves, the larvae feeding on bat droppings.

rapidity of the metamorphic cycle, a cycle faster than in other environments because of the availability of food and the high temperatures. Some species require less than a month to pass from egg to adult, and generations succeed one another without a break. Unlike tundra and desert Lepidoptera those of the tropical wet forests therefore produce many generations annually and do not usually undergo diapause. All this makes for great evolutionary potential as a result of a higher probability of mutation and combination and a correspondingly higher rate of genetic change.

Like other specialized tropical-forest organisms the reproductive strategies of butterflies and moths serve to control population size; the strategies are the opposite of those of organisms adapted to the environments previously described, where the chief limiting role is played by physical factors. One such strategy, named "K" by ecologists in contrast to strategy "R," consists in low egg production accompanied by behaviour patterns that enhance the likelihood of survival (parental care). An example of this is the care taken in egg laying by females of *Heliconius* (see the chapter on ecological relations). Another important feature is that the adults of some species can be very long-lived, surviving as much as six months in *Holiconius* and as much as nine months in some species of *Morpho*.

The examples illustrated in Plates 87, 95, 96 and 108 to 110 give only a summary notion of the incredible diversity of shapes and adaptations in the Lepidoptera of the tropical forests. Perhaps some figures taken from a recent work by G. Legg, concentrating exclusively on butterflies, may be more indicative. Thus single tropical biotopes such as Rangoon (Burma), Kaieteur Falls (Guyana), and Guayatebal (Colombia) contain, respectively, 180, 250, and 350 species, whereas such temperate areas as the New Forest in Britain or Fairfax County, Virginia, have 32 and 89 species. When larger areas are examined, the difference in numbers of butterflies found in tropical forests and temperate zones of equivalent size is even greater, with 500 species in Ghana, 1,300 in Guyana, and 69 in the United Kingdom.

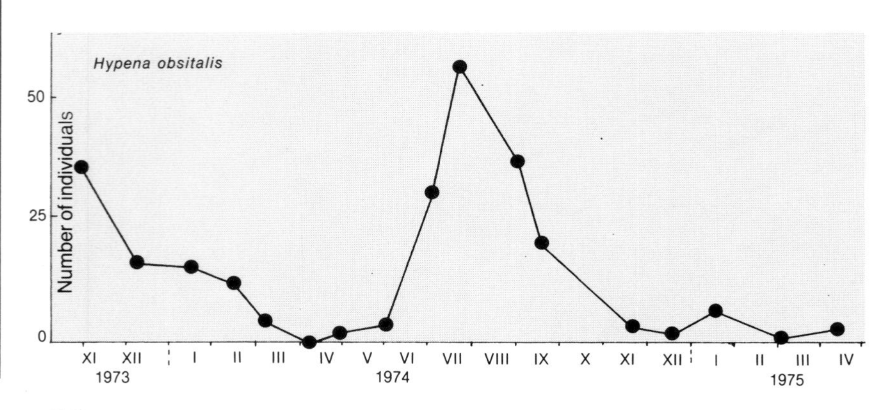

Geographical Distribution

Zoogeography

Zoogeography is the scientific study of the geographical distribution of animals. The starting point of zoogeographical analysis is a description of the distribution of species and an examination of the causes of this distribution; it takes into account both evolutionary history and ecological factors. To a large extent the geographical distribution of species appears to have been determined by extrinsic factors, such as climatic conditions and geographical barriers, and by intrinsic factors, such as the ability of individual species to disperse and colonize new areas. This means that it is possible — in theory at least — to explain the geographical distribution of a taxon in terms of its evolution during the appropriate period of geological history. A change in the geographical distribution of species often reflects an evolutionary change. It is no coincidence that the two founders of the theory of evolution by natural selection were both particularly interested in the problem of animal geography: Darwin assigned two chapters of *The Origin of Species* (1859) to the subject and Wallace an entire volume, *The Geographical Distribution of Animals* (1876), which remains a classic study in this field. Generally speaking, the study of geographical distribution can be conducted at three distinct levels: The first is the chorological level, which is concerned with the populations of a species or of a group of related species; the second is concerned with the qualitative and quantitative analysis of the populations of selected areas (especially in relation to ecological and dynamic factors); and the third is concerned with the detailed description of the faunas of the great geographical regions.
The first approach, study of the geographical distribution of populations, is probably the best way of achieving a better understanding of the small-scale processes of evolution because it enables us to examine the connections between geographical isolation of populations and the development of new species (see also the chapter Origin of Species). The second level of analysis is used when we want to know, for example, how the population of an island has been formed. In this case we investigate the places of origin of the various species making up the fauna and examine their relative antiquity, their dispersive mobility, their ability to form new colonies, the physical barriers they have had to surmount, and so forth. These methods of analysis are relatively modern and are characterized by a dynamic view of geographical distribution. The zoogeographical study of areas is the oldest of the three approaches, its traditional aim being the identification and delimitation of regions, subregions, and provinces in terms of the distribution of animal species. This type of study was originally purely descriptive, but today it has been expanded to cover new problems and new methodologies. It should be noted, finally, that zoogeographical phenomena are generally due to the joint effects of many historical and dynamic factors, which are interlinked in an extremely complex

Up to this point, butterflies and moths have been described in terms of their morphological, genetic, taxonomic, behavioural, and ecological diversity. We have also seen that certain characteristics and certain functions (colour, perception of the environment, feeding habits, sexual activity, and relationships within a community) are expressed by these insects in highly specific ways — so much so that they provide a perfect definition of the zone of the biosphere to which the species have become adapted. It has also emerged that the diversity of these creatures can be understood only in the light of the modern biological theory of organic evolution, which is based on the view that every lepidopteran species is the result of a historical process that takes place not only in time but also in space. This chapter, devoted to geographical diversity, will deal with the effects of space on the evolutionary process; in other words we shall consider the geographical distribution of a species.

For the sake of simplicity the chapter is divided into three distinct sections: In the first the distribution of lepidopterans is illustrated in terms of their geographical ranges; the second describes in wide terms the factors and processes, such as dispersal, colonization, and extinction, that govern the dynamic development of a species range; and the third part is devoted to a survey of the zoogeographic regions and their faunas.

Geographical Ranges

Butterflies populate the entire land area of the earth except for the polar regions and the most arid deserts. Each species occupies a definable geographical area, which is known as its *area of distribution* or, more simply, its *range*. Some species have ranges that cover a very small area. The Avalon hairstreak (*Strymon avalona*), for example, is restricted to southern California's Santa Catalina Island, only about 100 sq km in area.

The ranges of most species are, however, much larger than this. The dark-green fritillary (*Mesoacidalia aglaja*) covers the whole of Europe and Morocco and extends eastwards across central Asia, Siberia, and China to Japan. This is described as a "palaearctic" range because it covers a large part of the Palaearctic region. The long-tailed blue (*Lampides boeticus*), capable of tolerating widely differing climates, has an enormous range, which includes southern Europe, Africa, Asia, Australia, and the islands of Hawaii. There are some species, such as the migratory painted lady (*Cynthia cardui*), that are found on all continents and are consequently described as "cosmopolitan." The range of a species is described as "continuous" if every suitable area within the range is occupied. If only some of the suitable areas are occupied, the range is said to be "discontinuous." Most butterflies and moths have discontinuous ranges. It is easy to see how discontinuities arise. Originally, when a species consists of a single population, its range is necessarily continuous. Later a geographical or ecological barrier may arise and divide the species into two or more populations inhabiting separate areas; or the species may colonize territory at a distance from its original habitat. For an example of the first of these two mechanisms we may turn to the alpine argus (*Albulina orbitulus*), a European lycaenid adapted to live in cold climates (Plate 99). During the last ice age this insect probably had a continuous range to the south of the ice sheet; but during the present interglacial period, as the climate grew warmer and the populations of this insect were unable to adapt to the changed conditions, they either died out or moved to regions better suited to their requirements. Those in the southern part of the original range took refuge in the mountain massifs, while those in the northern part moved farther north to colonize the areas left free by the retreat of the glaciers. The present range of this species is thus separated manner. The definition of the relationship between these two factors is dependent on data of many different kinds (systematic, phytogeographic, ecological, paleogeographical, paleontological, and so on) that are often difficult to obtain. It is therefore very seldom — only in the simplest cases, when the data are clear and plentiful — that a powerful hypothesis can be formulated to explain the geographical distribution of a species or the origin of the population of a certain area.

into two parts: a southern one high in the Alps and a smaller northern one, at a lower altitude, in the mountains of Norway and Sweden. The geographical distribution of the large white (*Pieris brassicae*) is the result of a different mechanism. This insect was originally found only in the Palaearctic region, but it has recently increased its range by founding a colony in South America — probably with unintentional human assistance. This is a case of discontinuity arising from overcoming a geographical barrier. Butterflies with a wide and discontinuous geographical distribution are often polytypical, or divided into subspecies. This is not surprising; for we have already seen (at the beginning of the chapter on species origin) how physical barriers can isolate one population from another and thus greatly reduce opportunities for genetic exchange. A prolonged discontinuity of this kind will often favour the formation of new species. Numerous examples of these phenomena will be found if we examine the Lepidoptera that have *holarctic* distribution, a taxon being said to have such a distribution when it is present both in North America and in Eurasia. A holarctic distribution is therefore discontinuous by definition and may be possessed by a single species, a genus, or a taxon of higher grade. The holarctic range is one of the principal types of geographical range applicable to the creatures of the middle and high latitudes of the northern hemisphere. It reveals connections between the general fauna as well as the Lepidoptera of North America, northern Asia, and Europe. As we shall shortly see, many cases of holarctic distribution can be understood only in historical terms, and most of them are connected with the climatic and paleogeographical changes that took place in the Pleistocene. The description of this geographical category will give us an opportunity to introduce two other examples of discontinuous distribution, namely, the *circumpolar* and the *boreoalpine*, both of which occur in many species of holarctic Lepidoptera. We shall also mention some examples of vicarious geographical distribution and of zoogeographical relicts, two phenomena commonly found among the Lepidoptera of the world and indeed among other animals and plants in general.

Many butterfly genera are exclusively or mainly holarctic, including *Callophrys*, *Colias*, *Erebia*, *Anthocharis*, *Oeneis*, *Euphydryas*, *Boloria*, *Coenonympha*, and *Parnassius*. *Colias* contains two closely related and ecologically similar species: *C. chrysotheme*, which occurs from eastern Europe to the Altai Mountains, and *C. eurytheme*, which has a very wide distribution in North America. These two species derive from a common ancestor and occupy similar ecological niches but are resident in different areas. In each of the two regions mentioned one species replaces the other, which is an example of vicarious geographical distribution. In this instance the phenomenon is of recent origin, having been caused by the formation of the physical barrier of the Bering Strait. As is well known, the strait and the surrounding region of Beringia were above sea level during the first glaciations of the Pleistocene. The area was not completely covered by ice and enjoyed a relatively mild, though perhaps rather dry, climate. This made possible the dispersal of palaearctic species into North America and a contrary flow of nearctic species into Eurasia. During this period the Asiatic ancestor of the two *Colias* species probably crossed the intercontinental land bridge, together with other butterflies and such mammals as the wolf, the elk, and the bison. During the succeeding interglacial period the level of the sea rose, and the Asiatic populations were completely isolated from the North American ones. The creation of this geographical barrier has deeply influenced the evolution and distribution of many holarctic Lepidoptera, resulting in their differentiation into subspecies or separate full species. There are many examples of this among the arctiids, such as *Phragmatobia fuliginosa* and

Plate 99 **Ranges of distribution: arctic-alpine butterflies**

The arctic-alpine type of discontinuous geographical distribution is the result of the Pleistocene glaciations and their effects observed in lepidopterans and other animals. The arctic-alpine species are distributed more or less extensively in the most northerly regions of Eurasia and North America and also appear, much further to the south, as relict populations high on the slopes of the great mountain masses of the two continents, such as the Alps, the Pyrenees, the Apennines, the Caucasus, the Carpathians, the Rocky Mountains, the Sierra Nevada, the Cascades, and the Altai Mountains. Only a few such species have reached the Pamirs or the Himalayas. Not all montane butterflies with discontinuous distribution are to be included in the arctic-alpine category. One exception is the Apollo (*Parnassius apollo*, 1), which completely lacks the continuous circumpolar arctic distribution that characterizes true arctic-alpine species. The plate illustrates a few species of arctic-alpine butterflies, with maps showing the European or western Palaearctic parts of their respective ranges: (2) dewy ringlet (*Erebia pandrose*, Nymphalidae, Satyrinae); (3) *Pyrgus andromedae*, Hesperiidae; (4) shepherd's fritillary (*Boloria pales*, Nymphalidae, Nymphalinae); (5) mountain ringlet (*E. epiphron*, Nymphalidae, Satyrinae); (6) alpine argus (*Albulina orbitulus*, Lycaenidae); (7) bog fritillary (*Proclossiana eunomia*, Nymphalidae, Nymphalinae); (8) peak white (*Pontia callidice*, Pieridae).

Acerbia alpina, among the nymphalids *Boloria*, *Erebia*, and *Oeneis*, and among the parnassians. These four genera contain, respectively, 7, 7, 4, and 2 holarctic species with highly discontinuous distributions and at least two subspecies each. Quite a number of holarctic Lepidoptera have a circumpolar distribution. They include the arctiid *Acerbia alpina* (found in Scandinavia, Alaska, and northern Canada), various tundra-dwelling butterflies, such as the arctic fritillary (*Boloria chariclea*, found in Lapland and Greenland and in North America from Labrador to British Columbia and Alaska), the polar fritillary (*B. polaris*, found in Finland and Scandinavia above 68°N, northern Asia, northern Canada, Labrador, and Greenland), the dusky-winged fritillary (*B. improba*, Scandinavia above 66°N, Novaya Zemlya, and Siberia and from Alaska to Baffin Island), and the banded alpine (*Erebia fasciata*, arctic regions of Asia and from northern Alaska to Hudson Bay).

Other species of the northern hemisphere polar regions exhibit a different type of discontinuous geographical distribution, known as arctic-alpine, or boreoalpine, in which a northern population found at low altitudes is completely isolated from a southern population found higher up in the mountains. We have already seen how a geographical range of this type can arise in the case of the alpine argus (*Albulina orbitulus*); see also Plates 99 and 100. Like the circumpolar species the arctic-alpine butterflies also form subspecies. Thor's fritillary (*Boloria thore*), for example, has two subspecies. *B. thore thore* is found at Canazei in the Dolomites; the upper wing surface is very dark in colour, and the ventral side of the hind wings is traversed by a clearly defined bright-yellow discal band. *B. thore borealis*, on the other hand, is found at Abisko in Sweden, above 68°N latitude, and is characterized by the pale background colour of its wings and by an ill-defined muted discal band.

The arctic-alpine type of geological distribution can also be regarded as an instance of zoogeographical relictualism. A species whose range is very much smaller than it used to be often includes populations geographically isolated from the main center of distribution. Such populations are known as *relict populations*, and the areas they occupy are known as *relict areas*. *Zygaena exulans* has populations in the Balkans and on the slopes of the Gran Sasso mountains in Italy; these are excellent examples of relict populations dating from the Pleistocene (see the map showing its distribution, Plate 100). All the types of geographical distribution that have been mentioned are exemplified by the palaeno sulfur, or moorland clouded yellow (*Colias palaeno*). This insect has a discontinuous distribution of the holarctic type — mainly circumpolar since it extends from central to northern Europe across Siberia to Japan and from Alaska to the Yukon, the Baffin Islands, Labrador, and subarctic Canada. In Europe, however, its distribution is arctic-alpine, with subspecies *palaeno* in Finland and Scandinavia and subspecies *europome* in the mountains of central Europe. It also has relict populations in the Vosges, Bavaria, the Alps, and the Carpathians. As we have seen, geographical distribution is normally defined in terms of the area or areas in which the species is found, examples being palaearctic, holarctic, circumpolar, Eurasian, Euro-Siberian, Caspian, Sino-Tibetan, Holomediterranean, Apennine, Sardo-Corsican, tropical African, Sundan, Papuan, palaeotropical, and pantropical. These terms are, however, seldom defined in a clear and precise manner. We shall conclude this review with some other terms much used in zoogeography: The term *autochthonous* indicates a taxon that originated in the area where it is found today; *endemic* describes a species whose range does not extend outside the species's area. For example, *Acanthobrahmaea europaea*, which we discussed at the beginning of this chapter, is a species endemic to Basilicata and Apulia, while the arctiid *Ochnogyna corsicum* is endemic to Sardinia and Corsica.

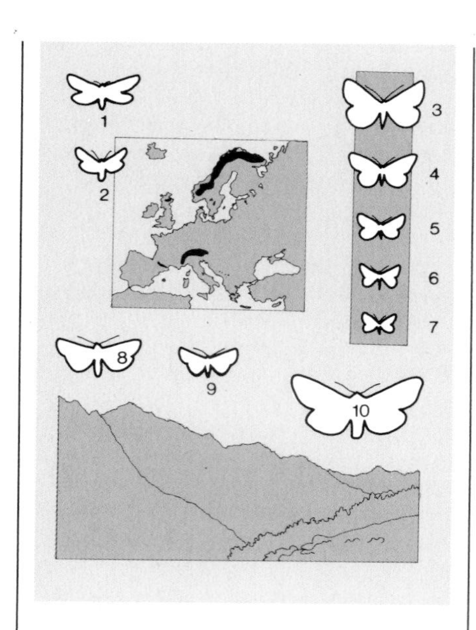

Plate 100 **Ranges of distribution: the arctic-alpine moths**

More than 50 European Lepidoptera have arctic-alpine distributions. Of these 23 are Macrolepidoptera, 22 are Microlepidoptera, and 10 are butterflies. The number of arctic-alpine Lepidoptera in palaearctic Asia has been less thoroughly studied but is known to include at least 24 species. In North America only 7 arctic-alpine Lepidoptera have been described, of which three have holarctic distribution. Among the moths the families most widely represented are the Noctuidae, Olethreutidae, and the Geometridae. All the species illustrated are found in Europe. (1) *Hepialus ganna* (Hepialidae); (2) *Zygaena exulans* (Zygaenidae), with ranges; (3) *Gnophos myrtillata*, Geometridae); (4) *Entephria flavicinctata* (Geometridae); (5) *Isturgia carbonaria* (Geometridae); (6) *Psodos coracina* (Geometridae); (7) *Pygmaena fusca* (Geometridae); (8) *Scotia fatidica* (Noctuidae); (9) *Caloplusia hochenwartii* (Noctuidae); (10) *Arctia flavia* (Arctiidae).

Dynamic Aspects of Geographical Range

From the geographical point of view, the distributional area of a species delimits a certain part of the earth's surface. As such it is subject to the action of the geological and climatic forces that slowly change the face of the earth and are also the principal factors tending to change ecosystems and thus promote the evolution of their component species. The species disperse and break up into many local populations, and these changes are reflected in their geographical range. For this reason the geographical range is considered a dynamic characteristic of the species. Changes of geographical range are generally of two types: expansion and reduction.

Expansion of the Geographical Range

Rapid expansions of geographical range occur quite frequently among butterflies and moths. In Italy, for example, the noctuid *Mythimna unipuncta* was a rare, localized species up to twenty or thirty years ago, but it is now rapidly spreading westwards, and in some years is so common as to be almost a pest. This species is also expanding strongly in Great Britain. It was reported for the first time in the Scilly Isles in 1957, and fifteen years later it had colonized the south coast of England as far east as Dorset. In the last ten years it is calculated that at least eighteen new species of Lepidoptera have established themselves in Sweden. In central-northern Europe the geometrid *Eupithecia sinuosaria* has moved the boundary of its range westwards by almost 1,600 km (900 mi) in about seventy years. It has been observed that there are certain years in which the number of new arrivals in a given area is particularly high, and it may be that this is related to a general improvement in the climate, of which the colonizing butterflies and moths take advantage. The phenomenon is an important one, and a reader of specialist periodicals will frequently come across lists of additional lepidopteran species appearing even in areas that have been well known for a long time. Some examples of expansion of geographical range will be found in Plate 101.

Geographical Barriers and Climates

The distribution of Lepidoptera and other terrestrial creatures depends in large part on climate; for butterflies and moths, which are coldblooded, temperature is particularly important. In many cases the distributional boundaries of individual species coincide more or less closely with winter or summer isotherms. In North America, for example, the northern boundary of winter survival for the Panamerican corn earworm (*Heliothis zea*) does not exceed a certain winter isotherm. Certain species, such as the European pine-shoot moth (*Rhyacionia buoliana*), succeed in maintaining their populations in areas where the minimum winter temperature is as low as −28°C. Tests involving the exposure of eggs, larvae, and pupae of different Lepidoptera to various temperatures for a period of twenty-four hours demonstrate great differences in resistance to cold. The eggs of the North American fall cankerworm (*Alsophila pometaria*, Geometridae) cannot endure temperatures below freezing; the larvae of the noctuid *Amathes ditrapezium* tolerate −15°C; the larvae of the arsid *Spilosoma nivea* tolerate temperatures of −25°C; and the pupae of the cecropia moth (*Hyalophora cecropia*) can survive at the colder-than-polar temperature of −70°C. High temperatures are of course also unsuitable for certain species; for example, the large white (*Pieris brassicae*), which is widely distributed in Europe and northwest Africa, is not found farther south than Israel, where it breeds only in the winter. This suggests that the factor determining its southern boundary is the high summer temperatures experienced farther south. Generally speaking, Lepidoptera with large geographical ranges covering a wide variety of

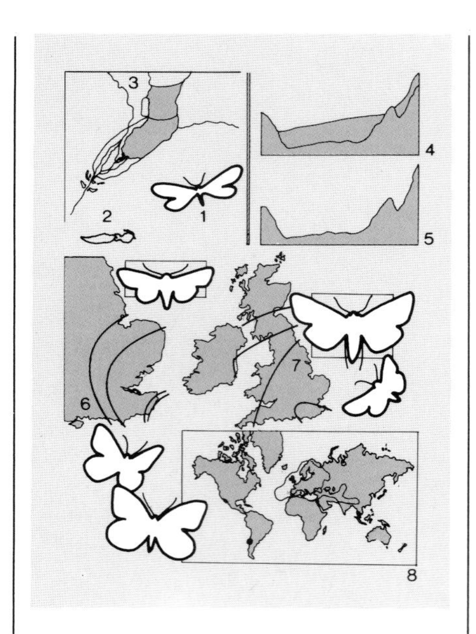

Plate 101 **Dispersion and colonization**

The geographical distribution of Lepidoptera changes under the influence of various factors. One example of this is the psychid *Solenobia triquetrella*, of which the adult male form is illustrated in Figure 1 and the case of the larva or of the adult in Figure 2. This insect is at present progressively working its way up the valleys in the Alps as the glaciers retreat. Figure 3 illustrates the retreat of the Rhône glacier between the years 1602 and 1961 and also indicates the places where *Solenobia* was found in 1961. J. Seiler has demonstrated that the ability of this insect to establish new colonies depends on its cytogenetic condition and on its method of reproduction. It is the tetraploid parthenogenetic females (indicated by grey dots in Figure 5) that lead the way into new territory at higher levels. Glacier valleys at lower levels, on the other hand, are colonized by diploid parthenogenetic females (indicated by white dots); after a variable time these are replaced by amphigonic diploid populations (black dots), which consolidate the new colony. Figure 4 illustrates the distribution of relict populations (always amphigonic) of *Solenobia* in *nunataks* (isolated refuges) at the peak of the last glaciation; Figure 5 shows the present position. Figures 6 and 7 illustrate two examples of rapid active dispersal by noctuid moths in Great Britain. In both cases progress is indicated by the series of boundary lines for different years. Figure 6 relates to the varied coronet (*Hadena compta*), and Figure 7 to *Polychrisia moneta*.
Figure 8 illustrates a recent case of passive dispersal. The large white butterfly (*Pieris brassicae*, Pieridae) has been spreading rapidly in Chile during the last few years. It has presumably been accidentally introduced by man from Europe or Asia, where it is native.

different climates are characterized by exceptional tolerance of extremes of temperature. (Similar considerations apply to other ecological factors.) The long-tailed blue (*Lampides boeticus*), for example, has an enormous geographical range, extending from Europe, Africa, Asia, and Australia to Oceania, and it flourishes equally well in Hawaii and at a height of 3,000 m (10,000 ft) in the Himalayas. A much narrower range of temperature tolerance is shown by species typical of high latitudes and those of alpine or tropical regions.

Climate is not, however, the only factor that controls the geographical distribution of butterflies. *Heliconius* populations, for example, are found in the American tropical forests but not in those of Africa or Asia. This restricted range can be explained if we assume that the evolution of this taxon (like that of most Lepidoptera) took place in an area isolated from the rest of the world by geographical barriers. The diffusion of animal species has always been, and still is, hindered by the presence of such barriers, to a varying extent depending on the mobility of the species under consideration. The ability to overcome quite formidable geographical barriers is an important characteristic of many lepidopterans, and enables them to expand their geographical ranges. This expansion, which is a continuous process, can be divided into the two successive phases of dispersal and colonization.

Dispersal

Generally speaking, the dispersal of Lepidoptera is favoured by their excellent mobility. Migratory butterflies can cover enormous distances, and nonmigratory species often travel some hundreds of kilometers. The sphingids show a high degree of mobility, as do some noctuids, many of whom have very large geographical ranges. The geometrids are weak fliers and hence have a limited capacity for dispersal. It has been calculated that the maximum cruising speed of Lepidoptera is about 25 km/hr (16 mi/hr) and that even strong fliers, such as the lady butterflies (*Vanessa*) and monarch and relatives (*Danaus*), seldom cover a distance of more than 1,000 km (625 mi); there are exceptions: one monarch (*D. plexippus*) is known to have travelled 2,112 km (1,320 mi) in forty six days. The implication is that distances of more than 1,000 km (625 mi) can be covered only with the help of a following wind. Dispersal depends to a large extent on chance, and the probability that a butterfly will reach an area very distant from its place of origin is extremely low. This helps to explain why the fauna of oceanic islands is so poor in Lepidoptera, typically comprising a very small number of species, most of which are from mobile groups.

The active process of dispersal is more rapid in some species than in others. In the bogworms (Psychidae), for example, it is very slow: The females are wingless, and the males move very little from their place of origin; so dispersal is left to the caterpillars. Many Lepidoptera have caterpillars that mine the roots of plants and consequently have very poor mobility, whereas leaf-eating caterpillars may be passively dispersed over long distance in a short period. The wandering tendencies of certain individuals are genetically determined, and dispersal may be activated by changes in the environment, such as a general deterioration of the climate. Dispersal is thus often an adaptive response to unfavourable environmental conditions. Human beings also sometimes help to enlarge the range of Lepidoptera, transporting them from one place to another, whether inadvertently or in the hope of gaining some benefit, as in the case of the pyralid *Cactoblastis cactorum* in Australia and the various species that produce silk. The most important of these is the silkworm.

Inadvertent human transport has quite often given rise to geographical distributions of an abnormal and apparently inexplicable nature, as can be

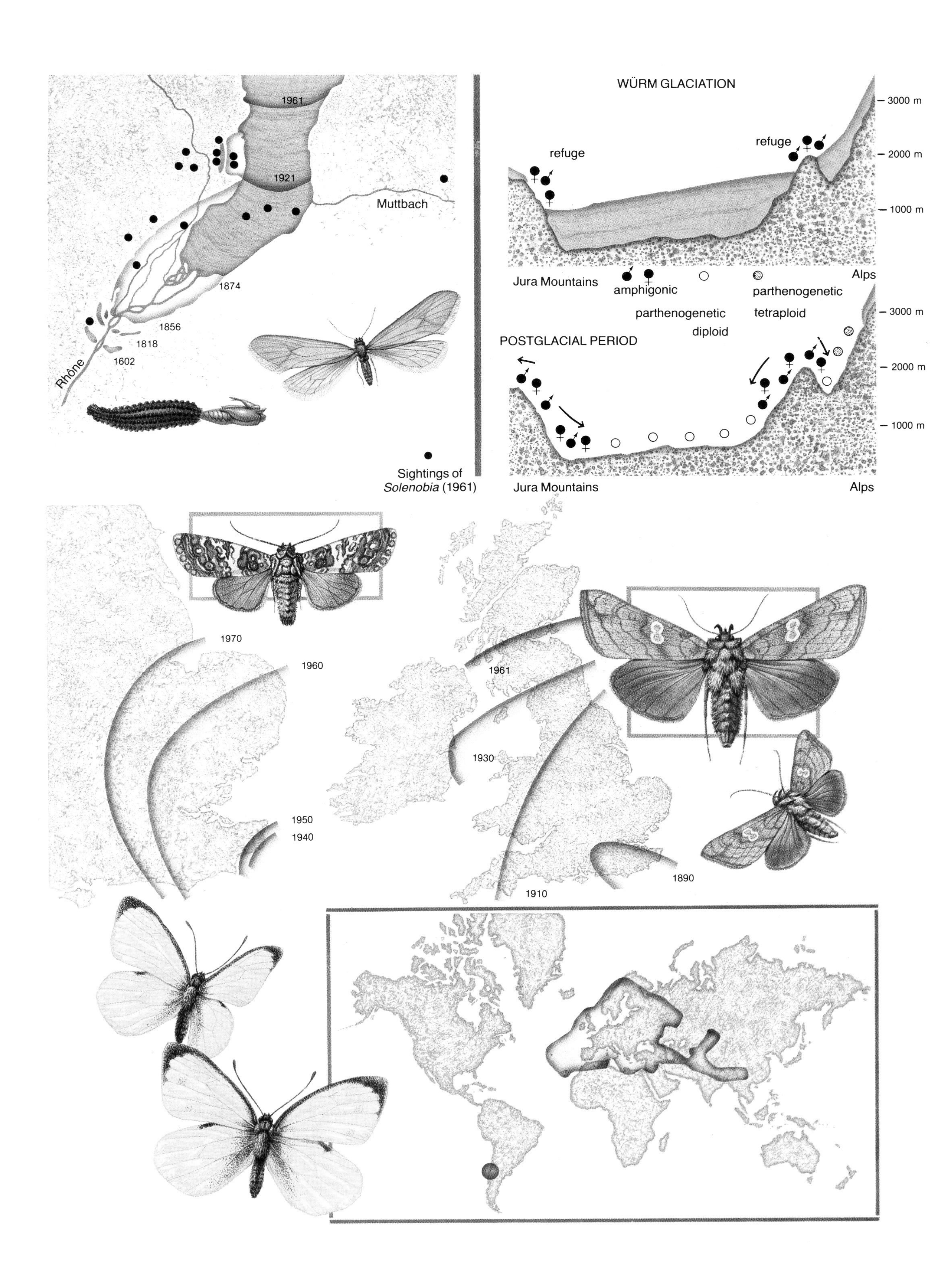
1961
1921
Muttbach
1874
1856
1818
1602
Rhône
Sightings of
Solenobia (1961)
WÜRM GLACIATION
refuge
refuge
3000 m
2000 m
1000 m
Jura Mountains
amphigonic
parthenogenetic
diploid
parthenogenetic
tetraploid
Alps
POSTGLACIAL PERIOD
3000 m
2000 m
1000 m
Jura Mountains
Alps
1970
1960
1950
1940
1961
1930
1910
1890

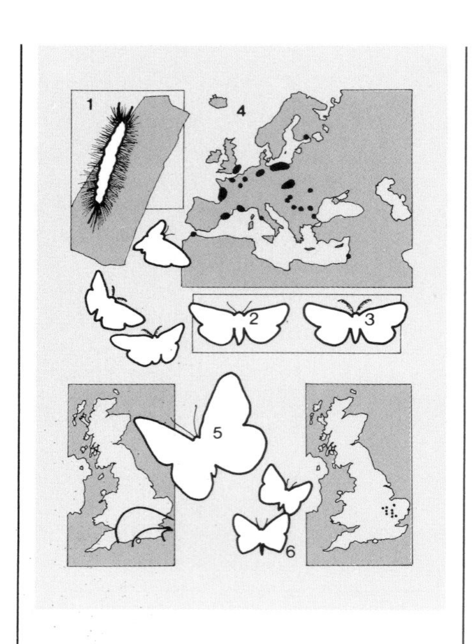

Plate 102A **Contraction of geographical range**

Over time the distributional range of a species expands or contracts. Contraction may be the result of many causes, including the demographic characteristics of the species itself, short-term and medium-term climatic changes, and artificial environmental changes. The lymantriid *Laelia coenosa* (1, caterpillar on *Phragmites*, 2, adult female, 3, adult male) is distributed discontinuously from Morocco to Japan and is highly dependent on damp environments, such as swamps, peat bogs, and brackish lagoons. Certain populations in central Europe are now in marked decline, and the English population became extinct in 1879 (Figure 4).
The date of extinction of many other species is known, especially in Great Britain (Plate 102B). Examples are the black-veined white (*Aporia crataegi*, Pieridae, 5), which became extinct in 1925, and the large copper (*Lycaena dispar*, Lycaenidae, 6), which disappeared definitively from Great Britain in 1848. Individuals of the first species, however, occasionally still reach England from Europe, and the second was reintroduced from the Netherlands in 1927.

seen from the following example: The thyridid *Banisia myrsusalis* is a sedentary species with a highly discontinuous tropical range, consisting of Central America and tropical Africa. Since cross breeding between the American and African populations is prevented by the geographical barrier of the Atlantic Ocean, one might expect that the populations of the two continents would have developed distinct subspecies. In fact, however, all the individuals resemble each other as closely as if they formed part of a single population. This similarity was unexplained until P. Whalley put forward a plausible hypothesis based on the fact that the geographical distribution of the insect is linked to that of its food plant. The caterpillars of *B. myrsusalis* feed on Sapotaceae, with a preference for the Central American *Achras sapote*, which is the tree from whose latex chewing gum was formerly extracted. Various attempts have been made to cultivate this tree in other tropical regions, including Africa. It accordingly seems probable that the eggs or caterpillars of *myrsusalis* have been accidentally transported to Africa with its food plant.

Colonization

Once the obstacles to dispersal have been overcome, the phase of colonization follows, and a new area is occupied. When an invading species has succeeded in establishing itself in a new environment, colonization is complete, and the species becomes a permanent resident. In many cases this outcome is accompanied by relative demographic stability so that the number of individuals that attain the adult state remains fairly constant from year to year. The process of colonization is delicate, and its success depends on the interaction of many factors. These include individual characteristics (such as ability to live on new food plants, tolerance of different climates, and fertility), characteristics of the entire group of the colonizing species (such as total numbers, numerical proportion between males and females, and genetic variability and mutational adaptability — see Plate 101), and equally the characteristics of the new environment (such as climatic factors, type of vegetation, and the number and effectiveness of competitive, predatory, and parasitic species). The success or failure of an attempted colonization is almost always unpredictable, and direct observation of the establishment of a lepidopteran in a new area is very seldom possible. However, an excellent opportunity for direct observation arises when Lepidoptera are artificially introduced into a country to control invasive plants. An interesting example is the colonization of a large part of Australia by the tiny pyralid moth *Cactoblastis cactorum*. This is a species of South American origin that has been used to stop the spread of two species of cactus, *Opuntia inermis* and *O. stricta*, introduced into Australia from the New World tropics during the nineteenth century. In their places of origin these two plants are kept under control by natural parasites; but in Australia no such parasites were available, and consequently they multiplied and spread rapidly, occupying many thousands of square kilometers in a period of eighty years. Queensland and New South Wales were particularly affected, livestock raising being the principal source of income in these states, and vast areas became unusable for this purpose. The situation was so serious that in 1920 the Australian government set up a special committee of experts to solve the problem. Various expedients were tried and failed, but finally a possible solution was proposed [in 1925, when the aggressive advance of the two cacti had reached its maximum point, having rendered useless some 243,000 sq km (95,000 sq mi)]: *Cactoblastis cactorum* was introduced from Argentina; 2,750 eggs were dispatched by sea, the caterpillars were unloaded, and the breeding process began. A few months later 2 million eggs of the second generation were spread over the plants, the caterpillars hatched out, and began to feed. The results

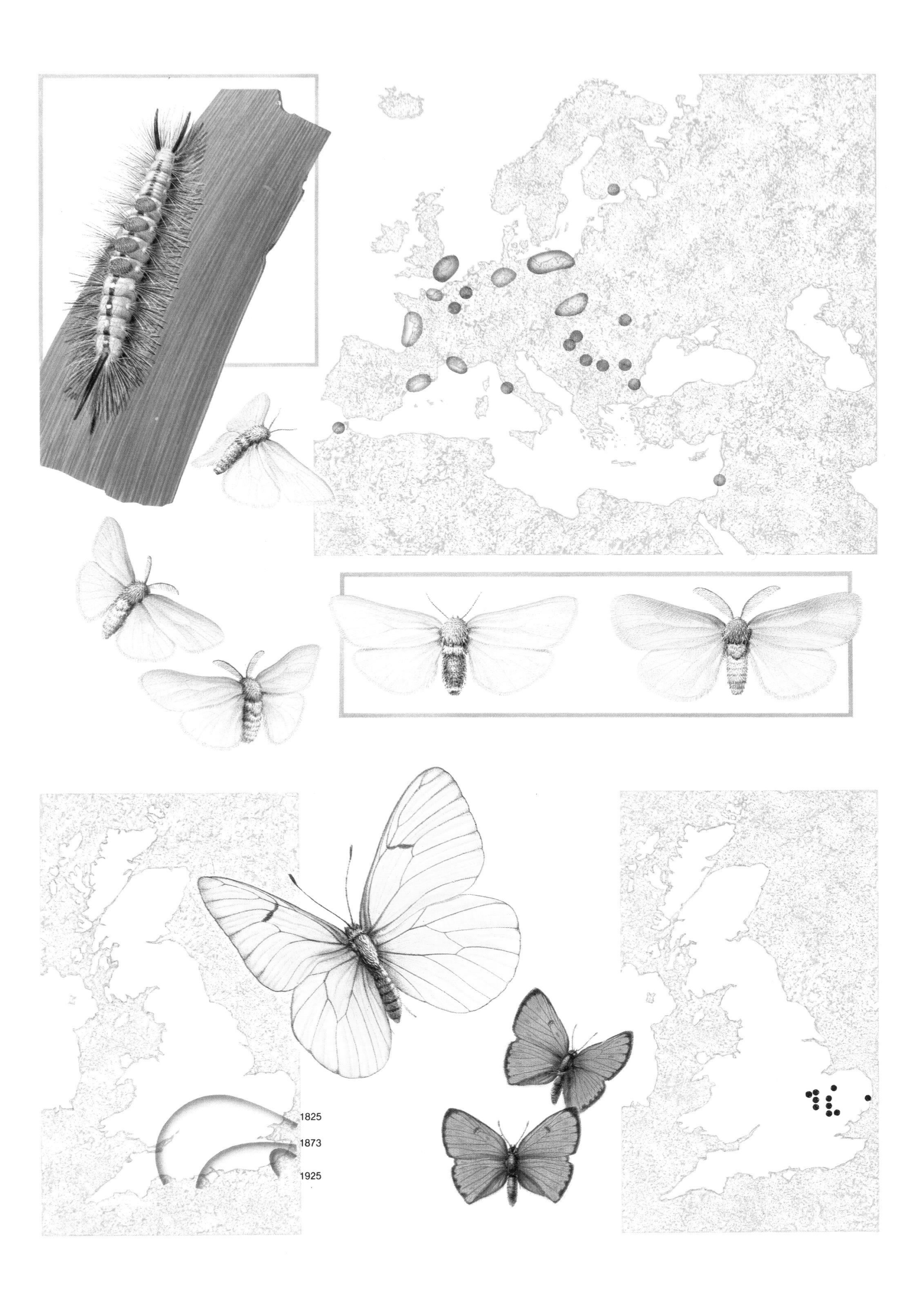
1825
1873
1925

Plate 102B **Contraction of geographical range**

In Great Britain the maintenance of records regarding the distribution and abundance of Lepidoptera has made it possible to confirm that during the last century at least thirty Macrolepidoptera have suffered severe loss or actual extinction. They include both species that were formerly widely distributed, such as the small rununculus (*Hecatera dysodea*), extinct in 1918, and *Aporia crataegi*, extinct in 1925, as well as very local species such as *Zygaena viciae* and the white prominent (*Leucodonta bicoloria*), which disappeared in 1927 and 1965, respectively. The causes leading to contraction of a range or total extinction include climatic change (as in the case of *Endromis versicolora*) and such artificial changes in the environment as reclamation of swamps [which has affected, among others, the gypsy moth (*Lymantria dispar*) and *Trachea atriplicis*] and alterations to the nature of woodlands [broad-bordered bee hawkmoth (*Hemaris fuciformis*)] and agricultural land [narrow-bordered bee hawkmoth (*H. tityus*)].
The upper half of the plate shows five extinct species, and the lower half shows five species whose numbers are in marked decline: (1) *Trachea atriplicis* (Noctuidae); (2) *Acronicta strigosa* (Noctuidae); (3) *Zygaena viciae* (Zygaenidae); (4) *Leucodonta bicoloria* (Notodontidae); (5) gypsy moth *Lymantria dispar* (Lymantriidae); (6) *Eriogaster lanestris* (Lasiocampidae); (7) *Hemaris tityus* (Sphingidae); (8) arctic skipper, chequered skipper *Carterocephalus palaemon* (Hesperiidae); (9) *Endromis versicolora* (Endromidae); (10) *H. fuciformis* (Sphingidae).

were astonishing. After only five years the cactus population collapsed, and ten years later, in 1935, both species were practically under ecological control. In the following years the reduction in the numbers of cacti caused a falling off in the population of *C. cactorum*. In due course, however, equilibrium was established between the host plant and its parasite.

Space does not permit a detailed study of all the factors that influence colonization; so we shall confine ourselves to commenting on the role of diet, which is certainly one of the most important. In the case of monophagous species it is obvious that their geographical distribution and the possibilities for colonization are closely linked to the distribution of the food plant. The European range of the two-tailed pasha (*Charaxes jasius*) is wholly contained within the range of the strawberry tree, on which its caterpillars feed. However, given similar conditions, monophagous species are at a disadvantage compared with oligophagous ones, and the latter in turn are at a disadvantage compared with polyphagous species. It should be noted that in certain cases colonization is accompanied by a change of diet. The European satin moth (*Stilpnotia salicis*) originally fed solely on the leaves of *Populus italica*. The insect was accidentally introduced into North America, where this tree is not found; the satin moth was on the point of dying out completely, but its caterpillars that hatched out on the leaves of the black cottonwood (*Populus trichocarpa*) began to feed on this new food plant. This change in eating habits was accompanied for several generations by very high mortality in the early stages of larval development, but in the long run it enabled the colonizing population to survive the critical period and to establish itself successfully in the new area. Similar cases have been observed in Germany, where several species [the lymantriid *Orgyia ericae*, the zygaenid *Rhagades pruni*, the arctiid *Rhyparia purpurata* and the emperor moth (*Eudia pavonia*)] arrived as colonists in the same area; all of them began to feed on *Calluna* although each of them was originally linked to a different food plant (respectively, *Salix*, *Prunus*, *Sarothamnus*, and *Rubus*). As in the case of *C. cactorum* two phases of colonization can be distinguished: In the first a small population is created, which, once it has adapted itself to local conditions, increases in numbers until it has saturated the environment; there may then be a decrease in the population (perhaps due to the beginning of a new stage of dispersal), and finally a state of more or less complete equilibrium is reached.

Reductions in Geographical Range

A progressive reduction of geographical range is a common event with many Lepidoptera; in extreme cases it may lead to species extinction. The contraction of geographical ranges is due to various causes — notably changes of climate and the effect of human activity, which is increasing all the time especially in temperate and tropical regions. A well-known case of range reduction due to climatic change is that of the large copper (*Lycaena dispar*), a species with wide distribution in moist habitats of the Palaearctic region. As early as the nineteenth century, a contraction of this insect's range in western Europe had left isolated populations in southern England, Holland, France, Germany, and northern Italy. In England the species became extinct owing to the reclamation of marshland and overcollecting. Range reduction and extinction are illustrated in Plates 102A and B; the relationship between environmental conservation and endangered species is discussed in the last chapter.

The Zoogeographic Regions

On the basis of the geographical distribution of the world's animals the

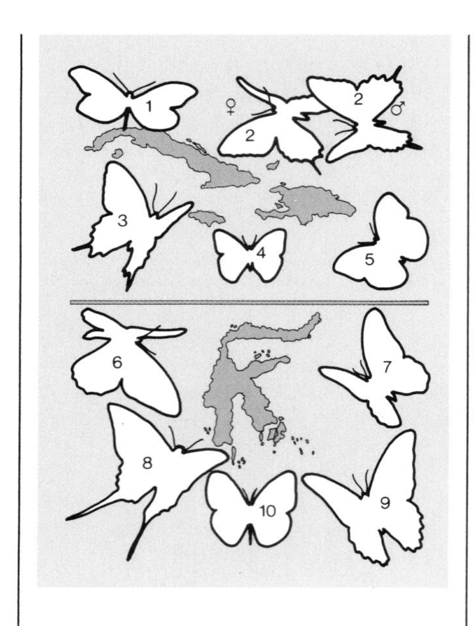

Plate 103 **Zoogeography of islands: endemicity**

Islands tend to be rich in endemic species, which have often originated by geographical speciation from related continental forms. The Antilles are a good example: 45 percent of their diurnal Lepidoptera (285 in all) are endemic. Cuba has 168 species in all, of which 11 are endemic to the island and 51 to the Antilles. Hispaniola is even richer, with 72 species endemic to the Antilles and 40 endemic to the island (which has a total of 175 species). Finally Jamaica, which is the smallest of the three major Antillean islands, has 120 butterfly species, of which 32 are endemic to the Antilles and 14 to the island itself.
(1) Cuban mimic (*Dismorphia cubana*, Pieridae), Cuba; (2) Poey's black swallowtail (*Papilio caiguanabus*, Papilionidae), left, female, right male, Cuba; (3) Gundlach's swallowtail (*Parides gundalachianus*, Papilionidae), Cuba; (4) Jamaican ringlet (*Calisto zangis*, Nymphalidae, Satyrinae), Jamaica; (5) red-splashed sulphur (*Phoebis avellaneda*, Pieridae), Cuba and Hispaniola.
In Wallacea the island of Celebes is particularly rich in cases of endemicity; of 26 swallowtails, for example, no fewer than 15 are endemic: (6) *Papilio velovis* (Papilionidae); (7) *Graphium encelades* (Papilionidae); (8) *G. androcles* (Papilionidae), (9) *G. milon* (Papilionidae); (10) Euploea mniszecki (Nymphalidae, Danainae).

land masses of our planet have been divided into a number of large territorial units, known as *zoogeographic regions*. These regions are distinguished from each other by each region's having its own characteristic fauna, that is, a complex of animals, all of which are subject to the geographical, ecological, and evolutionary factors that have arisen in the area in question. A meaningful description of the fauna of a region must include not only the exclusive taxa (its endemic species, genera, or families) but also the large groups, containing many species, and those with a wide distribution, found in the predominant biomes. Cosmopolitan species, for obvious reasons, cannot be used to distinguish one region from another. The regions are often delimited geographically by natural barriers such as oceans, major mountain ranges, or deserts; in many cases, however, they are separated not by a clear-cut boundary line but by a transition zone, across which some species may wander, leading to a certain degree of reciprocal exchange of fauna. Each is in fact a statistical entity based on average figures derived from a sample of species. Hence to the need to analyze the fauna quantitatively by means of indices such as the rate of endemicity or the coefficient of dissimilarity between different areas. In the nineteenth century A. R. Wallace defined six zoogeographic regions on the basis of the geographical distribution of vertebrates (mainly birds and mammals): the Palaearctic, the Nearctic, the Ethiopian, the Oriental, the Australian, and the Neotropical. Butterflies and moths are strong fliers, but their dispersion is not a matter of pure chance and is subject to various limitations: It is conditioned by climatic and biological factors and by historical factors such as continental drift, which has had so marked an effect on those species that are geographically specialized and consequently strongly influenced by geological developments in their natural habitat. Each zoogeographical region covers many millions of square kilometers and therefore cannot be a homogeneous area from the point of view of either geography or ecology; this means that the evolution of Lepidoptera will follow different lines in different parts of the region. It is thus necessary to make further use of the criteria employed to distinguish one region from another, to subdivide them, first, into subregions and, second, where required, into provinces on the basis of naturally occurring biogeographical discontinuities. In characterizing the lepidopteran fauna of a given region, we must rely on groups of species whose taxonomic structure and geographical microdistribution are very well known. Obviously not all families of Lepidoptera satisfy these requirements. We must therefore generally rely on two superfamilies of the butterflies, namely, the Hesperioidea and the Papilionoidea, which are incomparably better known both in terms of taxonomy and of their distribution. Some of the regions are also less well known than others. For example, the Neotropical, the richest of all, is largely unexplored from the entomological point of view. It is also the case that each region presents special aspects or zoogeographic problems that tend to predominate over the other distinguishing features. The special characteristics may concern the transition zones (this is true of the Oriental and Australian regions), the effects of glaciation on forest Lepidoptera (the Neotropical region), the abundance of boreoalpine species (for example in the Palaearctic and Nearctic regions), or the undersaturation of certain areas (as in the Ethiopian region). In the following pages we attempt a very brief summary of the Lepidoptera of each region.

Palaearctic Region

The Palaearctic region is the largest. It includes Europe, Africa north of the Sahara, the Near East, the Middle East, the northern part of the Arabian Peninsula, Asia north of the Himalayas, Korea, Japan, the Azores, and the Canary Islands.

Plate 104 **Island biogeography: number of species and geographical factors**

According to the theory of insular equilibrium put forward by Robert MacArthur and Edward Wilson, the number of species of Lepidoptera (or of any other kind of animal) present on a given island is determined by an equilibrium among its size, the immigration rate, the extinction rate, and its distance from a continental source area. This relationship is expressed in the graph (top right). The probability of extinction is linked to the area of the island. The larger the island and the more spatially diversified, the easier it is for the species to maintain a numerous population there. In the case of Lepidoptera there is plenty of evidence of the relationship between species richness and island area. The lower graph represents, on a logarithmic scale, the relationship between the number of species present in each of the main Mediterranean islands and their areas. The plate shows a selection of swallowtails typical of these islands (1–6) and a small sample of Mediterranean island species (all of which are endemic except for 12 and 13): (1) *Papilio toboroi* (Papilionidae); (2) pink-bodied swallowtail (*Pachliopta polydorous*, Papilionidae); (3) *Papilio bridgei* (Papilionidae); (4) *P. oberon* (Papilionidae); (5) *Graphium mendana* (Papilionidae); (6) *Ornithoptera victoriae reginae* (Papilionidae); (7) Corsican swallowtail (*P. hospiton*, Papilionidae); (8) Corsican grayling (*Hipparchia neomiris*, Nymphalidae, Satyrinae); (9) Corsican heath (*Coenonympha corinna*, Nymphalidae, Satyrinae); (10) Corsican fritillary (*Fabriciana elisa*, Nymphalidae, Nymphalinae); (11) Sardinian meadow brown (*Maniola nurag*, Nymphalidae, Satyrinae); (12) Eastern orange tip (*Anthocharis damone*, Pieridae); (13) western marbled white (*Malanargia occitanica*, Nymphalidae, Satyrinae); (14) Cretan argus (*Kretania psylorita*, Lycaenidae).

Climates within the region range from the arctic to the subtropical although temperate conditions predominate. Temperature and rainfall are consequently markedly seasonal. There is a latitudinal succession of vegetation types extending from Mediterranean scrubland or subtropical forest such as that found in central Japan to Arctic tundra, passing through areas of broad-leaved forests, mixed forests, and the vast expanse of the coniferous taiga. We must also note the alpine meadows, which shelter a rich community of Lepidoptera. The great plains of Asia stretch from 35 to 50°N latitude, from the Caspian Sea to the eastern boundary of Mongolia. They are characterized by various types of woodlands, steppes, and grasslands and also by deserts, which are linked with the Sahara via Iran, Iraq, and the northern parts of the Arabian Peninsula. In biogeographic terms the Palaearctic region is very similar to the Nearctic of North America. The two are in fact sometimes grouped together into a single Holarctic region. The relative poverty of the Palaearctic fauna is explained by its recent climatic history. During the Pleistocene ice extended no fewer than four times over the greater part of Eurasia, exterminating most Tertiary species. The effects of the Pleistocene on the evolution of palaearctic Lepidoptera are perfectly exemplified by mountain butterflies such as the Apollo (*Parnassius apollo*). After the last glaciation this butterfly, which is adapted to living in cold climates, probably avoided the increase in temperature in the southern part of its range by moving higher and higher up into the mountains, creating a number of small populations ecologically isolated from each other. This isolation prevented the exchange of genes between the populations, producing such a marked differentiation that almost all could be regarded as separate subspecies. This is probably why many of the most diversified taxa among the palaearctic Lepidoptera belong to the mountain fauna. In this habitat we find the greatest number of endemic species. At the family level endemicity is very rare. Apart from the endromids (which are represented by a single species) the only examples are the axiids and the heteroginids, the latter being confined to the Mediterranean area; the saturnids of the Agliinae subfamily are also endemic. Certain groups with mining caterpillars are important because of the large number and wide diffusion of their constituent species. Examples are the eriocraniids, the nepticulids, the tischeriids, the tortricids, the phaloniids, and the coleophorids — not to mention the geometrids, with some 3,000 palaearctic species. The Brephinae (endemic to the Holarctic), the Larentiinae, the Ennominae, and the noctuids (about 2,000 species) are of special interest, together with the Acronyctinae and the Amphipyrinae. All these groups of Lepidoptera are very well represented in the Nearctic region as well. Apart from the typically temperate taxa the palaearctic fauna includes often-endemic genera that have become highly diversified in the tropics. Among the Zygaenoidea and the Bombycoidea we may note *Zygaena*, *Procris*, *Somabrachys*, *Lemonia*, *Graellsia*, *Acanthobrahmaea*, and *Brahmaea* which are all genera centered in the Near East or in southern Europe. Also prevalent in southern Europe are the Scythridae.

According to a widely based study by A. S. Kostrowicki (1969), it appears that the Palaearctic region contains about 1,370 species of Papilionoidea, comprising 154 genera and divided among papilionids (85 species), pierids (161), nymphalids (over 720, of which 420 are Satyrinae), lycaenids (375), riodinids (17), and libytheids (1). Among the larger genera we may note the lycaenids *Polyommatus* (166 species), *Thecla* (48), and *Lycaena* (30) and the nymphalids *Satyrus*, in the broad sense (97 species), *Erebia* (83), *Melitaea* (61), *Argynnis* (59), *Lethe* (35), *Limenitis* (26), *Oeneis* (24), *Melanargia* (19), and *Pararge* (17). Important pierid genera are *Colias* (47), *Aporia* (24), *Pieris* (23), and *Anthocharis* (11). The papilionids include *Parnassius* (30) and *Papilio* (24). *Parnassius*

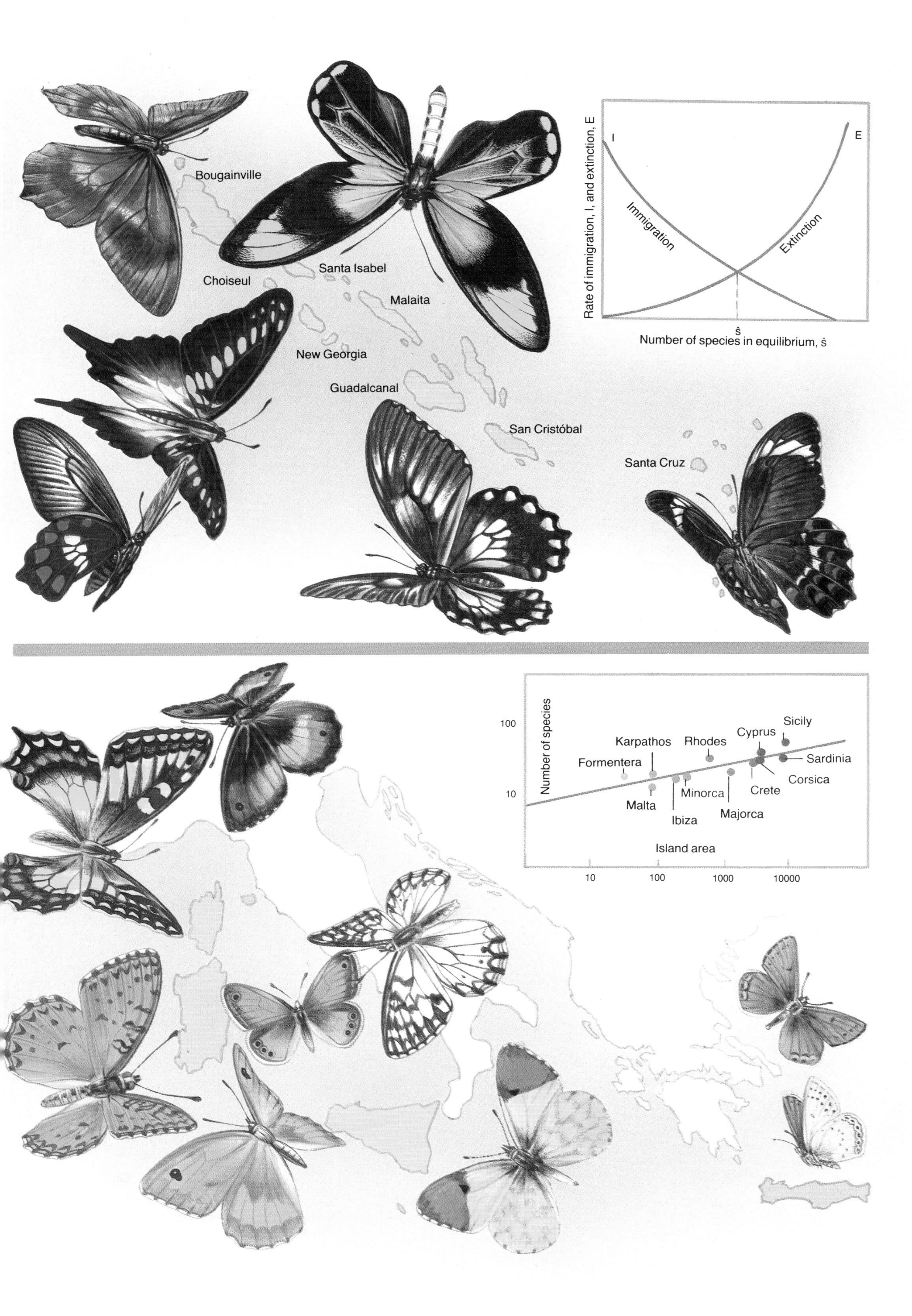
Bougainville
Choiseul
Santa Isabel
Malaita
New Georgia
Guadalcanal
San Cristóbal
Santa Cruz
Rate of immigration, I, and extinction, E
I
E
Immigration
Extinction
ŝ
Number of species in equilibrium, ŝ
Number of species
100
10
Karpathos
Rhodes
Cyprus
Sicily
Formentera
Sardinia
Corsica
Minorca
Crete
Malta
Ibiza
Majorca
Island area
10
100
1000
10000

Islands

Islands may be natural laboratories of evolution, where species often undergo the strangest modifications in terms of physical adaptation to environment and in behaviour. This phenomenon is principally due to the fact that, when animals and plants reach an island, they are relatively free from certain environmental constrictions or selective agencies, such as predation and — above all — competition, that retard the progress of evolution on the continents. Isolation and especially the extreme isolation found in oceanic islands also facilitate the rapid differentiation of insular populations and give maximum scope to the adaptive capabilities of the species, to the point of permitting the extreme expression of such biological characters as dwarfism, gigantism, loss of wings (aptery), and exceptional longevity. Like other insects and birds certain insular Lepidoptera have rudimentary wings and are incapable of flight. Examples are the noctuids *Brachypteragrotis patricei* of Amsterdam Island, *Dismorphinoctua cunhaensis* of Tristan da Cunha, and *Peridroma goughi* in the neighbouring Gough Island. Cases of gigantism are also common, especially among the Sphingidae. Certain taxa also present deviant adaptations: *Eupithecia* is a large genus of Geometridae with worldwide distribution; its caterpillars are normally vegetarian, but in Hawaii no fewer than six species have carnivorous caterpillars, which live on small insects. In certain cases a single ancestral genus or species may, in an island environment, give rise to an extraordinary evolutionary radiation. In the Hawaiian Islands, with their total area of about 16,700 sq km (6,500 sq mi), we find at least 26 endemic *Agrotis* noctuids. For purposes of comparison the Italian region of Lazio, with an area of 17,200 sq km (6,700 sq mi), has only 10 *Agrotis*, none of which is endemic. In addition to favouring species formation, the island environment favours the perpetuation of species that have become extinct elsewhere and the development of numerous endemic species. Hawaii, for example, is the home of approximately 1,000 endemic Lepidoptera. Darwin and Wallace had both noticed that island faunas differ from those of continents both in quality and quantity. The relative poverty and imbalance of island faunas are particularly marked on oceanic islands. The Krakatoa disaster gave a tremendous new impetus to the study of the quantitative aspect of island faunas, and to interest in the processes of species colonization and dispersal. In 1883 the volcano of Krakatoa, an island situated between Java and Sumatra, erupted catastrophically, halving its surface area and destroying all animal and plant life on the half that remained. This event provided a unique opportunity to study the colonization of an empty island. Research showed that within the next fourteen years Krakatoa was colonized by 61 plants and 132 insects. After thirty-six years there were 27 birds, a number that remained constant for many years thereafter.

is one of the most distinctive genera of the Palaearctic region, and includes a total of 31 montane species, 28 of which are endemic to the Palaearctic, while 2 (*P. eversmanni* and *P. phoebus*) have circumpolar holarctic distributions, and only one (*P. clodius*) is endemic to the Nearctic region. Most of the palaearctic *Parnassius* are concentrated in the area to the south of the mountain ranges that run from the Pamir to Bhutan, with a northward extension into Tibet and Mongolia (they are particularly abundant in the Altai Mountains). Only three *Parnassius* are found in Europe (*P. apollo*, *P. phoebus*, and *P. mnemosyne*). Of the six provinces into which the Palaearctic region is subdivided, the richest is the Tibetan, with 638 species of Papilionoidea, of which 82 are endemic; it is followed by Central Asia, with 488 species, 104 of which are endemic, the Euro-Siberian province, with 477 species, 71 of which are endemic, the eastern palaearctic, with 372 species, 56 of which are endemic, western Asia, with 342 species, 55 of which are endemic, and finally North Africa, with 141 species, 23 of which are endemic.

The Tibetan province stretches from southwest China to the Nan Shan Mountains and from Tibet to Kashmir; its wealth of Lepidoptera is due to the penetration of many taxa from the neighbouring Oriental region. (Of the 132 genera found in the Tibetan province, 71 are of oriental origin.) The species that have dispersed widely across a large part of the Palaearctic region include the pierids *Pieris napi*, *Anthocharis cardamines*, and *Gonepteryx rhamni*, the nymphalids *Argynnis paphia*, *Apatura ilia*, and *Araschnia levana*, the lycaenids *Lycaena phlaeas* and *Celastrina argiolus*, and the libytheid *Libythea celtis*.

Nearctic Region

The Nearctic region includes North America and the northern part of Mexico above 20°N, the latitude that marks the approximate boundary with the Neotropical region. In both climatic and zoogeographic terms the Nearctic region is very similar to the Palaearctic, and hence Lepidoptera of the two regions are also similar although the Nearctic has relatively fewer species in all. This impoverishment is particularly marked at high altitudes and in certain temperate areas; it is largely the result of the Pleistocene glaciations.

However, the resemblances between the Nearctic and Palaearctic regions are more important than the differences, and some zoogeographers prefer to combine them into a single, enormous Holarctic region. In our discussion of the geographical ranges of species we mentioned many examples of holarctic and/or circumpolar Lepidoptera. Here we need only mention a few groups of noctuids, such as *Plusia*, *Lithacodia*, *Amphipyra*, and *Apamea*, all of which have holarctic distributions; *Anarta cordigera*, *A. melanopa*, and the pyralid *Crambus perlella* all have typically circumpolar ranges. Certain families of Microlepidoptera that are well established in northern areas consequently have many nearctic species; they include the eriocraniids, the nepticulids, the tischeriids, the phaloniids, and the coleophorids. Finally we must note the presence of species such as the European cabbage white (*Pieris rapae*), the European skipper (*Thymelicus lineola*), and the gypsy moth (*Lymantria dispar*); these three are of European origin but have been introduced into North America by man. A peculiarity of nearctic Lepidoptera is their wealth of neotropical species, which detracts from their holarctic character to an appreciable degree. The pierids *Eurema proterpia* and *E. mexicana*, the papilionids *Papilio cresphontes* and *Battus polydamas*, the nymphalids *Danaus plexippus*, *D. gilippus*, *D. eresimus*, *Dryas iulia*, *Heliconius charitonius*, *H. erato*, and *Phyciodes vesta*, and the lycaenids *Cyanophrys miserabilis* and *Calycopis isobeon* are some of the typically neotropical taxa that have come from Mexico or the Antilles and

Observations from many other island situations gradually brought to light the ways in which certain variables exercise a quantitative control over the development of island fauna. As shown by Robert MacArthur and Edward Wilson, the number of species on an island is closely related to its area. If we express this relationship on a double logarithmic scale (as the lower graph in Plate 104), the points plotted form an almost straight line. The analysis of the relationships among area, number of species, and degree of isolation has led to the formulation of an important quantitative theory of the population of islands. This theory, which we owe largely to MacArthur and Wilson, is known as "the model of island biogeography." The model establishes that the number of species on an island is in equilibrium (which means that it will not change with the passage of time) when the immigration rate, or number of new species arriving on the island in a given unit of time, equals the extinction rate, or number of species already established on the island that become extinct in the same unit of time. (See Plate 104.) The island's area and its distance from the nearest continent are the two main geographical factors that control the quantitative equilibrium of insular species. The surface area influences the extinction rate (which is higher on small islands than it is on large ones), while the distance affects the immigration rate (a distant island will receive fewer immigrants in a given unit of time than will one relatively close to the continent).
If we compare two islands that differ only in area, we find the following: The extinction rate is higher for the smaller island; the immigration rate is greater for the large island; and the equilibrium species number is higher for the larger island.
If we compare two islands that differ only in their distance from a continent, we find the following: The immigration rate is higher for the nearer island; the rate of extinction is the same for both islands; and the equilibrium species number is greater for the nearer island. The island-biogeography model, which we have presented in a much simplified form, has been confirmed by numerous observations, including those relating to Lepidoptera (Plate 104). The model is very useful for purposes of quantitative zoogeographical analysis both of real islands and of areas such as host ranges or mountain tops, which are ecological islands.

have established themselves in the Nearctic region, especially in the southern part extending from California to Florida. Other noticeable invading species from the Neotropical region are the Megalopygids and the Mimallonids. Both families also contain numerous species endemic to the Nearctic region. Some other striking examples may be found among the saturnids and the noctuids. The saturnids, which include many very large endemic species, are represented by species such as luna moth (*Actias luna*), promethea moth (*Callosamia promethea*), royal walnut moth (*Citheronia regalis*), and polyphemus moth (*Antheraea polyphemus*). As well as important noctuid genera such as *Acronicta*, *Hadena*, *Oncocnemis*, *Euxoa*, *Heliothis*, and *Mamestra*, we also find *Catocala*, *Autographa*, and *Schinia*, which contain many more nearctic than palaearctic species.

The butterflies of North America amount to about 700 species. The Hesperiidae are most numerous with 250 species, followed by the Nymphalidae with over 200, the Lycaenidae with about 100, the Pieridae with over 50, the Papilionidae with 25, the Riodinidae with 24, the Megathymidae with 20, and the Libytheidae with 2 species. Additional species are found in Mexico north of 20°N latitude.

The western states of North America are noticeably richer in species than are the eastern: Colorado has more than 250 species, whereas Pennsylvania has only 142 and Florida about 185. In the same way the south of the country is richer than the north.

Ethiopian Region

The Ethiopian region comprises Africa to the south of the Sahara, the southern part of the Arabian Peninsula, and a number of islands: Madagascar, the Comoros, the Mascarenes, and the Seychelles. Owing to the peculiar nature of their fauna, these islands are treated as forming a separate Malagasy subregion, which is quite distinct from the continent of Africa in zoogeographical terms. The great deserts of the Sahara and of Arabia form a very definite barrier between the Ethiopian region and the Palaearctic. These deserts have very few resident butterflies and moths. The Sahara is inhabited by only about 20 butterfly species; they include palaearctic forms such as the Bath white (*Pontia daplidice*) and the clouded yellow (*Colias croceus*) and also tropical African species typical of arid savannah country, such as the smoky orange tip (*Colotis euippe*), the African migrant (*Catopsilia florella*), and the broad-bordered grass yellow (*Eurema brigitta*). Africa has a fairly high average altitude because of the presence of many tablelands, normally above 1,000 m (3,300 ft), rising to 2,000 m (6,600 ft) in Ethiopia. As we travel away from the Equator, we pass from a humid tropical climate to a dry tropical climate. Along the line of the Equator itself we may note a gradient of rainfall, with a maximum value in the west of the continent and a minimum value in the east. The distribution of the principal types of vegetation (rain forests, savannah, and grasslands) is very clearly defined geographically. Generally speaking, the tropical forests, which extend from Sierra Leone to the Rift Valley, are surrounded by expanses of moist, grassy savannah, which are in turn surrounded by dry steppelike plains, which are more extensive to the north of the Equator.

Savannah (known locally as "bushveld"), extensive dry woodlands ("karoo"), and a sort of Mediterranean scrubland are typical of southern Africa. The slopes of the Rift Valley and of the tablelands of Ethiopia and the Cameroons are characterized by vertical vegetation zones, beginning with the savannah (dominated by *Themeda* grasses) and followed by mountain forests of juniper and podocarpus, the bamboo forest (genus *Arundinaria*), a zone dominated by heathers, and finally the extraordinary African alpine zone, which is dominated by giant lobelias and giant

groundsels. It must be said that the butterflies living at high altitudes in the African mountains are of small zoogeographic interest. The holarctic *Erebia* and the other Satyrinae, which characterize high-altitude lepidopteran fauna in almost all other parts of the world, are completely missing. Many species present on the mountains (such as *Cupido aequatorialis*, *Colias electo*, *Papilio demodocus*, *Danaus chrysippus*, and *Mycalesis dentata* and also numerous *Acraea* and *Catopsilia florella*) are in fact also found at low altitudes on the steppes and are very widely distributed in Africa. Moths are much less numerous and seem to include some that are either autochthonous or at least restricted to high altitudes. Examples are *Phryganopsis elongata* and *Episilia rhodopea*.

In Madagascar the principal types of vegetation succeed one another from west to east in accordance with the increase in annual rainfall. The island is roughly divisible into three longitudinal zones, characterized respectively by xerophyllous vegetation, moist savannah, and rain forests. The main climatic changes are linked to variations in rainfall, which is highly seasonal over large areas of Africa. The Lepidoptera in such areas display conspicuous differences in coloration between the generations of the dry season and those of the rainy season (Plate 17). In addition it is probable that variations in rainfall have determined the geographical distribution of most African butterflies. From a historical point of view it should be noted that Africa, like the other major tropical areas, has been affected by the Pleistocene glaciations, which produced alternating pluvial and interpluvial ages, causing the area of low-altitude tropical forests to expand and contract correspondingly. The fragmentation of the rain forest, which is still the home of the richest and most diversified communities of African Lepidoptera, certainly caused alterations in the ranges of many forest butterflies.

Reliable information is lacking about the number of taxa in Africa; many areas have been inadequately explored, and we can present only a superficial picture of the position, especially as regards the moths. The following families are endemic to the Ethiopian region: the Anomologidae (South Africa), the Chrysopolomidae, and the Thyretidae, together with the subfamilies of the Pompostolinae (Zygaenidae), the Ludinae (Uraniidae), and the Lipteninae (Lycaenidae). The following groups are

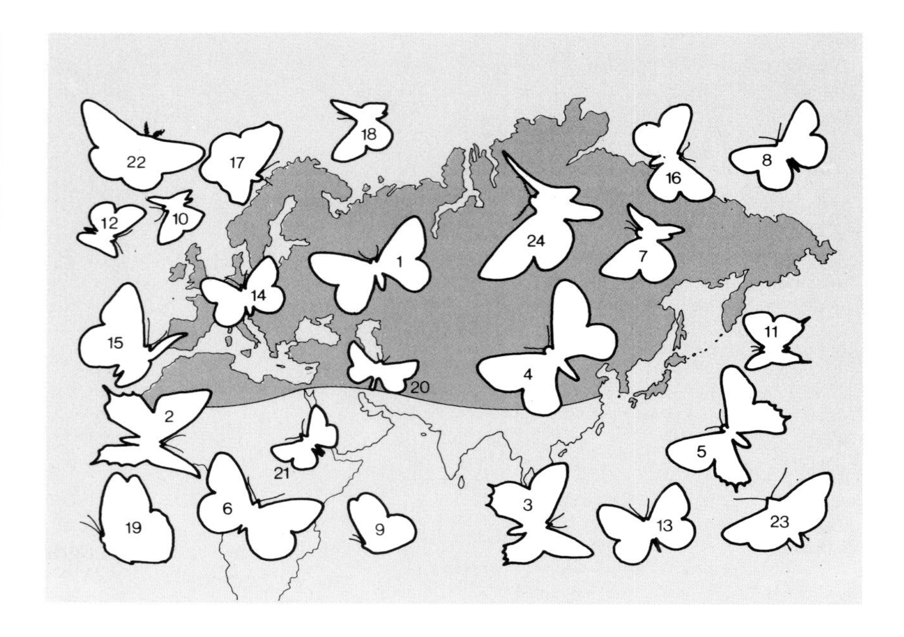

Plate 105 **Palaearctic region**

This is the largest of all the regions in terms of area. It is situated in the northern hemisphere, and is the home of a very large number of Lepidoptera adapted to temperate or cold climates. Its most characteristic Lepidoptera are therefore found in mountainous areas or at high altitudes; tropical elements are relatively scarce. The plate shows a selection of Palaearctic Macrolepidoptera, many of which occur in Europe: (1) Eversmann's parnassian (*Parnassius eversmanni*, Papilionidae); (2) southern swallowtail (*Papilio alexanor*, Papilionidae); (3) *Allancastria cerisyi* (Papilionidae); (4) *Parnassius charltonius* (Papilionidae); (5) *Luehdorfia puziloi* (Papilionidae); (6) black-veined white (*Aporia crataegi*, Pieridae); (7) palaeno sulphur, moorland clouded yellow (*Colias palaeno*, Pieridae); (8) *C. wiskottii* (Pieridae); (9) eastern orange tip (*Anthocharis damone*, Pieridae); (10) purple-shot copper (*Heodes alciphron*, Lycaenidae); (11) *Japonica saepestriata* (Lycaenidae); (12) chalk-hill blue (*Lysandra coridon*, Lycaenidae); (13) *Melanargia titea* (Nymphalidae, Satyrinae); (14) Scotch argus (*Erebia aethiops*, Nymphalidae, Satyrinae); (15) great banded grayling (*Brintesia circe*, Nymphalidae, Satyrinae); (16) *Oeneis mongolica* (Nymphalidae, Satyrinae); (17) red admiral (*Vanessa atalanta*, Nymphalidae, Nymphalinae); (18) Cynthia's fritillary (*Euphydryas cynthia*, Nymphalidae, Nymphalinae); (19) cardinal (*Pandoriana pandora*, Nymphalidae, Nymphalinae); (20) arctic skipper, checkered skipper (*Carterocephalus palaemon*, Hesperiidae); (21) dingy skipper (*Erynnis tages*, Hesperiidae); (22) emperor moth (*Saturnia pavonia*, Saturniidae); (23) large yellow underwing (*Noctua pronuba*, Noctuidae); (24) puss moth (*Cerura vinula*, Notodontidae).

Past history

Over 200 million years ago the ancient supercontinent of Pangaea broke up, and its fragments began to drift apart. These events have played an important role in the evolutionary history of the flora and fauna of our planet, influencing the differentiation and the geographical distribution of the various species and so conditioning the evolution of the ecological communities and the competition of the various faunas that we can see today. The evolutionary history of the Lepidoptera began about 140 million years ago and runs parallel with the geological and geographical changes due to continental drift. In spite of the lack of a satisfactory fossil record for Lepidoptera it is reasonable to suppose that the evolutionary history and the present distribution of these insects have been more or less closely linked with the history of the development of the continents, in keeping with the distribution of the groups of flowering plants that provide immature Lepidoptera with their main source of food. Similarities between the faunas of areas now far apart (such as South America and Africa or South America and Australia) can be at least partly explained in terms of continental drift, and so can the complex patterns of distribution that arise in certain physical areas such as Wallacea. However, the connection between continental drift and the biogeography of Lepidoptera has not yet been thoroughly investigated, and the dispersal mobility of these insects remains one of the most important factors in explaining most of the present pattern of geographical distribution. In the Mesozoic the landmass on our planet was divided into two supercontinents, Laurasia to the north and Gondwana to the south; between them lay the Tethys Sea, of which the present Mediterranean is a remnant. Laurasia included the areas now known as Eurasia, Greenland, and North America; Gondwana included the Antarctic, Australia, New Guinea, New Zealand, Madagascar, Arabia, Africa, and South America. It is thought that the breakup of Gondwana took place before that of Laurasia; Gondwana certainly gave rise to the greater number of continental islands. Below is a chronological table showing some of the main geological events that are of special biogeographical interest. It concentrates on the southern regions, which are now the richest in Lepidoptera.

Millions of years ago	
180	Opening up of the central Atlantic between Africa and North America
180 160	Beginning of the separation of Africa and South America
140	India and Madagascar begin to detach themselves from the rest of Gondwana
120	India and Madagascar completely detached from the rest of Gondwana
100	India separates from Madagascar and begins to migrate rapidly towards the north. Africa and South America continue to move apart
70 65	Africa now completely separated from South America
55	Australia and New Guinea begin to detach themselves from Antarctica
40	Australia and New Guinea are separated from Antarctica. India collides with Asia
20	Australia and New Guinea approach Asia
15 10	Beginning of the collision between New Guinea and Australia and the lands of Asia in the area of Wallacea. In this first phase the Peninsula of Sula (a western portion of the New Guinea platform, from which islands such as Sula and Obi emerge today) attached itself to Celebes, forming the eastern limb of the island in its present form. Other parts of Wallacea were also affected by the collision
5–3	First Ceram and then Timor collide with the arc of volcanic islands in the Banda Sea
2	Pleistocene

very well represented in terms of both quantity and quality: Metarbelidae, Xyloryctidae, Zygaenidae, Drepanidae, Lemoniidae, Lasiocampidae, Hypsidae, and among the Nymphalidae the Acraeinae. This region undoubtedly has its closest affinity with the Oriental region in terms both of families (Methacandidae and Hyblaeidae) and of genera (which are numerous among the Papilionoidea), the many examples including *Belanois*, *Colotis*, *Melanitis*, *Ypthima*, *Acraea*, *Charaxes*, *Kallima*, *Neptis*, *Cyrestis*, *Hypolycaena*, *Castalius*, *Azanus*, and *Abisara*. Certain neotropical species and genera (such as *Hypanartia paullua*, *Actinote*, *Apaustus*, *Carystus*, *Eunica*, and *Megalura*) have counterparts in the African continent (respectively, *Antanartia delius*, *Acraea*, *Acleros*, *Ploetzia*, *Asterope*, and *Cyrestis*); the genera *Brephidium* and *Zizula* are common to both regions. Affinities with the Palaearctic region involve only a few taxa, most of which are of Palaearctic origin: *Pieris*, *Pontia*, *Colias*, *Euchloe*, *Pararge*, *Issoria*, and *Melitaea*. A noteworthy exception is the two-tailed pasha (*Charaxes jasius*), a tropical species found in the Mediterranean area, which is present also in Africa and belongs to one of the largest genera of Ethiopian Nymphalidae. The most important recent contribution to the zoogeography of the African butterflies has been made by R. H. Carcasson (1964). From this work and from D. F. Owen's splendid *Tropical Butterflies* (1971), which contains a wealth of analysis of the evolutionary, ecological, and geographical aspects of the African butterflies, we have taken the numerical data and the brief zoogeographical notes that follow.

Where we give a parenthetical pair of figures, the first indicates the number of forest species found in the African continent, and the second indicates the number of species found in open country. The Ethiopian region is the home of more than 2,600 species of butterflies. The Lycaenidae (1,092 species) are represented by genera such as *Anthene* (100/22), *Epitola* (72/6), *Liptena* (65/2), *Telipha* (28/0), *Alaena* (0/17), *Virachola* (10/11), and *Spindasis* (7/20). The Nymphalidae have 938 species divided as follows: Nymphalinae, 500, Satyrinae, 245, Acraeinae, 168, and Danainae, 25. They also comprise some twenty important genera, including *Amauris* (14/0), *Bicyclus* (70/9), *Ypthima* (2/10), *Neocoenyra* (0/13), *Charaxes* (65/15), *Cimothoe* (62/0), *Euryphura* (10/0), *Euriphene* (55/1), *Bebearia* (53/0), *Najas* (40/0), *Neptis* (36/4), *Precis* (11/14), *Bematistes* (20/0), and *Acraea* (82/54). Among the Hesperiidae (423 species) we may note *Colaenorrhinus* (18/0), *Gorgyra* (20/0), *Kedestes* (0/11), and *Coeliades* (9/4). Among the Pieridae (150 species) we may note the genera *Belenois* (15/8), *Mylothris* (18/3), and *Colotis* (1/35). The Riodinidae (12 species) include the *Abisara* (9/0) and the endemic Malagasy *Saribia*. The Papilionidae have 85 species; of those belonging to *Papilio* (36/2) we must mention the enormous African giant swallowtail (*P. antimachus*) and the other magnificent swallowtails, *zalmoxis*, *rex*, *phocas*, *hesperus*, *zenobia*, and *nireus*; noteworthy species among the *Graphium* (28/3) are the species *G. ridleyanus*, *leonidas*, and *illyris*. There are many butterflies with mimetic colouring, including the famous mocker swallowtail (*P. dardanus*), the white-barred acraea (*Acraea encedon*), the plain tiger (*Danaus chrysippus*), the mimic (*Hypolimnas misippus*), and *Bematistes* (see plate 81). The richest geographical areas are the Cameroons (1,150 species), Zaïre (1,000), the Ivory Coast (750), and Ghana (500); Madagascar (250) and South Africa (200) are relatively poor. The areas with the most numerous and diversified Lepidoptera are the low-altitude tropical rain forests, but there is a marked difference between those in West Africa and East Africa: The ratio in terms of numbers of species is 12.5:1. The eastern forests are today, as they were in the past, more fragmented than are those in the west, and this may explain both the relative poverty of species (98) and the presence of many endemic

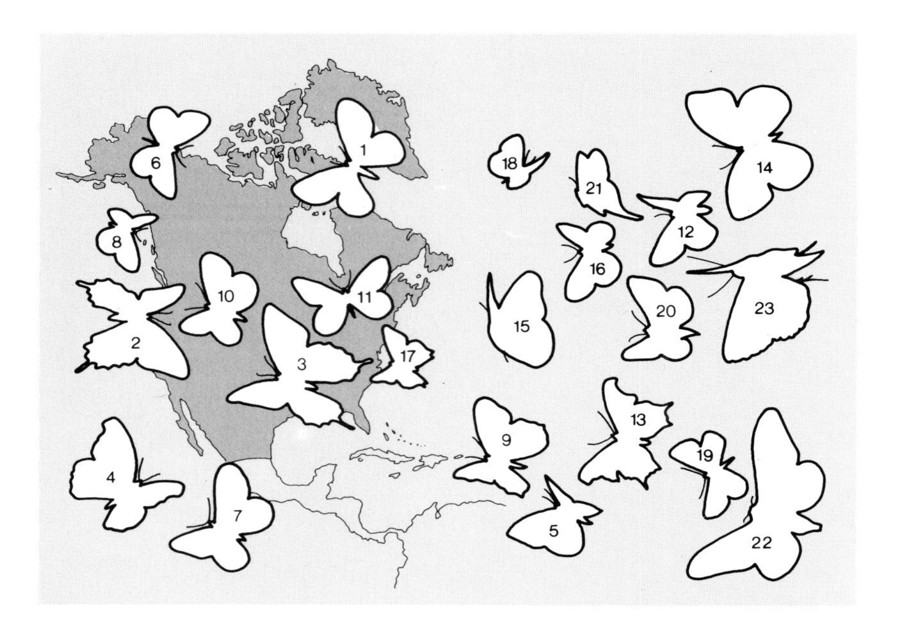

Plate 106 **Nearctic region**

The Lepidoptera illustrated in this plate belong to the Nearctic region, whose climate and fauna are very similar to those of Eurasia. The northern part of the region is covered by huge, cold expanses of tundra and taiga, where there are relatively few butterflies or moths. As we move further south, however, the number of species found increases rapidly. In the strip of territory that runs from southern California to Texas and in the southern tip of Florida we find some typically tropical species, which come from Mexico or from the Antilles.
(1) Clodius parnassian (*Parnassius clodius*, Papilionidae); (2) spicebush swallowtail (*Papilio troilus*, Papilionidae); (3) western tiger swallowtail (*P. rutulus*, Papilionidae); (4) southern snout (*Libytheana carinenta*, Libytheidae); (5) clouded sulphur (*Colias philodice*, Pieridae); (6) Sara orange tip (*Anthocharis sara*, Pieridae); (7) pine white (*Neophasia menapia*, Pieridae); (8) California ringlet (*Coenonympha california*, Nymphalidae, Satyrinae); (9) pearly eye (*Enodia portlandia*, Nymphalidae, Satyrinae); (10) theano alpine (*Erebia theana*, Nymphalidae, Satyrinae); (11) California arctic (*Oeneis chryxus ivallda*, Nymphalidae, Satyrinae); (12) chalcedon checkerspot (*Euphydryas chalcedona*, Nymphalidae, Nymphalinae); (13) question mark (*Polygonia interrogationis*, Nymphalidae, Nymphalinae); (14) great spangled fritillary (*Speyeria cybele*, Nymphalidae, Nymphalinae); (15) American painted lady (*Vanessa virginiensis*, Nymphalidae, Nymphalinae); (16) atala (*Eumaeus atala*, Lycaenidae); (17) tailed copper (*Lycaena arota virginiensis*, Lycaenidae); (18) Miami blue (*Hemiargus thomasi*, Lycaenidae); (19) Mormon metalmark (*Apodemia mormo*, Riodinidae); (20) Strecker's giant skipper (*Megathymus streckeri*, Megathymidae); (21) *Codactractus alcaeus* (Hesperiidae); (22) royal walnut moth (*Citheronia regalis*, Saturnidae); (23) *Catocala ilia* (Noctuidae).

butterflies. Moreover, some typically forest genera, such as *Najas*, *Cymothoe*, *Bebearia*, and *Bematistes*, are poorly represented in the eastern forests, whereas others, such as *Kallima* and *Ariadne*, are missing altogether. The Malagasy subregion contains 301 butterfly species, of which 233 are endemic; there are 81 genera, of which 12 (including 7 Hesperiidae) are endemic; and 2 (*Euploea* and *Atrophaneura*) are absent from the African continent. The Lycaenidae are few in number (the Lipteninae are absent), but the Satyrinae are plentiful (no fewer than 54 distinct species of *Henotesia*), as are the Hesperiidae. There are many vicarious species or subspecies of those living on the African continent, and most of the species found on the island are clearly of African derivation (see the example in Plate 20B). Among the moths we may note two very beautiful species of *Chrysiridia* (Uraniidae), namely, *C. ripheus*, which is endemic to Madagascar, and *C. croesus*, which is found in East Africa and Zanzibar.

Oriental Region

The Oriental region lies almost entirely between 30°N latitude and 10°S latitude. It is the only one to be almost entirely tropical in character. Bounded to the north by the Himalayas, it extends southwards in the form of a series of warm, moist peninsulas, some parts of which are subject to monsoons and consequently characterized by deciduous tropical forests; vast areas of India and of the inland parts of southeast Asia have this type of vegetation. Other parts have an equatorial climate and are dominated by rain forests, such as in the Western Ghats (India), the coastal areas of southeast Asia, and much of the Malay Peninsula. The southeastern boundary is formed by the shallow Sunda Sea, from which emerge the great islands of Sumatra, Java, and Borneo, covered with a rich, dense, tropical vegetation, as well as a mosaic of smaller islands (such as Palawan and the Philippines), which extend northwards from Borneo or eastwards from Java towards the transitional zone known as *Wallacea*, which separates the Oriental and Australian regions. The eastern part of the continental boundary with the Palaearctic region is somewhat ill defined; there is in fact a zone of integration between the faunae of the two regions in south-central China, where the spurs of the Himalayas sweep down towards southwest China and Burma. The western part of the frontier with the Palaearctic region, on the other hand, is clearly marked by the Hindu Kush and by the mountains of south Pakistan. The Oriental region is very rich in Lepidoptera, being second in total number of species only to the Neotropical region. The principal families are all well represented. Families concentrated in the Oriental region include the Hyblaeidae, the Drepanidae, the Callidulidae, the Eupterotidae, and the Brahmaeidae. *Attacus* is an important genus among the Eranidae. A characteristic group of the Lepidoptera of this region is made up of species that are exclusively or mainly Indo-Australian, such as the Palaesetidae, the Dudgeoneidae, the Agonoxenidae, the Timyridae, the Tineodidae, and the Oxychirotidae. Butterflies are extremely numerous — more than 1,400 species on the Indian subcontinent, more than 1,000 in Malacca, 850 in Thailand, about 240 in Sri Lanka, and more than 370 in the tiny area of Singapore. The more important genera include the following: among the Hesperiidae *Suastus*, *Tagiades*, *Baracus*, and *Plastingia*; among the Papilionidae *Papilio* (with species such as *memnon*, *hector*, *fuscus*, *iswara*, *castor*, and *paradoxa*), *Graphium* (with *doson* and *antiphates*), *Troides* (with *helena* and *mirandus*), and species such as *Atrophaneura coon*, *Teinopalpus imperialis*, and the magnificent *Trogonoptera brookiana*; and among the Pieridae *Prioneris*, *Valeria*, and *Eurema*. The very numerous Nymphalidae include the following genera: *Amathusia*, *Faunis*, *Stichopthalma*, and *Thauria* in the Amathusiinae; *Euploea*, *Idea*, and *Ideopsis* in the Daninae; *Tethosia*, *Euripus*, *Kalima*, *Pantoporia*, *Polyura*, and *Tanaecia* in the Nymphalinae; and *Mycalesis*, *Elymnias*, *Lethe*, and *Ypthima* in the Satyrinae. There is a very conspicuous group of Lycaenidae, including *Allotinus*, *Jamides*, *Nacaduba*, *Narathura*, *Simiskina*, *Sinthusa*, *Talicada*, and *Tajura*. Finally we must mention *Dodona* and *Laxita* among the Riodinidae. The process by which the Oriental region was populated with Lepidoptera is a very complicated one because of the geological history of the Indian subcontinent and the geographical changes that overtook the Greater Sunda Islands during the Pleistocene. Several times these islands became linked together with the Malay Peninsula to form a vast territory known as *Sundaland*. Rises and falls in sea level created alternating situations of insularity and continentality, which in turn connected or isolated the various populations of butterflies. This caused a lively development of subspecies and species, the effects of which we can observe today in the large number of pairs of related species and genera, such as the lycaenids *Poritia* and *Arhopala* or the nymphalids *Amathusia* and *Tanaecia*. There are also a number of instances in which one member of a pair of closely related species has a relatively modest geographical range, while the other has a very large one. *Eurema lacteola* is confined to the Sunda, while *E. andersonii* ranges from the Sunda right across to India and Sri Lanka. Another example is *Papilio iswaroides* and *P. helenus*. It is very difficult to make a full assessment of the effect of ancient changes in the geography of the region on the fauna of today, with its wealth of species of very different origins, including East-Palaearctic, Himalayan, Sino-Tibetan, Sundan, African, Central Asian, and Indo-Australian. J. G. Holloway, a recognized authority on the Lepidoptera of the Oriental region, has made a statistical analysis of the geographical distribution of a highly significant sample of the Southeast Asian Macrolepidoptera and of Indian butterflies. Here we can give only a schematic exposition of the interesting results obtained by Holloway, treating separately the Indian subcontinent (from Kashmir to Assam and Sri Lanka) — which we shall refer to by the convenient name of India — and Southeast Asia (from Burma to northern Vietnam and the Sunda). For India we must distinguish two principal zoogoegraphical elements:

Plate 107 **Ethiopian region**

Of all the tropical regions this is the poorest in Lepidoptera. Certain groups are, however, particularly well represented, such as the lycaenids (this family accounts for almost half of the true butterflies); and certain areas, such as Madagascar and its neighbouring islands, are very rich in endemics (over three-quarters of the Malagasy butterflies are restricted to this insular area). The species selected exemplify the rather dull and monotonous coloration of tropical African Lepidoptera, with the one magnificent exception of Figure 16: (1) narrow blue-banded swallowtail (*Papilio nireus*, Papilionidae); (2) citrus swallowtail (*P. demodocus*, Papilionidae); (3) purple tip (*Colotis ione*, Pieridae), left, male, right, female; (4) orange-patch white (*C. halimede*, Pieridae); (5) *Mylothris erlangeri* (Pieridae); (6) pearl charaxes (*Charaxes varanes*, Nymphalidae, Charaxinae); (7) club-tailed charaxes (*C. zoolina*, Nymphalidae, Charaxinae), male, wet-season form; (8) blue temora (*Salamis temora*, Nymphalidae, Nymphalinae); (9) African map butterfly (*Cyrestis camillus*, Nymphalidae, Nymphalinae); (10) *Hypolimnes dexithea* (Nymphalidae, Nymphalinae); (11) plain tiger (*Danaus chrysippus, dorippus* form, Nymphalidae, Danainae); (12) neobule acraea (*Acraea terpsicore*, Nymphalidae, Acraeinae); (13) figtree blue (*Myrina silenus*, Lycaenidae); (14) striped policeman (*Coeliades forestan*, Hesperiidae); (15) sunset moth (*Chrysiridia madagascarensis*, Uraniidae); (16) *Eudaemonia brachyura* (Saturniidae); (17) *Dasiothia medea* (Sphingidae); (18) *Amphicallia tigris* (Arctiidae).

Tropical Lepidoptera and glaciations

In the chapter entitled Origin of Species we have already seen how climatic changes during the Pleistocene influenced the evolution of montane Lepidoptera, isolating populations and thus favouring the development of montane species on the mountain masses of North America, Europe (Plate 22), and South America (Plate 24). In the present chapter we have also shown that certain discontinuous geographical ranges, such as the arctic-alpine, are due to the effects of the Quaternary glaciations. These few examples illustrate a phenomenon well known to biogeographers, namely, the influence of the Pleistocene climate on the evolution and the geographical distribution of temperate species. Until recently it was so firmly believed that the equatorial biomes were not affected by the dramatic changes that took place in the climate of temperate areas that the supposedly unchanging nature of the tropical climate during the Pleistocene was put forward as one of the primary explanations of the great wealth and variety of species in the tropical communities. More recent research has modified this interpretation by showing that the successive deteriorations of the climate during the Pleistocene have in fact profoundly

One is formed of species belonging to genera whose distribution is mainly concentrated in the zone that extends from Assam to the Sunda; the other is made up of species belonging to genera with ranges centered in the Palaearctic region to the north of tropical Asia. The first, or eastern, element has the larger number of species (more than 300 in all, divided among more than 70 genera), with many low-altitude butterflies typical of the rain-forest biotopes. Examples are *Papilio paradoxa*, *P. slateri*, *Graphium macareus*, *G. doson*, and *Appias indra*. Endemic species are numerous in the rain forests. This contingent of species presumably colonized southern India during the Pleistocene. Some of these eastern species are also found at middle and high altitudes in the eastern Himalayas and include *P. castor*, *P. arcturus*, *P. bootes*, *Delias atostina*, *G. glycerion*, and *Ochodes brahma*. A smaller group is found in the dry habitats and the plains of southern India (*G. nomius*). The palaearctic element of the Indian lepidopteran fauna is mainly made up of typically alpine species (such as *Parnassius*, *Colias*, and *Baltia*), which are commonest in the western Himalayas. A group of species that is more modest in size but has much wider ecological and geographical distribution is characteristic of northwestern India, examples being *P. alexanor*, *Aporia leucodica*, and *Melitaea persea*. As regards Southeast Asia, we must consider first a group of highly mobile species (such as *P. demoleus*, *P. agamemnon*, *Danaus genutia*, *Hypolimnas misippus*, and *Euploea mulciber*) and a small group of Lepidoptera peculiar to the small islands of the Sunda, which is widely diffused in areas of secondary vegetation. In addition to the two groups just mentioned there are two principal zoogeographical categories in the region. The first is known as the "Melanesian" group and is formed of species belonging to genera whose range is centered in New Guinea. Examples include many high-altitude moths on the slopes of Mount Kinabalu in northern Borneo, one of which is *Horisme labeculata*. The other group, which is known as "Sunda-Oriental," is larger than the Melanesian and may be divided into predominantly continental species (such as *D. eryx* and *E. doubledayi*) and species living on the islands or the Malay Peninsula (such as *G. empedevana*, *Callidula sumatrensis*, *Chalcosia zehma*, and *Cossus kinabaluensis*). The area to the north of the River Kedah forms the climatic barrier that separates these two subgroups.

influenced the tropical regions, causing cases of ecological insularity to arise in the rain forests, just as they have in temperate zones. Although the first data to provide evidence for this came from observations made in Australia and Africa, it is only with the results of the South American studies of many authors (such as K. S. Brown, J. Haffer, J. R. G. Turner, and P. E. Vanzolini) that a well-structured theory of forest refuges as ecological islands has emerged. Put very simply, the theory is as follows: The climatic cycles of the Pleistocene generated periodical conditions of aridity, during which the ancient rain forests must have been split up into a number of much smaller, fragmented woodlands separated by large areas of savannah as a result of the retreat of the vegetation towards moister areas. During this arid phase the biotypes of the rain forest must have functioned as islands of refuge for many animals and plants, whose populations will, therefore, have evolved allopatrically. The return of a period with a wet climate would have favoured the expansion of the forests, in many cases reconnecting the Lepidoptera populations that had previously been cut off in the moist refuges. A process of this kind, probably repeated several times during the last million years, would have caused the appearance of numerous subspecies, semispecies, or new species, according to the degree of differentiation achieved by the individual isolated populations. From Brown's very extensive observations of Heliconiinae and Ithomiinae, two groups of mimetic Lepidoptera typical of the neotropical forests, it seems that there is a great abundance of subspecies and semispecies that are still capable of producing hybrids (see Plate 20A: *2a*). This suggests that the genetic divergence between these populations, which are now living together, is of fairly recent origin. The plants, birds, and lepidopterans of the Amazon forests all exhibit many cases of pairs of closely related species living in the same area, of endemic species concentrated in certain areas, and of hybridization zones that are often the same for creatures of widely different kinds — all of which can be explained by postulating that the populations now coexisting in ecological and geographical continuity were formerly separated by climatic barriers.

Australian Region

The Australian region comprises Australia, New Guinea, Tasmania, New Zealand, and the numerous islands and archipelagoes that extend from the revised Weber line eastwards towards Polynesia. It can be roughly divided into three subregions: The Australian subregion includes Australia itself, Tasmania and neighbouring islands, and New Zealand; the Papuan subregion comprises New Guinea and island groups such as the Bismarck Archipelago and the Solomon Islands; the Oceanic subregion contains thousands of volcanic or coral islands, many of them very small, where the number of species and higher taxa of Lepidoptera is relatively very low. We shall therefore discuss only the first two subregions, concentrating on Australia itself and New Guinea, which are the largest constituent areas and the richest in terms of Lepidoptera. Very isolated islands tend to have few species even if they are quite large in area; an example is New Zealand, which has only 17 species of butterflies, 6 or 7 of which are migratory. New Zealand has, however, some interesting examples of endemicity, such as the entire family of the Mnesarchaeidae. The Australian region adjoins another zoogeographic region only along part of the Wallace line, which separates it from the Oriental region. The main factors determining the character of the Australian fauna are probably to be found in the peculiar paleogeographical history of the continent and in the present bioclimatic conditions, which we shall now consider briefly. From the Miocene onwards the progressive desiccation of many areas favoured the development of xerophyllous vegetation. The vegetation diversified greatly and was enriched with new species during the Pliocene, with its alternation of wet and dry periods. Fluctuations in sea level during the Pleistocene provided an intermittent connection with New Guinea, which brought many new species of Lepidoptera. At present two-thirds of the surface of Australia has an arid or semiarid climate. The vegetation of this huge area (known as the Eirean province) is dominated by grasses such as *Triodia* and by trees and bushes of the *Acacia* group (600 species). The northern and eastern coastal zones of Australia (known as the Torres province) have a tropical or subtropical climate, with many extensive areas of tropical rain forest similar to that of Southeast Asia. In the Bass province (consisting of the southern coastal belt, part of the eastern coastal belt, and the island of Tasmania) the climate is cooler, permitting the development of wet temperate forests of trees such as the southern beech (*Nothofagus*). Inland regions with annual rainfall of over 380 mm (15 in) have sclerophyllous vegetation, forming savannahs dominated by over 600 species of gum trees (*Eucalyptus*). The geographical distribution and the basic composition of the Australian Lepidoptera is strongly influenced by these highly varied environmental conditions. The butterflies and moths of the Australian rain forests are largely of oriental or Papuan origin, with numerous common genera. The Lepidoptera found in the areas dominated by *Eucalyptus* and *Acacia* (the two plant genera with which most Australian butterflies are associated) are, generally speaking, much more original in character. A striking example of a high degree of evolutionary success among the Lepidoptera found on *Eucalyptus* trees is the Oecophoridae, which account for about 2,500 out of the 5,000 species of known Australian Lepidoptera. There are also 360 Tortricinae (the Tortricidae of Australia amount to about 800). There are other important groups of Australian Lepidoptera rich in forms associated with the *Eucalyptus*; examples are the Incurvariidae, the Cossidae, the Anthelidae, and the Notodontidae. Among the south Australian fauna we find endemic forms (such as the Satyrinae) belonging to genera such as *Geitoneura*, *Heteronympha*, and *Oreixenica* (found also in Tasmania), which are probably derived from ancestors originating in the Antarctic continent when it enjoyed a temperate climate. The total number of

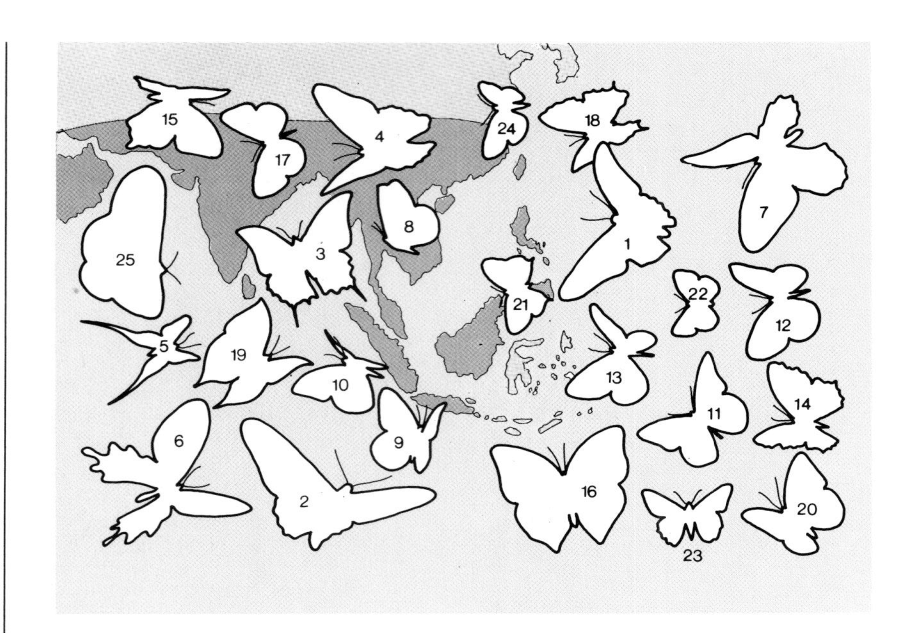

Plate 108 Oriental region

Made up of a great continental landmass and many islands, the Oriental region is a tropical area with a very rich fauna, which is the home of some of the most beautiful and best known butterflies and moths in the world. Plate 108 shows a small but important sample of oriental species, which brings out their great aspect diversity. The line separating the large island of Borneo and the smaller island of Celebes (which is shown in white) corresponds to the imaginary zoogeographical watershed between the Oriental and Australian regions. (1) *Troides aeacus* (Papilionidae); (2) *Trogonoptera brookiana* (Papilionidae); (3) *Teinopalpus imperialis* (Papilionidae); (4) blue triangle (*Graphium sarpedon*, Papilionidae); (5) *Lamproptera curius* (Papilionidae); (6) *Atrophaneura latreillei* (Papilionidae); (7) *A. horishanus* (Papilionidae); (8) *Delias eucharis* (Pieridae); (9) *D. crithoe* (Pieridae); (10) *Hebomoia glaucippe aturia* (Pieridae); (11) *Appias nero* (Pieridae); (12) *Euploea dufresne* (Nymphalidae, Danainae); (13) *Ideopsis vitrea* (Nymphalidae, Danainae); (14) *Elymnias nesaea lioneli* (Nymphalidae, Satyrinae); (15) *Neorina krishna archaica* (Nymphalidae, Satyrinae); (16) *Zeuxidia aurelius* (Nymphalidae, Amathusiinae); (17) *Neptis nandina* (Nymphalidae, Nymphalinae); (18) map butterfly (*Cyrestis thyodamas formosana*, Nymphalidae, Nymphalinae);(19) dead-leaf butterfly (*Kalima horsfieldi philarchus*, Nymphalidae, Nymphalinae); (20) *Tanaecia julii* (Nymphalidae, Nymphalinae); (21) *Arhopala horsfieldi* (Lycaenidae); (22) *Poritia philota* (Lycaenidae); (23) *Satarupa gopala* (Hesperiidae); (24) *Erasmia pulchella* (Zygaenidae); (25) *Brahmaea wallichii* (Brahmaeidae).

Wallacea

The biogeographic differences between the Australian and Oriental regions are numerous and profound; some of them (such as the almost total absence of placental mammals in Australia) are extremely ancient and may be traced back to the time of Pangaea, when the landmasses of our planet were united in a single megacontinent and the block that later became Australia was attached to Antarctica. When Australia and New Guinea reached their present position on the globe, an exchange of species on a large scale took place — and indeed still does — with the Asiatic islands of the Sunda platform (Borneo, Java, and Sumatra) along the route passing through the long island chain of the Lesser Sunda, the Moluccas, and Celebes. The lowering of the sea level during the Pleistocene glaciations assisted the process of interaction between the two faunae. Moving away from Asia and Australia as we approach Celebes, we find a progressive weakening of zoogeographical contrasts until we reach a real no man's land of islands with a thoroughly mixed fauna. This zone of transition, Wallacea, is bounded to the west by the Wallace line and to the east by the

Australian Lepidoptera is more than 11,200, belonging to 177 families. The distribution of species and genera among the various families is very uneven. A quarter of the total is made up of Oecophoridae (2,500 species) and Xyloryctidae (420 species). Apart from the Oecophoridae and the Tortricidae there are three well-represented archaic families: the Hepialidae (109 species), the Cossidae (99), and the Castniidae (29), all with internal-feeding caterpillars well adapted to arid environments. Groups of more recent origin, found mainly in the north, are the Sesiidae (20 species), the Timyridae (87), the Papilionidae (18), the Pieridae (31), the Sphingidae (59), the Lymantriidae (74), the Hypsidae (8), and the Agaristidae (37). Cases of endemicity are extremely numerous, even at the family level; examples are the Lophocoroniidae (3), the Cyclotornidae (5), the Anthelidae (73), the Carthaeidae (1), and the Agathiphagidae (1, with a separate species in Fiji). The butterflies amount to 382 species in all, of which 174 are endemic and 112 have Australian distributions. An analysis of the butterflies by geographical distribution and by endemicity reveals the marked differences between the faunae of the various provinces of the Australian region. The Torres province has 320 species (84 percent), of which 112 (35 percent) are endemic. The Eirean province has 92 species (24 percent) of which 3 are endemic, 2 being skippers (Hesperiidae) and 1 a lycaenid. The Bass province has 195 species (51 percent), of which no fewer than 133 are endemic (68 percent, mostly belonging to the Trapezitinae and Satyrinae, the two groups of Australian butterflies that show the highest rate of endemicity). New Guinea is such a large island as to be almost continental in status. It is very mountainous, with many peaks higher than 4,000 m (13,000 ft) and one of 5,000 m (16,400 ft). The great mountain chains are separated by innumerable deep river valleys. Much of the interior is dominated by tropical rain forests although the island's wide range of altitude has permitted the formation of many other types of biome, such as temperate woodlands, forests of southern beech, mountain biomes, and even tundra. The geological and climatic history of New Guinea is complicated. In the Eocene the Australian landmass broke away from Antarctica and moved northwards. Between 15 and 5 million years ago New Guinea collided with an arc of islands of Tertiary origin, which became welded to New Guinea and now form an integral part of it. This

Weber line, these two imaginary lines marking the boundary between the faunae of the Oriental region and the Australian region. The area is named after A. R. Wallace, a great nineteenth-century naturalist who travelled extensively in these archipelagoes. Having observed a marked discontinuity of faunae between two islands as close to each other as Bali and Lombok, Wallace was the first to trace the boundary between the Australian region and the Oriental region. Naturalists later found between pairs of islands differences that did not agree with the Wallace line. Huxley moved the frontier to the west of the Philippines, while Weber moved it much farther to the east, including Celebes and the Moluccas in the Oriental region. Unlike these early naturalists, however, the zoogeographers of today prefer to treat Wallacea as a wide zone of transition, in which the two faunae intermingle in proportions that vary from island to island and depend on the group of organisms under consideration. There is some doubt about the zoogeographical character of the Lepidoptera of Wallacea. There seems, however, to be a predominance of oriental forms in certain groups, such as *Tridrepana* and *Canucha* among the Drepanidae and *Troides* among the Papilionidae. There are also groups of Wallacean origin, such as the exquisite *Ornithoptera*. Among the Pieridae we may note the *Delias* group, which is of Papuan origin (60 species in New Guinea and 8 species in Australia) but is widely distributed in Wallacea (7 species in Celebes, 8 in Halmahera, 11 in Ambon and 10 in Buru). This group has also penetrated into the continental part of the Oriental region, with 9 in the Malay Peninsula, 7 in Burma, and 7 in northern India. If we limit ourselves to examining the Papilionidae, we shall find 55 species in Wallacea; the island with most species is Celebes (26), followed by the Moluccas: Halmahera (21), Ceram (18), Buru (15), Flores (14), Sumbawa (13), and Sumba (12). Timor and Lombok, though very different in surface area, have the same number of species (11). Islands like Bali, which are small in area but situated on the boundary of the Oriental region, often have a number of species of Papilionidae (16 in Bali) that is comparable with that found in territories of far greater size, such as Australia (16), but remote from the main centres of distribution of these butterflies. In addition to such typically Wallacean species as the Ulysses butterfly (*Papilio ulysses*), *P. weiskei*, and Cairn's birdwing (*Ornithoptera priamus*) Wallacea also provides a home for species of undoubted oriental origin [such as *P. memnon*, *P. polytes*, the orchard butterfly (*P. aegeus*), and the green-spotted triangle (*Graphium agamemnon*)] and species of Australian origin [such as the big greasy (*Cressida cressida*) and *G. wallacei*)].

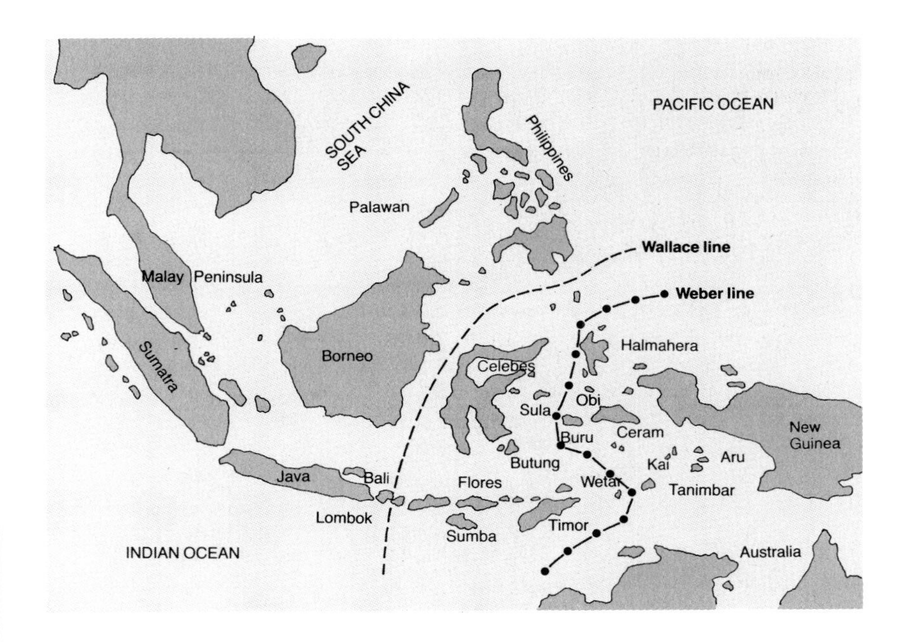

event, which took place at a short distance from Wallacea, brought the fauna of the Oriental region into contact with that of the Australian region for the first time. During the Pleistocene, New Guinea was in contact with Australia to the south and with the Moluccas to the north. The glaciations made New Guinea suitable for colonization by the ancestors of fauna such as the Satyrinae *Platypthima* and *Peridopsis*, which are adapted to an alpine environment. The alternation of periods of geographical connection and isolation favoured the immigration of additional species from Wallacea, resulting in a high concentration of different kinds of Lepidoptera in New Guinea. Other species probably emigrated from New Guinea during the same period, such as members of the *Delias* (which is the largest butterfly genus in the Australian region) and the members of *Taenaris*. These two groups, together with many other species, radiated outwards towards the Oriental region as well as towards nearby Australia. New Guinea is very rich in Lepidoptera, some of which are extraordinarily beautiful, notably the members of *Delias*, the *Ornithoptera*, and the *Milionia* (see Plate 9). Among the Papilionidae (39 species) we may note some endemic forms, such as *O. alexandrae*, *O. chimaera* and *O. rothschildi*, and among the Nymphalidae we must mention *Tellervo* (the only genus of the Ithomiinae to be found outside the tropics).

Neotropical Region

The Neotropical region extends from Mexico to Tierra del Fuego and takes in the island groups of the Antilles and the Galápagos. The long mountain chain of the Andes runs along the entire western coast of the South American continent. To the east of the northern Andes lies a huge area of tropical rain forest, drained by a vast network of rivers. This is known as the "Brazilian area" and corresponds roughly to the Amazon basin. It is characterized by conditions of great climatic stability, which have a marked influence on the flora and fauna, favouring a high degree of ecological diversity. Other, very different, climatic and biogeographical configurations occur high in the Andes and also, of course, further to the south — in the Chilean area, for example. The slopes of the Andes are divided by altitude into four zones (tropical, subtropical, temperate and

Plate 109 **Australian region**

Thanks to its geological past and its present outlying geographical position, the Australian region has an unusual fauna full of contrasting elements. The character of the Australian Lepidoptera is unique. It is marked by intense development of certain archaic groups and of species belonging to arid biomes; many important families are completely missing, whereas others are most generously represented. New Guinea is dominated by species characteristic of rain forests. The oceanic islands have relatively few Lepidoptera, but a high proportion of them are inevitably endemic. The exchange of species between this region and the Oriental region can be observed in the islands of Wallacea. (1) *Ornithoptera chimaera* (Papilionidae); (2) *Troides criton* (Papilionidae); (3) big greasy (*Cressida cressida*, Papilionidae); (4) *Papilio toboroi* (Papilionidae); (5) *Delias niepelti* (Pieridae); (6) wood white (*D. aganippe*, Pieridae); (7) *Taenaris catops* (Nymphalidae, Amathusiinae); (8) common silver xenica (*Oreixenica lathoniella*, Nymphalidae, Satyrinae); (9) Australian leafwing (*Doleschallia bisaltide australis*, Nymphalidae, Nymphalinae); (10) *Mynes woodfordi* (Nymphalidae, Nymphalinae) male; (11) *Danis schaeffera caesius* (Lycaenidae) female; (12) dark purple azure (*Ogyris abrota*, Lycaenidae), female above, male below; (13) common imperial blue (*Jalmenus evagoras*, Lycaenidae), male; (14) *Zelotypia staceyi* (Hepialidae); (15) *Milionia paradisea* (Geometridae).

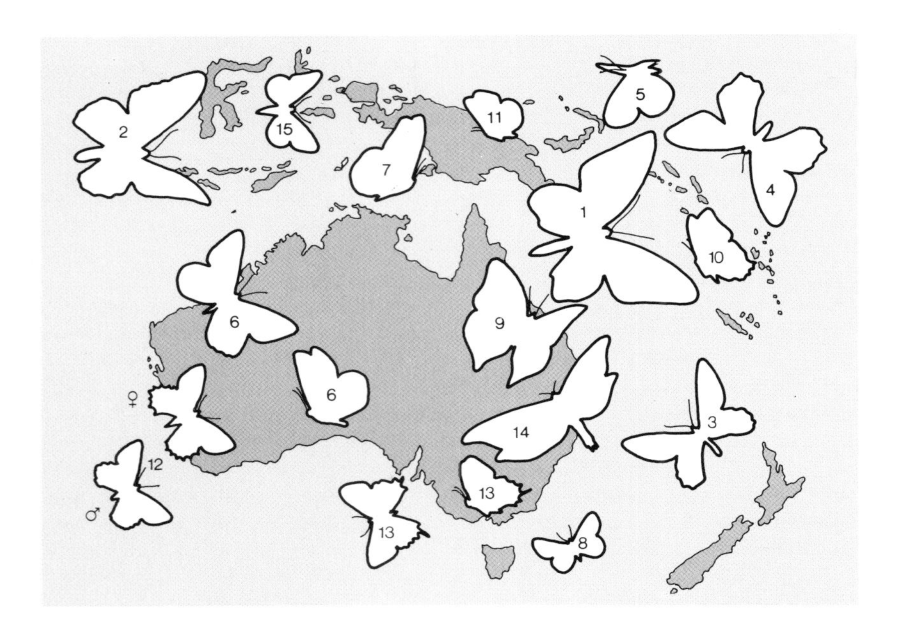

páramo). However, this zoning is also affected by latitude and exposure to winds. The areas immediately sheltered by the Andes chain are ecologically varied, consisting as they do of an unbroken succession of hills, valleys, and plains. They contain the richest variety of Lepidoptera to be found anywhere in the world. Up to 350 species of butterflies have been found in quite small districts in Colombia, while at least 4,000 species are known in Peru.

From the Tropic of Capricorn to Cape Horn and from the mountains of Chile to the Atlantic there is a predominance of pampas and steppes, which become tundra in Patagonia and Tierra del Fuego. In Central America (known as the "Mexican area") the mosaic of different climates and environments is even more complicated than in the Brazilian area. Considered as a whole, the Neotropical region is richer in Lepidoptera than any other, with at least 10,000 species of butterflies alone. However, given that much of South America is still little explored in entomological terms, there are no complete and detailed lists of species. We must therefore content ourselves with a brief qualitative review, limited to the most important taxa. One of the peculiarities of the neotropical Lepidoptera fauna is the high incidence of endemicity not only at the levels of species or genus but also at the levels of subfamily and family. Endemic subfamilies include the Rhescynthiinae (Saturnidae) and the Pyrrhopyginae (Hesperiidae); endemic families include the Arrenophanidae, Oxytenidae, Cercophanidae, Dioptidae, and Sematuriidae. The following families are largely concentrated in the Neotropical region: the Stenomidae, the Megalopygidae, the Mimallonidae, the Castniidae, the Ctenochidae, the Hypsidae, and two other subfamilies of Saturnidae, namely, the Hemileucinae and the Citheroniinae. Turning now to the butterflies, we find about 170 species of Papilionidae (of which at least 160 are endemic), belonging to six distinct genera: *Battus* (14 or 15 species, of which 3 are also nearctic), *Parides* (a genus also found in the Indo-Australian area and in Madagascar and having 44 neotropical species, all of which are endemic), *Euryades* (endemic with 2 species), *Papilio* (with over 50 species), *Eurytides* (a Pan-American taxon with about 50 endemic species and 2 also found in the Nearctic region), and finally *Baronia* with the Mexican species

Plate 110 **Neotropical region**

Central and South America have the richest Lepidoptera fauna of all the regions. The species that exemplify it best are to be found in the great tropical rain forests. Certain valleys in Colombia and Peru are famous for the variety of species that they contain; a few hectares of forest may be the home of literally hundreds of different species of Lepidoptera. There are many endemic and rare species in the region, and instances of mimicry are common.
(1) Homerus swallowtail (*Papilio homerus*, Papilionidae); (2) *Parides montezuma* (Papilionidae); (3) *P. childrenae* (Papilionidae); (4) lycidas (*Battus lycidas*, Papilionidae); (5) tiger pierid (*Dismorphia amphione*, Pieridae); (6) *Catasticta manco* (Pieridae); (7) tailed orange (*Eurema proterpia*, Pieridae); (8) ghost brimstone (*Anteos clorinde*, Pieridae); (9) *Pteronymia cotytto* (Nymphalidae, Ithomiinae); (10) *Dircenna klugii* (Nymphalidae, Ithomiinae); (11) *Bia actorion* (Nymphalidae, Satyrinae); (12) *Morpho patrocius* (Nymphalidae, Morphinae); (13) *Paulogramma pyracmon* (Nymphalidae, Nymphalinae); (14) *Napeocles jucunda* (Nymphalidae, Nymphalinae); (15) the Doris (*Heliconius doris doris*, Nymphalidae, Heliconiinae); (16) *Evenus coronata* (Lycaenidae); (17) *Mesosemia loruhama* (Riodinidae); (18) *Amenis pionia* (Hesperiidae); (19) *Pyrrhopyge spatiosa* (Hesperiidae); (20) *Castnia papilionaris* (Castniidae); (21) white-tailed page (*Urania leilus*, Uraniidae); (22) *Unidentified moth* (Hypsidae); (23) *Cosmosoma cardinale* (Ctenuchidae).

B. brevicornis. The Pieridae are represented by endemic or almost endemic taxa such as *Dismorphia*, *Pseudopieris*, *Catasticta*, *Ascia*, *Leptophobia*, *Eroessa*, *Andina*, and *Phalia*. Among the Nymphalidae the following are exclusively neotropical: the Brassolinae (*Caligo* and *Narope*), the Morphinae (with about 80 species of *Morpho*), almost all the Ithomiinae (including *Hypoleria*, *Mechanitis*, and *Pteronymia*), and almost all the Heliconiinae. Among the Nymphalidae we find some highly characteristic endemic groups such as the Charaxiinae *Prepona*, *Agrias*, and *Anaea*, the Nymphalinae *Nessaea*, *Callicore*, *Callithea*, *Dynamine*, *Anartia*, *Eunica*, *Catonephele*, and *Doxocopa*, the Danainae *Lycorea* and *Ituna*, and the Satyrinae *Pierella*, *Cithaerias*, *Calisto*, and *Pronophila*. One of the most famous Lepidoptera of the Neotropical region is the ghost moth (*Thysania agrippina*), a huge noctuid, with a wing span of up to 300 mm (12 in), whose range extends from Mexico to Brazil.

The fauna of the mountain areas are particularly interesting from a zoogeographical point of view. Their Lepidoptera are a mixture of species belonging to holarctic genera (such as *Pieris*, *Colias*, and *Argynnis*) and of species belonging to autochthonic or endemic genera (such as the pierid *Trifurcula* and the nymphalids *Pseudomaniola*, *Lymanopoda*, *Pedaliodes*, and *Steroma*). *Erebia* and *Oeneis* are not represented at all. The high-altitude species of the Andes include *Pieris xanthodice*, *Colias alticola*, *Piercolias andina*, *Phulia paranympha*, and *Argynnis inca*.

Butterflies and Man

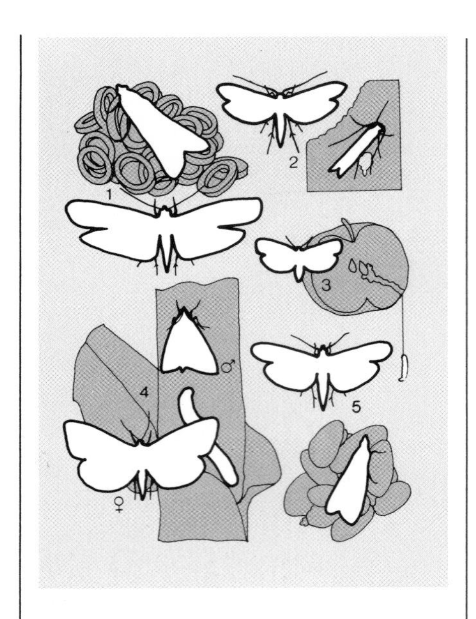

Plate 111 **Harmful Lepidoptera**

The five Lepidoptera illustrated on this plate are but a few of the most harmful moths from an economic point of view. Each of them has been able to modify or extend its ecological niche by adapting itself so as to exploit new resources provided by man, such as crops, foodstuffs, and articles manufactured from plant and animal materials. These species originated from relatively restricted geographical areas in the Palaearctic region but later, in historical times, spread abroad so widely as to become cosmopolitan: (1) Mediterranean flour moth (*Ephestia kuehniella*, Pyralidae), 20–25 mm (0.7–0.9 in); (2) webbing clothes moth (*Tineola bisselliella*, Tineidae), 11–15 mm (0.4–0.6 in); (3) codling moth [*Cydia (= Laspeyresia) pomonella*, Tortricidae], 15–22 mm (0.6–0.8 in); (4) European corn borer (*Ostrinia nubilalis*, Pyralidae), 26–34 mm (1.0–1.3 in); (5) Indian meal moth (*Plodia interpunctella*, Pyralidae), 13–18 mm (0.5–0.7 in).

Harmful Moths and Butterflies

Ecology teaches us that all organisms in the natural ecosystem are actually "useful," in the sense that every one fulfills a function in maintaining the balance of the ecosystem to which it belongs. However, as everyone knows, there are some species that we describe as "harmful" because, in a wide variety of ways, they are a threat to the interests of human commerce. It should be made clear, however, that the harmfulness of a species is not actually one of its biological features but is something that results from peculiar circumstances. No creatures are consistently harmful everywhere, but under particular conditions certain species compete with human beings for the enjoyment of resouces (space, food, manufactured articles, or health) and so are harmful from an economic or health point of view. Because the diet of caterpillars is almost wholly herbivorous, the Lepidoptera are the group potentially most damaging to crops. With advanced agricultural methods and the tendency to plant crops in large uniform blocks, in monoculture, human beings have given the herbivores a superabundant, predictable resource of food. In a relatively short time (a few thousand years) this has resulted in the adaptation of many Lepidoptera to new ecological niches in association with crops or to the extension of those that already exist. One after another, new species hitherto regarded as economically unimportant have become real plagues as they have moved from native vegetation to agricultural or commercially forested areas. This exploitation of new food resources is also true of those moths considered harmful to manufactured goods and foodstuffs (Plate 111).

The list of harmful Lepidoptera is remarkably long: In a recent treatise on entomology applied to agriculture Balachowsky catalogues about 400 economically important species in Europe alone. They belong to some thirty families, the majority being Tortricidae, Yponomeutidae, Pyralidae, and Noctuidae.

Control methods have undergone a remarkable evolution as the use of chemical means has been in part supplanted by biological techniques (using natural predators and parasites); and this in turn, by integrated pest management ("IPM"), in which use is often made of sterile males, of cultural methods, or of pheromones to capture or confuse the males so as to inhibit mating. Naturally efforts to combat destructive insects do not aim at the eradication of the species concerned — an aim that, as well as being virtually unattainable, is in any case undesirable. They are undertaken in the hope of "controlling" its damage within economically acceptable limits.

An outstanding example of a hostile relationship between Lepidoptera and man is that of the gypsy moth (*Lymantria dispar*), an omnivorous species of palaearctic origin that kills or damages many deciduous trees. The gypsy moth is subject to periodic demographic explosions, which cause a massive increase in the larval population, leading to the defoliation of huge areas (sometimes thousands of hectares) of deciduous forest. In Europe, particularly in the Mediterranean basin, where the gypsy moth mainly attacks the cork oak (*Quercus suber*, Fagaceae), this results in economic damage due to a reduction in the annual trunk-diameter increment and the loss of the year's crop and to difficulty in harvesting the cork in the legally prescribed time. In eastern Canada and the northeastern United States, where the species was introduced accidentally in 1869, the damage caused by the gypsy moth is far greater and is estimated to be in the neighbourhood of hundreds of millions of dollars a year. The spread of the gypsy moth in North America began with an accident at Medford, Massachusetts, when a crate containing caterpillars imported by a thoughtless French entomologist, Léopold Trouvelot, was

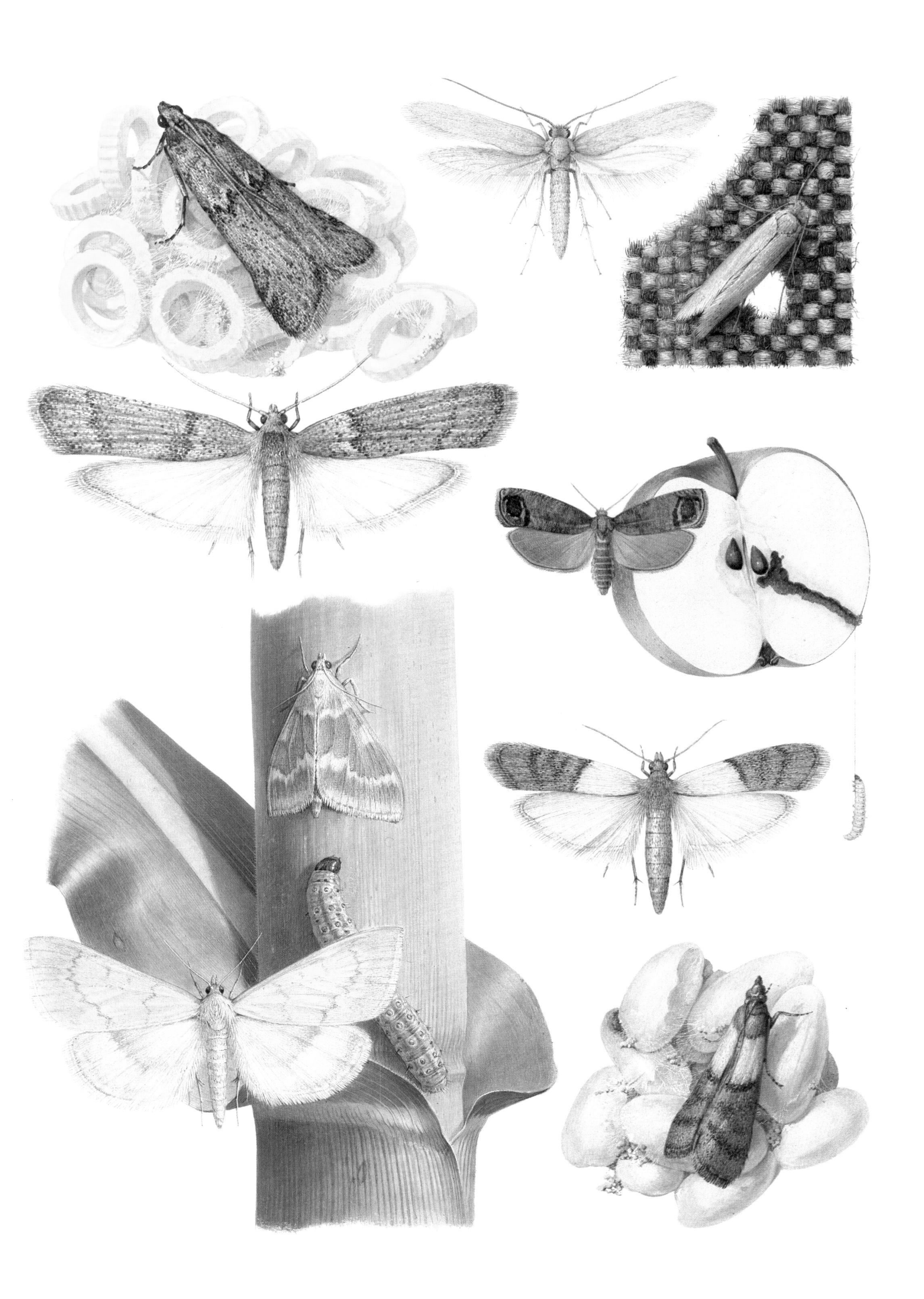

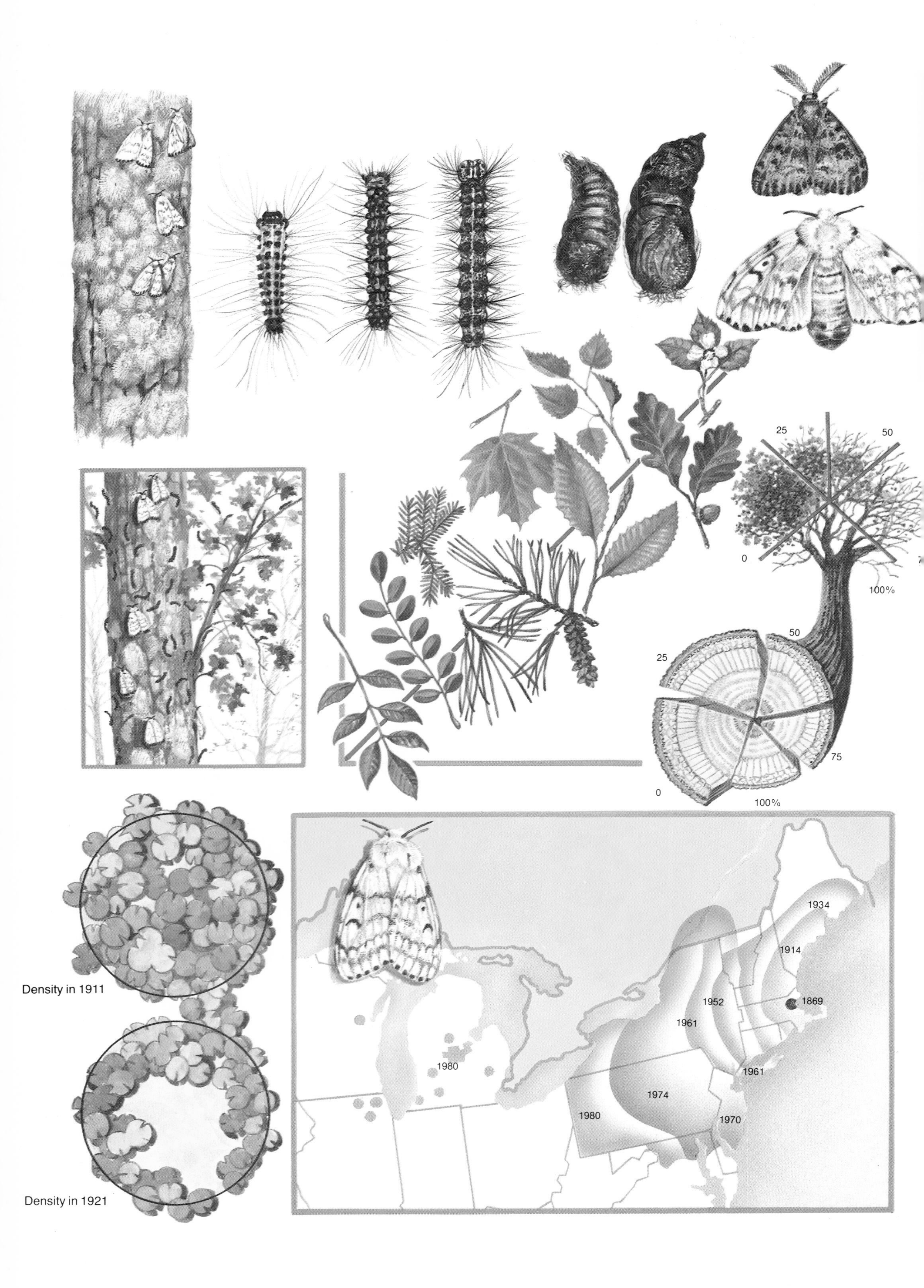

25
50
0
100%
50
25
75
0
100%
Density in 1911
Density in 1921
1934
1914
1952
1869
1961
1980
1961
1974
1980
1970

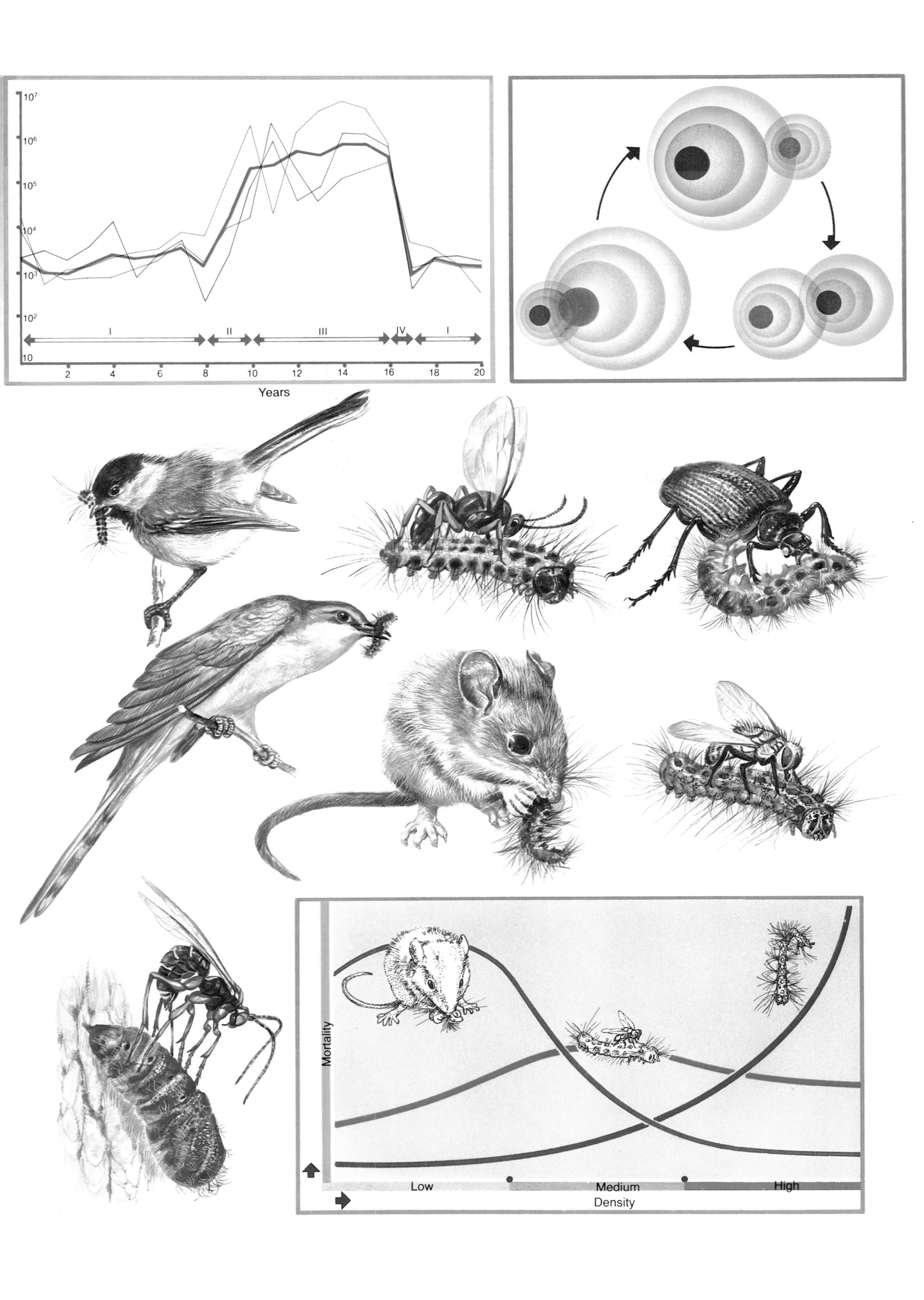

10⁷
10⁶
10⁵
10⁴
10³
10²
10
I
II
III
IV
I
2
4
6
8
10
12
14
16
18
20
Years
Mortality
Low
Medium
High
Density

Plate 112 **Natural history of the gypsy moth (*Lymantria dispar*) in North America**

Originally a European species, the gypsy moth (*Lymantria dispar*, Lymantriidae) was accidentally introduced into the United States in 1869 and in a few years became established there as the most harmful forest insect in the northeastern United States. The species produces one generation a year and overwinters in the egg stage. The eggs are laid in clusters of 80 to 800; each batch is protected by a dense layer of stinging hairs detached from the tip of the female's abdomen, which ensure that the eggs are insulated against the cold winter weather. The egg patches (1) are usually laid on tree trunks of a few species; but they are also laid on tree stumps, walls, rocks, wooden buildings, and other surfaces. Postembryonic development occurs in five larval stages in the male and six in the female. The eggs hatch at the end of April or beginning of May, and the first-instar larvae (2) move upward to tree foliage, from which they then disperse in the air by means of silk threads. Until the third instar (3) the caterpillars feed during the day, but from the fourth to the sixth instar (4) they are nocturnal. Pupation takes place in late June or early July. The male pupa (5) is considerably smaller than that of the female (6). The males emerge after about two weeks, the females a few days later. There is marked sexual dimorphism, as the Latin name of the species suggests: The male (7) is smaller [average wingspan, 40 mm (1.5 in)], and the female (8) larger [average wingspan, 60 mm (2.3 in)]. The females are sedentary and attract the males, which are active fliers, by a specific pheromone, known either as "gyplure" or "disparlure." The females lay their eggs shortly after mating, the number laid depending on the size of the moth.

The destructiveness of the gypsy moth is due to its immense numbers (9). In particular the species is subject to periodic demographic explosions, during which it can cause serious defoliation in several kinds of deciduous trees, especially oaks. Different kinds of tree suffer different degrees of defoliation, as the diagram in Figure 10 shows; it is relatively small for the ash and the false acacia but very high for oaks and apple. Other trees subject to attacks by gypsy-moth caterpillars are birch and, to a decreasing degree, beech, maple, and various conifers, such as pines and hemlock. The foliage's being attacked results in a diminished production of wood. It has been observed that a defoliation of less than 50 percent of the leaves has a limited effect, whereas more severe defoliation drastically reduces annual growth (11) and threatens the survival of the affected tree. After these attacks the density and indeed the composition of a forest may be substantially modified over a period of years, as has been shown by a research study carried out on a forest in New England (12). The map (13) illustrates the progressive spread of *Lymantria* in North America since its introduction into Massachusetts. The thick red line on the graph (14) shows the numerical

broken open; he was breeding the species in the hope of crossing it with *Bombyx mori* to obtain silk from caterpillars that could feed on oak leaves. After 1869 *Lymantria dispar* spread farther and farther to the southwest until by 1980 it had reached Ohio and West Virginia and in smaller colonies several of the southern states, as well as Wisconsin, Illinois, and Michigan. The natural history of *Lymantria dispar* has been closely studied to see what useful lessons may be learned from it: The population dynamics of this single moth have been dealt with in more than 600 scientific papers; but they also give us excellent reference material in which we can observe in an integrated way a whole set of biological phenomena that have been dealt with individually in the chapters of this book. The main aspects of the gypsy moth's biological cycle, the relationships with food plants, the history of its colonization in North America, and its demography and ecological interactions with parasites and predators are illustrated in Plate 112.

Useful Lepidoptera

The list of butterflies and moths that can be regarded as directly useful to man is very short — in no way comparable to the list of harmful species. It may be that certain Blastobasidae, Pyralidae, Lycaenidae, and Noctuidae, which have carnivorous larvae, contribute to the control of herbivorous insects such as aphids and scales, but their role (although insufficiently explored) appears to be modest. In contrast the use of the herbivorous caterpillars of the pyralid *Cactoblastis cactorum* from South America, to control certain invasive *Opuntia* cacti (Cactaceae) in Australia and South Africa, has been very effective. An interesting example of the practical use of certain moths is that of *Sitotroga cerealella*. The eggs of this small gelechiid, which is easy to raise in the laboratory, are used in the large-scale breeding of millions of parasitic *Trichogramma* wasps, which are released as adults and parasitize the eggs and caterpillars of many harmful Noctuidae. And of course there is the silkworm; the economic importance of this domestic moth is still considerable despite the progress over a period of twenty years of a hypothetical North American gypsy moth population in terms of the number of eggs per acre: The overall progress of the population can be divided into four phases, low density (I), demographic growth (II), high density (III), and collapse (IV). Phases I and III are relatively steady for periods of about eight to ten years, while the transitional phases II and IV are short. In practice the stability of the (harmful) phase of greatest abundance seems to be the result of the alternate rise and fall of individual local subpopulations (thin lines), as illustrated in Figure 15. As one subpopulation (blue circle) grows, another (red circle) diminishes, having reached its numerical peak the previous year, when it was subjected to severe attacks by predators and parasites. The various predators and parasites, together with diseases and starvation, have an uneven effect on the mortality of gypsy moth caterpillars, depending on their density (16). The main predators and parasites illustrated are the following: (17) black-capped chickadee (*Parus atricapillus*), (18) yellow-billed cuckoo (*Coccyzus erythrophthalmus*), (19) white-footed deer mouse (*Peromysus leucopus*), (20) predaceous ground beetle (*Calosoma sycophanta*), (21) braconid wasp (*Apanteles melanoscelus*), (22) ichneumon fly (*Itoplectis conquisitor*), and (23) tachinid fly (*Compsillura concinnata*). (Adapted and updated from publications by R. W. Campbell, D. R. Houston, M. R. McManus, R. J. Sloan, W. E. Wallner, and R. T. Zerillo, U.S. Department of Agriculture.)

production of synthetic textiles with similar properties (see the accompanying discussion). Lepidoptera can also contribute to the enrichment of human diet, especially in tropical areas traditionally short of readily available food resources: Several species of caterpillar are eaten by native populations in South America and New Guinea; in Madagascar the caterpillars and pupae of the Lasiocampid *Borocera madagascariensis* are eaten; in Mexico the giant yucca skipper (*Megathymus*) larvae are a delicacy; and Australian aborigines regularly eat "witchetty grub," which consists of the larvae of the cossid *Xyleutes leucomochla* gathered from the roots of *Acacia*. The caterpillars of sphingids and saturnids, living or dried, are sold in markets in Africa, and we can find the caterpillars of the silkworm sold in supermarkets or delicatessens. Such foods will perhaps seem less exotic if we recall that the larvae of *Cossus cossus*, served with honey, were greatly appreciated by bons vivants in ancient Rome.

The Butterfly As a Cultural Object

The human interest in the natural environment has always found expression in a variety of ways and on different levels. Butterflies and moths, which form such a conspicuous part of nature, are treated differently in different cultures. Since man makes virtually no practical use of these creatures (butterflies and moths are only marginal resources, unlike other animals and plants, and only recently have they become the object of scientific research, it appears that man's relationship with the butterfly has been played out — at any rate up to now — on the stage of the imaginary and the symbolic. We therefore consider the processes that promote these creatures from part of the natural world to objects of cultural interest. To assess exactly how the promotion comes about or why the Lepidoptera should be endowed with cultural significance above and beyond their ecological function would call for an analysis of the psychological processes and the anthropological and philosophical concepts involved in the interpretation of nature. We do not intend to go into such issues; we have the more modest aim of shedding light on a few of the cultural aspects of the relationship between man and the Lepidoptera and shall confine ourselves to two areas that give rise to interesting hypotheses: language and popular traditions.

Butterflies and language

If we accept the principle that understanding of the environment can be gained only through experience, that is, through culture, and if it is true that it is mainly through language that we gain access to culture, then language must consist of the codification of experience.

G. R. Cardona, philologist and ethnolinguist at La Sapienza University, Rome, with whom we have discussed some aspects of the relationship between man and butterflies, has made a preliminary examination of certain significant points, which we summarize here. First and rather surprisingly we find that — despite the striking appearance of the Lepidoptera, their diversity of shape and colour, and their coexistence with human habitats — the world of butterflies is not classified linguistically. In other words the variety to be seen in nature is not reflected in language; it is very rare for a culture to have more than one word to describe all the various species of Lepidoptera, and indeed it is comparatively rare for any language, especially an ethnic language, to have a single word that applies solely to Lepidoptera. A good example is a language like Huave (state of Oaxaca, Mexico), in which the same word, *miek*, covers all species that we should classify as "butterflies," "moths," "dragonflies," and "bats." And yet Huave has words for the other insects

The conquest of the silkworm

Sericulture, which began in China 4,000 years ago, is one of the few examples of successful economic exploitation of a butterfly or moth. The origins of silkworm cultivation have come down to us as legends. But whatever the origins of this profitable industrial undertaking may have been, for more than twenty centuries China jealously kept secret the methods of silkworm culture. The production of silk was an imperial monopoly, and the death penalty awaited anyone who revealed the secret or produced silk on his or her own account. The Greeks, the Romans, and the other peoples of western classical antiquity were not acquainted with the silkworm; they imported small quantities of the precious thread or material from China. In the time of Augustus China and silk were equivalent: The Chinese were called Seres (from *se*, "silk"), and the information about them essentially related to silk, the silk trade, and the geographical knowledge that it gave rise to. From the first century B.C. and for some 600 years thereafter silk was imported into Europe along the many paths of what was called the "silk road." In imperial Rome it was exchanged for gold and was in such great demand that Tiberius was forced to ban silk clothing to stop the export of so much gold. In the Byzantine era Justinian installed silk looms in the imperial palace at Constantinople and decreed a government monopoly on weaving and trading in silk. Meanwhile the continual armed conflicts among the peoples of the Middle East made it increasingly difficult and expensive to replenish the stocks of raw material, and it seems that Justinian actually sent two men to China with the task of discovering the secret of silk. In A.D. 552 the silkworm was finally introduced into Europe, and the Chinese hegemony over sericulture came to an end. The most plausible version of this historic event — a real case of industrial espionage — recounts that two Nestorian monks, Persians by birth, brought silkworm eggs out of Sinkiang concealed in their bamboo staves. Breeding was established first in Lydia and Syria; then in the eleventh century *Bombyx mori* reached Spain thanks to the Arabs, and Basilian monks brought silkworms to Calabria from Greece and to Sicily in 1146. For some 500 years, from the beginning of the thirteenth century through the seventeenth, the foremost producer of silk was Italy: The principal centers grew up first in the south at such places as Catanzaro, Palermo, and Catania (there is a village in Calabria called Buonvicino — originally Bombicino, a name derived from the Latin for "silkworm" — reflecting the importance of sericulture in southern Italy). Sericulture later flourished in many northern cities — Genoa, Verona, Venice, and others — and from there it spread to France and the rest of Europe; it did not reach the United States until 1619. The end of the nineteenth century saw the decline of sericulture in the West and the resumption of silk imports from the East.

and for birds. This instance is not unique: Preuss, for example, says that Cora speakers (from Jalisco and Nayarit states in Mexico) use the same word for butterflies and moths as they do for birds. Having noted how little attention is given to multiplicity, let us go on to our second point and see how the names of Lepidoptera (when there are any) are constructed. In some cultures the butterfly or moth has a symbolic connection with the soul: For instance the Greek word ψυχή means both "soul" and "butterfly," and in Russian *bábochka* literally means "grannie" (*bábka* is "grandmother"), and the dialect word *dúshichka* "little soul" (*dushá*, "soul"). The words used in other languages reveal different associations: "Butterfly" in English may reflect the old Anglo-Saxon belief that cream and butter were stolen by fairies and witches disguised as butterflies; originally the word probably referred only to the common yellow species, such as some of the pierids. The vocabulary of the European languages contains many mythical and legendary references, but this is true of specific names not applicable to all Lepidoptera. Moreover these names are normally translations into the national language of scientific names coined by lepidopterists. However, there are many instances of popular names unrelated to the official nomenclature. In a dialect of northeastern Calabria, for instance the *Amata*, a common and abundant moth in that area, is known as the *previtariello*, the "young priest" or "seminarist," a name that seems justified by its characteristic black and red colouring.

But let us look at the words used for Lepidoptera in general. In most languages the words have no special significance in themselves but are made up of sounds that create a sensory impression. It is not so much the colours of a butterfly that tend to suggest the word for it as its silent flight, unique among flying creatures in that the opening and closing of its wings is clearly visible with the naked eye. Hence it is its behaviour rather than appearance that suggests the name. The German word *Schmetterling* comes from the verb *schmettern*, "to throw down," "to crash." The butterfly's motion in flight tends to be reproduced linguistically in a double-movement pattern: the repetition of a single or double syllable. This syllable very commonly consists of a labial sound (*p*, *b*, or *f*) followed by a vowel and then a liquid consonant (*l* or *r*). An example of the pattern is the Latin *papilio*, derived from the doubling of a root *peb*, "fly," "swim," as if it were *palpelio*. The word *papilio* is preserved in the French *papillon* and in the name of the tent that is derived from it, *pavillon* in French, or "pavilion" in English (because of the association between the raised flaps of the tent and the wings of a butterfly — an association that was made as long ago as the late Roman era). The Italian word *farfalla* follows the same pattern, as does the Portuguese *borboleta*, although both of these are fairly far removed from the Latin. The name of a butterfly or moth is reinvented in much the same way even in different geolinguistic areas. Thus the pattern described above (repetition of a single or double syllable) is also found in some Dravidian languages (spoken in southern India), such as Tamil and Malayalam, in the word *pāppātti*, in Koda with *papili*, and in Kurukh with *papla*. The Balinese *kupukupu* is also based on imitative sounds although its component phonic units are different from those of other languages.

Popular Traditions

The Angamis (a Naga tribe) are extremely careful not to hurt certain species of butterfly. In the Solomon Islands the magical connection between the soul and the butterfly is experienced personally by those on the point of death, who gather their relatives around them and tell them into what animal, bird, or butterfly they have decided to transmigrate. From then on the family treats the chosen species as sacred. On meeting the creature, a son will say, "That is my father!" and offer it a coconut.

Butterflies and Man

Butterflies in art

Butterflies are the insects found more often than any others in the fauna of figurative art. Probably the oldest representations are in Egyptian wall pictures at Thebes and those on certain Mycenaean seals. In China we find the silhouette of the silkworm on jade amulets of the Chou dynasty; in pre-Columbian Mexico the butterfly is a frequent motif in ceramic art. The artists of the Alexandrian period also decorated sarcophagi and low reliefs with butterfly images, and the statue of Psyche (the maiden loved by Eros) in the Capitoline Museum has butterfly wings. More recently *art nouveau* often drew inspiration from these insects, making good use of their pictorial impact. The attitude of the great tradition of oriental art is quite different: Chinese and Japanese artists, so sensitive in their treatment of natural subjects, depict butterflies in all their mobile delicacy.
The examples are too numerous to mention: The fascination butterflies arouse has found expression in an enormous variety of works of art and in many different art forms: painting, sculpture, engraving, gem carving, pottery, textiles, metalwork, glass, enamels, and drawings. In lyric poetry too butterflies have enjoyed a place of great importance. The butterfly is surely the poetic creature *par excellence*, which can be associated with such antithetical human emotions as joy and sorrow. Endowed with great symbolic significance, a symbol of the passage of time, the butterfly embodies two opposites, eternal life and transience: The human taste for the ephemeral has curiously consecrated the immortality of the butterfly.

Some European peoples believe that witches or fairies often turn themselves into butterflies or moths and so get into houses, causing all kinds of mischief. In Westphalia an expulsion ceremony was practiced regularly. On 22 February the children would go from house to house carrying hammers, chanting gay songs, or reciting rhymes telling all the butterflies to fly away. A similar ceremony could be carried out by the residents of the houses, who would knock on the doors of all the rooms to drive out the witches. If this rite were not performed, there would be disastrous consequences: The house would be invaded by countless little parasites and rats, and butterflies and moths would swarm in great numbers round the milk bowls. In Serbia it is believed that butterflies are the souls of witches; the method of getting rid of them is as follows: If you find a witch asleep, you should stop her mouth so that her soul-butterfly cannot reenter her body through the mouth and she will die. Elsewhere butterflies are considered signs of good or bad luck. In Brunswick the nature of the luck is determined by the colour of the first butterfly seen when summer comes: If the butterfly is white, is presages bad luck; if it is yellow, the birth of a baby; if multicoloured, a wedding. In the pre-Columbian cultures of Central America butterflies were very important. Religious and mythical tradition prescribed the greatest respect for them, and they were depicted everywhere in the decoration of temples, statues, liturgical objects, and ornaments. The symbolic references seem to vary, but in fact they all refer to life in its different forms and to the gods (often cruel gods) who embody it. Butterflies are associated with the souls of the dead, new plant growth, the heat of fire, the light of the sun, and relevant sacrificial rites. This close relationship between butterflies and the endless transformations of nature is also the source of the symbolism that gives rise to the butterfly-soul identity; it is typical of a magical conception of nature, according to which nature is a continuum of things and processes in reciprocal transformation. This conception is given poetic expression in one of the most famous passages in the Chinese philosophical tradition. In the work named after him the great Taoist Chuang-tze tells how he dreamed that he was a butterfly. On waking and pondering upon his dream, he came to doubt his own identity. "Who am I? Chuang-tze, who sleeps and dreams that he is a butterfly, or a butterfly that is awake and imagines it is Chuang-tze?"

Now if it is true, as it seems to be from these brief considerations, that our interest in the magical world of butterflies and moths does not focus upon their diversity of form and colour but rather concentrates on a single activity, flight, and a single process, metamorphosis, this interest may well be due to these two attributes, of the Lepidoptera having greater evocative and symbolic power. On the other hand to subsume Lepidoptera under the categories of movement and change may be a way of avoiding conflict with the mythical notion of nature as a sequence of multiple states of the same single reality, of which we are an integral part. It is worth emphasizing that this complete involvement has produced the respect for the environment that has allowed our species to survive.

Classification of Lepidoptera

Classification and Nomenclature

The clarity of a biological classification is one of its most important attributes and must be based on an unambiguous system of nomenclature. In essence a classification system consists in a series of names arranged according to certain criteria; it is essential that it should enable us to discover unequivocally the taxon to which a given organism belongs. Notwithstanding these premises, however, the official systems of nomenclature are not so perfect as they might be, and for some groups, of which the Lepidoptera are one, the confusion among names of both genera and species is extremely awkward. As has been mentioned in the chapter on systematics, there are rules governing the usage and priority of the scientific names of the Lepidoptera. The norms were adopted in 1961 and, to clear up the countless points not in agreement with the new code, have required the reexamination of the whole of the enormous literature from the time of Linnaeus to the 1960s. This positively titanic task has been undertaken by a group of experts at the Natural History Museum of London, who with the aid of lepidopterists from all over the world are working on the indexing and critical examination of all the names of genera of Lepidoptera and of their species, with the aim (1) of establishing what the valid name of each genus should be and (2) of linking the species typical of that genus to the name. When this great *catalogue raisonné* of the generic names of the Lepidoptera (over 26,000 names at a recent count) has been compiled, it will make the classifiers' work easier and will clarify their decision.

Traditional Classifications

From Linnaeus onwards the criteria on which past writers have based the systematic division of Lepidoptera have been numerous and heterogeneous. Consequently the old systems of classification differ even from each other and are difficult to compare with modern systems. However, although they are now regarded as too arbitrary and crude and are no longer used in modern scientific literature, it is opportune to mention them briefly; for use is made of them almost every day. In the first volume of *Systema Naturae* Linnaeus divided the order of Lepidoptera into three groups, *Papilio*, *Sphinx*, and *Phalaena*, which differed from each other in the form of their antennae and the position of their wings at rest. This classification more or less coincides with the distinctions of the great French entomologist P. A. Latreille (1762–1833): "diurnal," "crepuscular," and "nocturnal" groups. Again, these three groups correspond (roughly speaking) to the groups called Rhopalocera (see below), Sphingidae, and a third group comprising all the other families. Another type of distinction, based on the size of the adult, is that between Microlepidoptera and Macrolepidoptera. Although this division has no scientific basis, it is still sufficiently popular to be worth mentioning. Specialists agree that all forms in the first three of the modern suborders of Lepidoptera (Zeugloptera, Dachnonypha and Monotrysia) fall into the category of Microlepidoptera although some collectors usually exclude the large Hepialidae. As regards the families of the fourth suborder (Ditrysia), however, there is no consensus. In some cases the Ditrysia regarded as Microlepidoptera include the Pyralidae and all the families of the Tortricoidea, Tineoidea, and Pterophoroidea. Other classifications are more comprehensive and include in the Microlepidoptera all the families of Ditrysia from the Cossoidea to the Pterophoroidea. All the other families of the Ditrysia belong to the Macrolepidoptera. A distinction still widely used today is that between Rhopalocera — butterflies with clubbed antennae — and Heterocera — moths with variously shaped antennae. However, this distinction, which we owe to J. B. A. Boisduval (1801–1879), is also one that cannot really be justified; for while it is true that the Rhopalocera (that is, the superfamilies Papilionoidea and Hesperioidea) form a homogeneous group, the Heterocera include moths that are so diverse in morphology and biology as to make the group entirely artificial. Table 1 shows how the three early classifications described above correspond to each other.

Table 1

Linnaeus	Latreille	Boisduval
Papilio	Diurnal	Rhopalocera
Sphinx	Crepuscular }	Heterocera
Phalaena	Nocturnal }	

Table 1
Comparison of the Classifications of C. Linnaeus, P. A. Latreille, and J. B. A. Boisduval.

Modern Scientific Classifications

Besides the early classifications mentioned above, there are many other classifications of a far more technical kind, covering the larger taxonomic categories (suborder and superfamily). As a result of strict selection of differentiating morphological features, they have all contributed to the current systematic arrangement; it is thus useful to mention at least the main ones and compare them with the classification used in this book. A comparison of the main classifications at the suborder level appears in Table 2.

Table 2

	Comstock	Tylliard	Börner	Hinton	Common
Micropterigoidea				(non Lepidoptera)	Zeugloptera
Eriocranioidea	Jugatae	Homoneura		Dachnonypha	Dachnonypha
Hepialoidea			Monotrysia		
Nepticuloidea				Montrysia	Monostrysia
Incurvarioidea	Frenatae	Heteroneura			
Other superfamilies			Ditrysia	Ditrysia	Ditrysia

Table 2
Comparison of the Classifications of Suborders Proposed by J. H. Comstock (1982), R. J. Tillyard (1918), C. Börner (1925), H. Hinton (1946), and I. F. B. Common (1970).

At the end of the nineteenth century the American entomologist J. H. Comstock (1892) divided the Lepidoptera into two suborders, Jugatae and Frenatae. He observed that the connection between forewing and hind wing could be made either by the jugum or by the frenulum (for a description of these two structures see the opening chapter). A few years later F. Karsch (1898) and R. J. Tillyard (1918) proposed a different classification, based on the wing venation. Two suborders were described, which Tillyard referred to as Homoneura and Heteroneura, depending on whether the fore- and hind wings had similar or different venation. It will be seen that Tillyard's suborder Homoneura corresponds to Comstock's Jugatae, to the present-day suborders Zeugloptera and Dachnonypha, and to the superfamily Hepialoidea of the Monotrysia. Tillyard's Heteroneura corresponds to Comstock's Frenatae and embraces the current suborder Ditrysia plus the superfamilies Nepticuloidea and Incurvarioidea of the Monotrysia.

A different classification was adopted by A. S. Packard (1895), based on the adult mouthparts. This writer distinguished two suborders, Laciniata and Haustellata, differentiated by whether the adult has developed, functional mandibles adapted to mastication or has mandibles that are atrophied or absent and replaced by maxillary lobes developed to form a proboscis. Compared with current usage, Packard's suborder Laciniata corresponds to the Zeugloptera, and the Haustellata to all the other Lepidoptera.

The next step was taken by C. Börner (1925), who in a study of the Lepidoptera of Germany divided his classification into Monotrysia and Ditrysia, two suborders distinguished by the presence in the female of a single reproductive aperture or of two. Börner's suborder Monotrysia corresponds to the suborders Zeugloptera, Dachnonypha, and of course Monotrysia. There is no difference between Börner's Ditrysia and the same suborder in today's classification. Börner's classification soon attracted the attention of scientists and provides the basis for all later systems; in broad outline it is still valid today.

The classification into Monotrysia and Ditrysia was later modified by an Englishman, H. E. Hinton (1946), who separated some primitive families of Eriocranioidea from Börner's Monotrysia and classified them as the present suborder Dachnonypha (which we have mentioned several times). Hinton also regarded the Micropterygoidea as so different from Lepidoptera that he put them in a separate order, the Zeugloptera. His classification of Lepidoptera thus has three suborders, Dachnonypha, Monotrysia, and Ditrysia. However, some authors have recently put the Zeugloptera back among the Lepidoptera, making a fourth suborder. The classification adopted in this book has in fact four suborders, Zeugloptera, Dachnonypha, Monotrysia, and Ditrysia; at the suborder, superfamily, and family levels it uses, apart from some slight modifications, the classification proposed by I. F. B. Common (1970) in *Insects of Australia*.

The most comprehensive of the four suborders is the Ditrysia, which

Table 3
Summary of the Number of Superfamilies and Families in Each of the Four Suborders of Lepidoptera.

includes about 98 percent of the species described; only 2 percent belong to the other three suborders. The number of superfamilies and families in each suborder is given in Table 3 and in the list that closes this chapter.

Table 3

Suborder	Superfamilies	Families
Zeugloptera	1	1
Dachnonypha	1	5
Monotrysia	3	9
Ditrysia	18	93
Total	23	108

For details of the principal taxonomic characters of the suborders, superfamilies and families, see the chapter Systematic Account of the Families of Lepidoptera.

Zeugloptera
MICROPTERIGOIDEA: Micropterigidae

Dachnonypha
ERIOCRANIOIDEA: Eriocraniidae, Agathiphagidae, Neopseustidae, Lophocoronidae, Mnesarchaeidae

Monotrysia
HEPIALOIDEA: Prototheoridae, Palaeosetidae, Hepialidae
NEPTICULOIDEA: Nepticulidae, Opostegidae
INCURVARIOIDEA: Incurvariidae, Prodoxidae, Heliozelidae, Tischeriidae

Ditrysia
COSSOIDEA: Cossidae, Dudgeoneidae, Compsoctenidae, Metarbelidae
TORTRICOIDEA: Tortricidae, Phaloniidae
TINEOIDEA: Pseudarbelidae, Arrhenophanidae, Psychidae, Tineidae, Lyonetiidae, Phyllocnistidae, Gracillariidae
YPONOMEUTOIDEA: Sesiidae, Glyphipterygidae, Douglasiidae, Heliodinidae, Yponomeutidae, Epermeniidae
GELECHIOIDEA: Coleophoridae, Agonoxenidae, Elachistidae, Scythridae, Stathmopodidae, Oecophoridae, Ethmiidae, Timyridae, Blastobasidae, Xyloryctidae, Stenomidae, Cosmopterigidae, Gelechiidae, Metachandidae, Anomologidae, Pterolonchidae
COPROMORPHOIDEA: Copromorphidae, Alucitidae, Carposinidae
CASTNIOIDEA: Castniidae
ZYGAENOIDEA: Heterogynidae, Zygaenidae, Chrysopolomidae, Megalopygidae, Cyclotornidae, Epipyropidae, Limacodidae
PYRALOIDEA: Hyblaeidae, Thyrididae, Oxychirotidae, Pyralidae
PTEROPHOROIDEA: Pterophoridae
HESPERIOIDEA: Hesperiidae, Megathymidae
PAPILIONOIDEA: Papilionidae, Pieridae, Nymphalidae, Libytheidae, Riodinidae, Lycaenidae
GEOMETROIDEA: Drepanidae, Thyatiridae, Geometridae, Uraniidae, Epiplemidae, Axiidae, Sematuridae
CALLIDULOIDEA: Callidulidae, Pterothysanidae
BOMBYCOIDEA: Endromidae, Lasiocampidae, Anthelidae, Eupterotidae, Mimallonidae, Bombycidae, Lemoniidae, Brahmaeidae, Carthaeidae, Oxytenidae, Cercophanidae, Saturniidae, Ratardidae
SPHINGOIDEA: Sphingidae
NOTODONTOIDEA: Notodontidae, Dioptidae, Thyretidae
NOCTUOIDEA: Ctenuchidae, Hypsidae, Nolidae, Arctiidae, Lymantriidae, Noctuidae, Agaristidae

Is Classification a Science?

It was said earlier that biological systematics is the study of the diversity of creatures in relation to their classification and phylogenetics. Biological classification in turn consists in the grouping of organisms on the basis of common characteristics. The result of this process is a formal classification; that is, the species are placed in increasingly comprehensive groups. There are of course several different formal classifications of the Lepidoptera (both at higher levels, like the suborders, and at the lower, like the subfamilies). In addition, as we have seen in the course of the review of the different families, there are vast numbers of groups whose systematic position is uncertain. This state of affairs suggests that biological classification systems are necessarily no more than highly approximate models of observed reality. They constitute attempts to interpret organic diversity, and while they are not wholly subjective, neither are they totally objective in that they reflect the systematic philosophy of their authors. A classification system does not therefore constitute a scientific law — or at any rate not in the sense in which we say that the law of universal gravity is a scientific law. A biological classification has in fact no necessary relationship with its object; it has no absolute normative value but consists rather of a provisionally valid model that incorporates and links many observed and interpreted facts in the light of theory but that usually gives rise to a number of exceptions. Generally speaking, the best classification systems make it possible to verify theories that may be formulated on the basis of the assumptions that underlie the model in question. However, although the scientific value of biological classifications has its limitations, it must nevertheless be said that we cannot do without them if we want to give meaning to the enormous diversity of living creatures and so to our knowledge of nature.

Catching and Preserving Lepidoptera

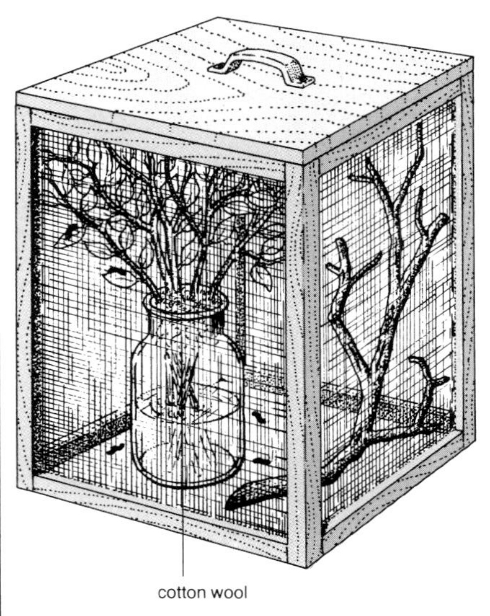

Figure A
A breeding cabinet

Breeding

The breeding, or rearing, of Lepidoptera provides one of the most convenient ways of carrying out many interesting observations on the biology of these insects. Lepidoptera can be bred during any stage of their development cycle. Eggs and caterpillars are not difficult to find; note and identify the food plant, and inspect it carefully, paying particular attention to the underside of the leaves, where the eggs are often laid. It is sometimes more difficult to find pupae because many species pupate under stones or bark or in the earth. When collected, the eggs are put into clean, well-aired containers, the bottom of which is covered with blotting paper; as the larvae hatch, it will be necessary to transfer them to larger containers, containing plenty of food and some twigs to serve as perches. It is always important to avoid overcrowding. Fresh leaves of the plant that the caterpillars feed on must always be available; a small supply kept in the refrigerator will come in useful. Every two or three days or more often if necessary replace the old leaves with new ones and clear the container of waste matter. After the second or third change the caterpillars may be transferred to a cage like the one illustrated in Figure A; it should be provided with branches for the pupae to attach themselves to. For species that pupate in the earth it will be necessary to put a layer of clean, not too fine sand on the floor of the cage. Containers must always be large enough to allow the adult butterflies or moths to fly after they emerge from the pupa. The temperature, humidity and light conditions will vary depending on the species being raised; it is only with experience and advice from more expert breeders that you will learn the tricks of the trade.

Preservation of specimens is different for the various preimaginal stages. Eggs may be frozen or killed in cyanide and then placed in small airtight vials. Preserve larvae by killing and distending them in boiling water or KAAD solution (kerosene, acetone, acetic acid, and dioxane) and then storing them in vials with 70 percent ethyl alcohol (ethanol). Larvae, including caterpillars, may be preserved for display by freeze drying, but individual collectors rarely have the proper equipment for this.

Every specimen in a collection should have a card recording the place and date of collection and the name of the collector. For a specimen that has been bred the card should record the place and date of collection of the initial stage and also the date of emergence from the pupa. It is recommended that a diary be kept of all information about the breeding.

Collecting Diurnal Lepidoptera

The methods of collecting adult specimens and the apparatus needed are different, depending on whether you are collecting butterflies or moths. The indispensable instrument for butterflies and for diurnal moths is an aerial-insect net, which you can either buy from a biological-supply house or make yourself. The basic design consists of a handle, a frame to which the net is attached, and the net itself. The handle may be made of wood, bamboo, rattan cane, aluminum, or glass fiber. In open areas with low vegetation, where it is possible to chase the butterflies, the handle need be only 60 to 80 cm (24 to 32 in) long, but to catch species that normally settle or fly very high, the handle will have to be proportionately longer — up to 4 to 5 m (12 to 15 ft) — and possibly telescopic. The frame to which the net is attached consists of a stainless-steel ring, with one or more joints and 30 to 50 cm (12 to 20 in) in diameter, attached to the handle. Figure B shows two jointed rings and three different ways of attaching the frame to the handle. The bag itself can be made of nylon netting (white, green, or black) or of some other translucent material, but it must be as strong as

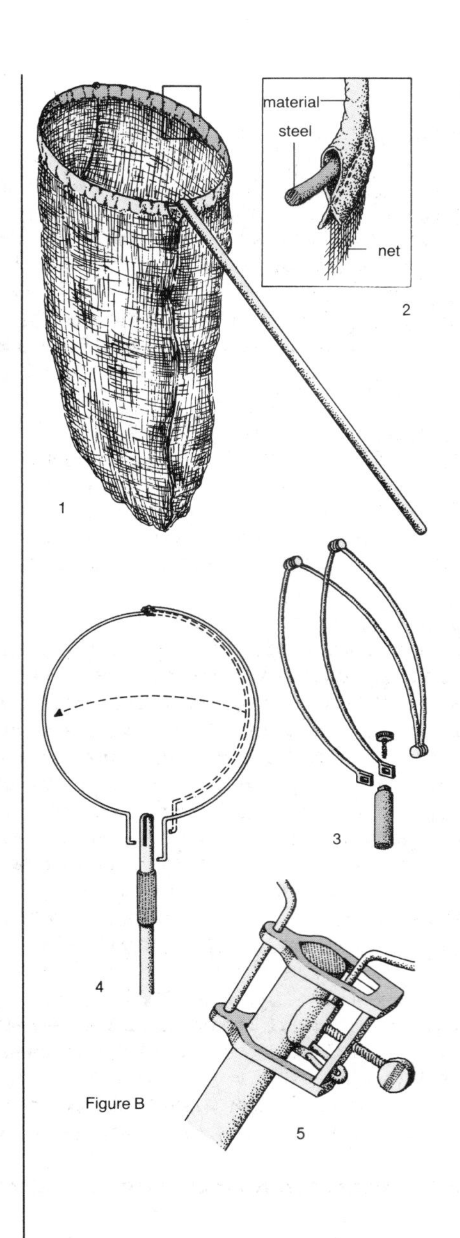

Figure B
(1) An insect net, with a detail showing how the net is attached to the ring (2).
(3–5) Various types of frame and attachment for detachable nets.

Figure C
A simple way to make an envelope for Lepidoptera.

Figure D
Trap baited with fermented fruit for catching diurnal Lepidoptera. This type of trap is particularly recommended for attracting and collecting Nymphalidae, Hesperiidae, and Riodinidae in tropical areas.

possible to withstand tearing by briars and branches. The bag, shaped and attached as shown in Figure B: 2, should be at least 70 to 80 cm (28 to 32 in) deep to prevent the insects from escaping. Once in the net the lepidopteran may be killed by gentle pressure on the thorax with the index finger and thumb or by the introduction of a jar containing ether, chloroform, or potassium cyanide (KCN) into the net. The specimen, if a butterfly, is then transferred to a paper envelope of appropriate size. Many collectors like to use square glassine envelopes; others prefer triangular envelopes made by folding a sheet of paper as shown in Figure C. While catching the insects, keep the envelopes in a fairly stiff container within easy reach in one of the many pockets necessary to the entomologist. If you go to several places on the same day, make sure that you keep the envelopes from each biotope separate, and mark each one in indelible ink. Most moths should never be placed in envelopes but should be transferred to tins containing slightly moistened tissue. If possible, they should be field pinned or spread on the same day as capture. Lycaenidae are best pinned in the field because they are very hard to set once they have dried. Zygaenidae have a highly elastic exoskeleton; so it is best to kill them by using pins previously dipped in a solution of nicotine or a tobacco extract, which can be made by soaking a cigar in a little water. Lycaenidae and Zygaenidae should be put in a small box after pinning.

Other indispensable items are entomological tweezers, a set of insect pins, and a notebook to record everything of interest in the behaviour of a given species, the food plants, and the composition of the fauna of a given biotope.

Many entomologists use bait to attract diurnal or noctural Lepidoptera that would otherwise be hard to catch. This is a particularly common practice in the tropics. Some butterflies, such as the Apaturinae and some of the Nymphalidae, are attracted to rotting organic matter or by the excrement or urine of animals. Many Lycaenidae (blues), especially in periods of drought, swarm on wet sand near streams or rivers. In the tropics large aggregations of Pieridae (*Phoebis*, *Eurema*) or Papilionidae (*Eurytides*, *Papilio*) can be seen on wet sand or along watercourses where the local people wash their clothes. Apparently many species of *Eurytides* are actually attracted by soapy water. To lure the Ithomiinae, some collectors use pieces of dried *Heliotropum indicum*, one of the Boraginaceae, since it contains attractive pyrolizidinic alkaloids. Other baits commonly used in the tropics are fermented bananas and pineapple, which attract *Morpho*, *Caligo*, and other Nymphalidae, Hesperiidae, and Riodinidae, or shrimps left in the sun, which lure the males of Charaxinae. Some butterflies are also attracted by brightly coloured objects: the Heliconiinae by red and *Morpho* by blue. Dead specimens may also act as visual baits, for example, in certain *Morpho*, *Ornithoptera*, and *Papilio*. All these baits can be used in traps like the one shown in Figure D. This type of trap has the advantage of catching a remarkable number of species, which can yield important information on the composition of the fauna of a community, seasonal variation, or altitudinal zonation of certain taxa. The use of baited traps often proves to be the quickest and most reliable way to catch some species. Another distinct advantage of such traps is that they do not damage the specimens caught, which can be released after they have been examined and possibly photographed. Among the many natural lures whereby butterflies and moths may be caught flowers are of course the commonest; with a little experience you will soon learn to recognize which flowers are most attractive to Lepidoptera. Butterflies seem to prefer small flowers of various configurations, not necessarily scented but often white, pink or purple. The flowers visited by crepuscular and nocturnal species are generally scented, pale, or white in colour.

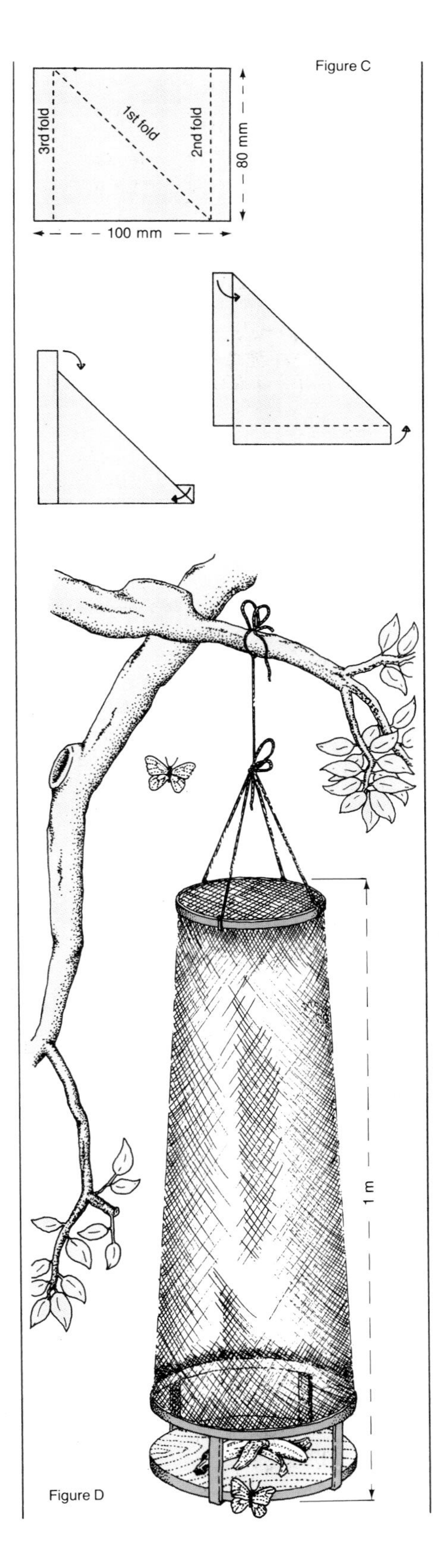

Collecting Nocturnal Moths

The best nights for collecting nocturnal moths are damp ones with no moon and no wind. Some collectors feel it advantageous to choose a spot high up to get a good view, but others find sheltered glades even better. To catch moths, a white sheet is used, stretched between two trees or two poles fixed in the ground, with a light shining on it. At one time the source of light was a kerosene or paraffin lamp; this has now been replaced by a filtered ultraviolet light or a mercury-vapour lamp of not more than 160 W placed above the sheet on a third pole about 1.5 or 2 m high. A motorcycle or automobile battery can be used as a source of power, or a small 200- to 300-W generator. When the moths reach the sheet, they are caught in a cork-stoppered wide-necked jar, inside which is a pad of cotton wool soaked in chloroform or ether; as soon as the moths are stupefied, they are to be transferred to a jar that contains KCN. Note, however, that this is an extremely powerful poison and that amateur collectors are not advised to use it; indeed the greatest care must be exercised in handling any of the chemicals for killing moths. You can kill big species, such as some Sphingidae and Saturniidae, by injecting a few drops of ammonia into the thorax.

An alternative means of catching moths, in which a sheet is not necessary, is the Rothamsted trap. When the moths come near its light, they spontaneously go through an opening into a container underneath the lamp. A series of baffles increases the internal surface area of the container so that there is room for many individuals. On the bottom of the container is a holder for ether and chloroform, which rapidly kills the moths. This trap catches and, if loaded with poisons, kills a great number of Lepidoptera and other insects indiscriminately. Its use is therefore limited to professional lepidopterists when they need very big samples, indispensable for certain types of research into the quantitative ecology of a community, such as estimates of abundance, or frequency, or diversity. Some nocturnal moths can also be attracted by scented baits; this is true of many *Catocala*, *Euxoa*, *Boarmia*, and so on. There are many recipes for such baits. Often sugary jam is mixed with sweet, perfumed wine or fermented beer; sponges are soaked in it and are threaded on pieces of wire. The sponges are hung up before dusk on the branches of bushes at regular intervals along paths or in the middle of clearings. Obviously these baits will be most successful when few flowers are out. Finally pheromones are extremely effective baits. Virgin females kept in cages attract males of their species; for species harmful to agriculture synthetic pheromones are used.

To catch nocturnal moths, lepidopterists have long made use of the moths' strong attraction to light. Although it is well known, this phenomenon has never been given an exhaustive scientific explanation, and it is only in the last few years that it has been properly investigated, thanks to the preliminary work of E. R. Laithwaite and P. S. Callahan. It has been shown that the infrared component of light stimulates, in many nocturnal moths, attraction responses similar to those induced by pheromones. It seems certain, moreover, that males are much more sensitive to this than are females; this difference does not seem to be a matter of chance, given that the females of some species emit not only pheromones but also infrared rays. Once it can be confirmed that there is an organ in the male that is sensitive to infrared rays, it will then be possible to deduce that such radiation exercises a sexual function; a receiver of this type has indeed been discovered for the first time in the male antennae of the lymantriid *Orgya antiqua*. It is interesting that a moth's flight towards a light is similar to the orienting flight towards a signalling female. In both cases the flight does not seem casual but is

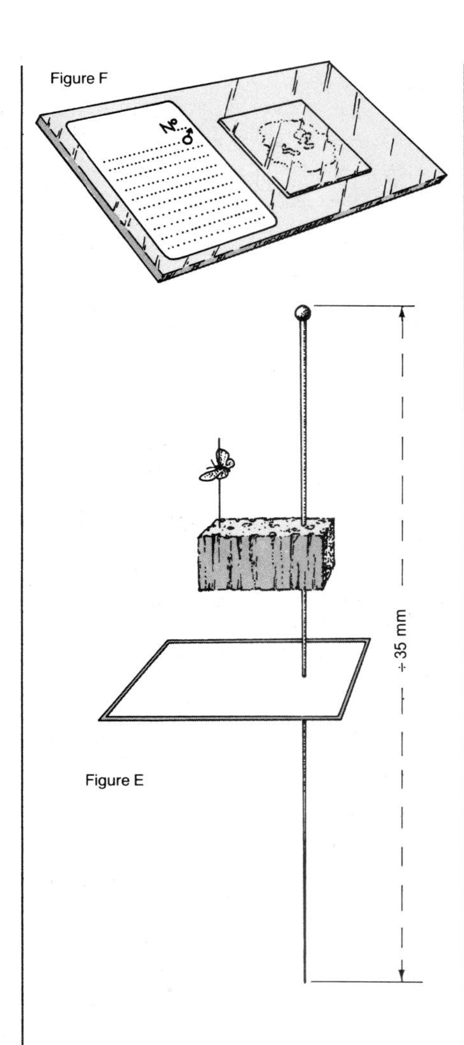

Figure E
Microlepidoptera set on a minutien.

Figure F
Slide mounted for storage of the genitalia and other anatomical features.

Figure G
(1) Butterfly on the spreading board: cross section showing the correct level to which the pin is inserted.
(2) General view of a spreading board.
(3–5) Sequence of operations for setting a butterfly with strips of paper and pins.

determined by the necessity of satisfying a need; that is, the male controls the path along which he flies. The path is actually defined by a gradient: in one case by the concentration of the pheromone, in the other by the intensity of the radiation. The path is probably plotted by the balance between the signals — pheromone or light wave — reaching each of the two antennae. The male in fact follows a path such that the strength of the signal (the gradient) emitted by the target (a virgin female or a light) is always maximized. The maximization is achieved by continuous comparison between the strengths of the signals — either spatially (as received by the right antenna compared with the left) or possibly temporally (the variations in strength from one instant to another). When the few facts available today are confirmed by a wider range of further observations, we shall know with certainty why moths flutter around a lamp.

Collecting Microlepidoptera

We need to deal separately with the Microlepidoptera: Because they are small — indeed often very small — there are special techniques for catching and preparing them. By day it is particularly easy to catch Pyralidae, Tortricidae, Incurvariidae, and Coleophoridae; many other species are caught in late afternoon or at dusk, others only at night. During the day beekeeper's bellows, attached to a receptacle filled with paraffin-soaked rags, may be used. When the rags are set alight, the smoke disturbs the Microlepidoptera, and instead of dropping to the ground, as they usually do, they fly up from the bushes. A very fine net is used to catch them. Having been caught, the specimen is transferred while still alive to a glass tube, which is immediately closed with a bung of plastic or cork. At home, the Microlepidoptera are killed with cyanide or ether and immediately set on setting boards about 5 cm (2 in) long. Alternatively they may be placed overnight in a moistened relaxing container. Strips of paper are not used as for the setting of Macrolepidoptera (see below), but with the aid of a lens a micropin (minutien) is inserted into the center of the thorax. Then, the insect turned over on its back, the wings are stretched open with two very light glass plates. When the specimen is dry, stick the minutien into a small block of polyporous pith or plastic, which is then placed on a standard entomological pin (see Figure E).

Setting and Storage of Lepidoptera

To set specimens their exoskeleton must be slightly elastic so that structures like the wings and the antennae can be arranged into the desired position, which will then be maintained when the exoskeleton dries and stiffens. Unless the specimens for setting are "fresh", that is, have just died or have been kept in a freezer and so are still soft, they will have to be dampened to restore a certain flexibility. A very easy method is to place the Lepidoptera overnight in a receptacle with humidity at saturation point; the inclusion of a few crystals of thymol or paradichlorbenzene (PDB) will prevent the formation of mould. Alternatively you can soften the specimens by injecting hot water or ammonia into the thorax, taking care that the liquid does not get into your eyes or damage the specimen. Normal humidification may not suffice for Lepidoptera with large bodies and powerful wing muscles; in such cases it is necessary to prick the joints of the wings with a pin or to make a small incision in the membranes of the mesothorax. To set Lepidoptera with their wings open, you will need spreading boards, insect pins, pins with glass heads, strips of parchment

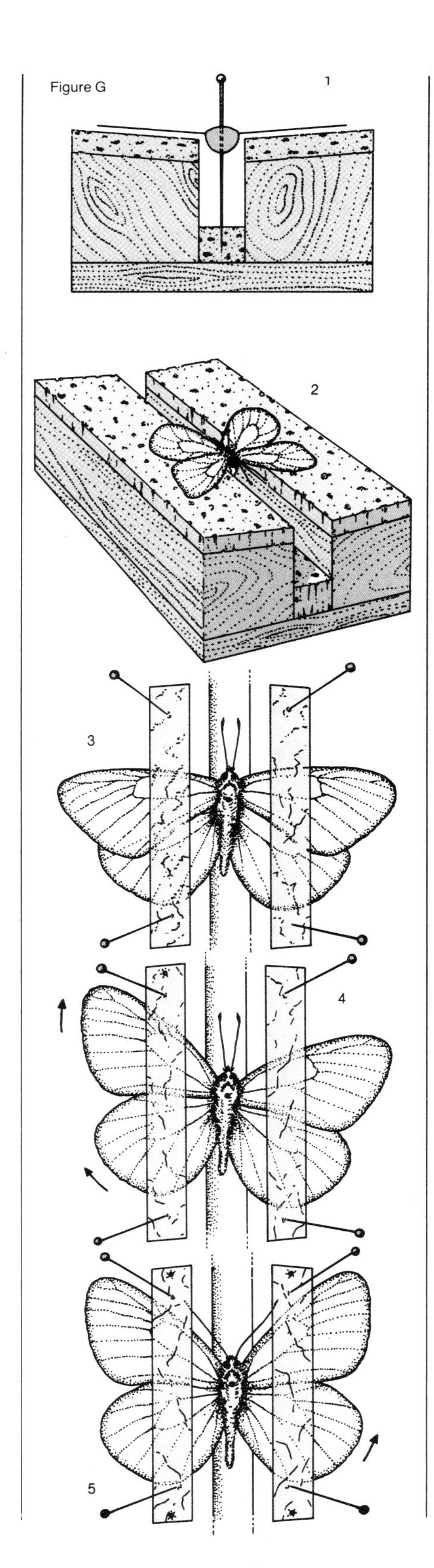

paper, setting needles mounted on handles, and pinning forceps. The mounting board consists of two parallel blocks of soft wood (birch, willow, lime, balsa, and the like) with a groove between them in which the specimen's body is placed. The base of the groove is a strip of material, such as cork or polystyrene, into which the pins are fixed. The specimen is held in position by a pin, the diameter of which should be appropriate for the size of the body. Two-thirds of the length of the pin is inserted vertically through the mesothorax. Insect pins have a standard length of 38 mm (1.5 in) but vary in diameter; the diameter is denoted by numbers from 000 to 7. Sizes 2 and 3 are used for most Lepidoptera. The setting of the wings may be carried out as follows: Turn the left forewing forward with a setting needle until its lower margin is at right angles to the main axis of the body; then turn the hind wing forward and fix it in position. Follow the same procedure for the wings on the right. The antennae are arranged parallel with the costal margins of the forewings. Strips of parchment paper and glass-headed pins are used to fasten the whole specimen firmly to the setting board. Every specimen on the setting board should have a label indicating the place where it was caught. The time taken for the specimens to dry after setting depends as much on the surrounding temperature and humidity as on their size. Generally they should be left for about three weeks in dry surroundings at a temperature of 25°C to guarantee perfect drying; the setting boards must be kept away from dust and out of the light. When the specimen is dry, take it off the setting board, put the label through the pin under the specimen, and place it in an insect storage box or glass-topped drawer. Because any zoological or botanical specimen in practice is useless without a precise account of the place and date it was collected, it is essential to take care to fill in this information on the label. The label should normally be smaller than the specimen and must have the following information in full if possible: precise details of the location where the insect was caught, altitude, date (with the month in Roman numerals), and the collector's name preceded or followed by "leg." (*legit*, "collected"); see example 1 below. Additional information, which is always useful, may be put on a second or third label. If the geographical coordinates of the place where the specimen was caught are not known, give its distance from a place or landmark (town, river, or mountain) that can be found in an atlas. In geographical areas for which we do not have accurate maps the place of collection may be described as in example 2 below:

1. COLO: Laramie Co.,
Fort Collins, 1510 m
VI-15-84 S. Jones leg.
2. SOLOMON ISDS.
Honiara, 35 km SE
Village 2 km inland,
100 m, I-2-83
S. Rossi leg.

If several labels for the same place are required, it may be worthwhile to have them photocopied or printed.

After being caught, Lepidoptera can be stored for several months in a deep freezer at −20 to −40°C. The specimens stay fresh and are as easy to set as if they had just been caught. Small species like the Lycaenidae will not keep long because the ice formed on the tissues soon sublimes at room temperature. If a deep freezer is not available, the envelopes with the dry specimens, with the locality duly marked, can be kept in boxes containing a little PDB. As soon as they are caught, nocturnal moths are pinned and kept in boxes containing a few grains of PDB until they can be set. Once set, the specimens are stored in insect boxes. Insect boxes vary in size from 43 by 48 by 6cm (17 by 19 by 2.5 in) to 30 by 33 by 6 cm (9 by 13 by 2.5 in); they must be of wood, the lids fitting perfectly to prevent carpet beetles

Studying Lepidoptera

Technically speaking, any collection is of value, provided the specimens are well preserved and perfectly labelled. But leaving technical aspects aside, the crux of the problem lies elsewhere; perhaps the best definition of a good collection is one that is intelligently put together. A "purely aesthetic" collection reveals a misunderstanding both of nature and of art. What is it really for? The collector usually aims at completeness, and his or her most pressing concern is to fill in any gaps. In addition to systematic collections, which include all the species, genera, and families of Lepidoptera in a given area, collections can be assembled in many different ways. Let us take just a few examples: collections covering an ecological field: Lepidoptera from wet lands or from a single habitat, Lepidoptera found in city gardens and parks, caterpillars, food plants, pupae, and adults of a small biotope, seasonal succession of the Lepidoptera of a certain community, vertical zonation of one or more families, and so on. Differently arranged collections might be polymorphic populations of Lepidoptera of a certain area, clines of geographical variation in one or more groups, or visual mimicry or cryptism in caterpillars or adults. Thus all sorts of collections of Lepidoptera can be made, and it is important to remember that every amateur's collection, if put together accurately, can make an important contribution to the study of these insects. When a collection of such a kind ceases to be of interest, it should be passed on to a university or natural-history-museum collection, which will look after it and make use of it. Some old collections of Lepidoptera, like those of Cramer, Fabricius, and Linnaeus, are still jealously and lovingly preserved, 200 years later. Of the great modern collections we can mention only the immensely rich and varied Rothschild collection (with two million specimens) in the Natural History Museum, London, and the great collections in the Paris Natural History Museum, the American Museum of Natural History in New York, and the Smithsonian Institution, Washington.
Where there is a large collection, there is usually also an equally important library of specialist literature and often, too, an association for the study of Lepidoptera. These exist all over the world, and some of them publish periodical reviews. In the United States there is the Lepidopterist's Society, which publishes the *Journal of the Lepidopterist's Society*, and the Lepidoptera Foundation publishes the *Journal of Research on the Lepidoptera*. The Xerxes Society, a society for the conservation of Lepidoptera, publishes *Atala*. In Europe the Societas Europaea Lepidopterologica publishes *Nota Lepidopterologica*; in Paris the Lepidoptera Department of the Natural History Museum publishes *Alexanor*.

(Dermestidae) or book lice (Procoptera) from getting in and damaging the specimens irreparably. The lid is made of glass, and the bottom must be made of something into which the pins may be stuck. Nowadays bottoms of high-density polyethelene covered with squared paper are used very successfully. In the past the bottoms were made of compressed cork, peat, or polystyrene, but these all had some serious disadvantages. The cases (in which the specimens have been displayed to advantage) are then put away like drawers in wooden or metal cabinets. Substances such as naphthalene or PDB, which will protect the collections from destruction by insects, are kept in suitable containers in both the boxes or drawers and the cabinets. Many different preserving preparations are used, but virtually all are poisonous to human beings: Some have poisonous fumes (PDB, naphthalene, carbolic acid, and so on); others are dangerous to touch (DDT, organophosphoric esters, and so on). An environment saturated with PDB may in time cause serious damage to the health of people working in it. Before an insect is added to a collection, it is a good rule to examine it carefully to make sure no parasites are present. If there are, it will be enough to put the specimen or the whole case into a deep freezer at −20 to −40°C for a few days.

Setting the Genitals and the Wings

Of the various techniques used to study particular anatomical structures of Lepidoptera we shall illustrate only those used to reveal the venation of the wings and to set the genital apparatus. The venation can be examined by pouring one or two drops of xylol or 90-degree ethyl alcohol onto the wings. The wings will become temporarily transparent with the veins clearly visible; the pattern will return to normal when the liquids evaporate. Another way is actually to remove the scales. To examine the genital apparatus, you will need a microscope. In some cases, with very big specimens that are still fresh, it is possible to squeeze the abdomen and with the aid of a powerful lens observe the genital apparatus as it protrudes. Otherwise proceed as follows: Cut off the last two or three segments of the abdomen (using very sharp scissors), and soften the abdominal tip by boiling it in a solution of distilled water and caustic potash (NaOH, 5 to 10 percent). The concentration and boiling time will depend on the size of the specimen and the degree of sclerification of the genitals. Then, using setting needles, you will have to remove any remains of the musculature, exoskeleton, scales, and skin from the section in the distilled water. When the section is perfectly clean, it must be dehydrated by immersion in alcohol for a few hours (at rising temperatures of 45, 75, 95, and 99°C) and then in xylol. After this place the prepared section on a slide, and cover it with a drop of Euparal or Canada balsam. Finally cover it with a cover glass, pressing down delicately to get rid of any air bubbles. The slides, prepared in this way, labelled, and marked with the reference numbers of the specimens in the collection and the name of the species, are left to dry for two or three days in a small stove with a thermostat set at 35 to 40°C. When they are ready, the slides may be kept in the appropriate cases. An alternative method of storage is to keep the male genital organs, especially large ones, in tiny glass or plastic test tubes containing 95-degree alcohol and a drop or two of glycerine. Always keep the proper reference label with them.

Conservation of Lepidoptera and Their Environment

The IUCN and WWF and Conservation of Lepidoptera

The International Union for the Conservation of Nature and Natural Resources (IUCN) and the World Wildlife Fund (WWF) are two of the principal international bodies working for the conservation of biological diversity on a world scale. Although they work closely together, they operate on quite distinct levels. Roughly speaking, the IUCN with its experts provides the technical support and advice needed by any governmental or public or private body on questions of environmental conservation, whereas the WWF, besides its public-relations activities, promotes and finances projects all over the world for the conservation of particular species or of entire communities and ecosystems. Despite the differences in their approach they work in close collaboration on problems of conservation. We shall see how they work together in the case of the world population of Lepidoptera: Since 1976 a Special Survival Commission has been formed by the IUCN with Butterfly and Moth Specialist groups reporting to it, whose expert lepidopterists, professional and amateur, given the IUCN the benefit of their valuable knowledge in their field. The groups maintain close contact with national associations for the protection of Lepidoptera and set out each year a "Programme of Priorities for the Conservation of Lepidoptera," which may be included in the IUCN's more comprehensive "World Conservation Programme." The latter identifies the areas most urgently needing intervention. The WWF in its turn bases its own financial undertakings on the proposals put forward by the IUCN. The intervention projects of the IUCN and the WWF for the protection of Lepidoptera are of necessity highly varied, with differing degrees of economic commitment. The projects may deal with the conservation of endemic species [for example, Wiest's sphinx moth (*Euproserpinus wiesti*) of North America] or that of genera occurring throughout the world (such as *Papilio*), or they may be designed to protect whole communities of Lepidoptera, which may be more or less widespread (for example, the community of alpine Lepidoptera in Korea or the Lepidoptera of the tropical cloud forest). In some cases their aim is to influence the ecology of rare species (for example, *Morpho*), and so on. Underlying this activity is the immense task of listing all the threatened species of Lepidoptera. This is in fact the primary and most far-reaching WWF project, which in due course will be published as the *Red Book of World Lepidoptera*.

Decline of Species

It is not surprising, given enormous changes in the environment, that even the insects that man loves best should be disappearing. No one today can say how many Lepidoptera species have become extinct in the last half century or how many are now threatened with extinction; it is probably on the order of thousands of species, most in the tropics. Yet it is certain that above all the deterioration of natural surroundings together with changes in climate and occasional instances of excessive collecting are responsible for the declining numbers and shrinking ranges of many Lepidoptera. In broad outline it is clear enough how these three causes are related to the disappearance of butterflies and moths, but it is much less clear when it comes to making accurate estimates, species by species, biotope by biotope. The extinction of many Lepidoptera that is taking place, both locally in terms of populations and globally in terms of species, is a complete anomaly in that it is totally disproportionate to the normal rates of extinction that we know take place in the course of immensely long periods of evolution; today the disappearance of a large number of populations or species is evident in the course of a human life span. The massive decline is due to the increased pace and intensity with which man is changing his environment, the control mechanisms of which are geared to a slower tempo of change in response to much slighter pressures than are now the norm.

We can see how the Lepidoptera in particular are involved in the changes in the balance of the environment merely by considering their principal ecological characteristic: They are phytophagous insects. As such, Lepidoptera are affected directly by damage to the earth's vegetation. Many species are monophagous or oligophagous and are consequently extremely sensitive to the diminution or disappearance of some of their food plants; other species are highly reluctant to leave their own habitat and so are not capable of dispersing in order to escape changes in their biotope. The ecology of many species is such that their survival depends on complex communal relationships or on delicate relationships that are easily changed by, for example, the introduction of new predators, the prevalence of a certain parasite, and the introduction of a new plant that competes with, and succeeds over, the caterpillars and their food plants. Moreover, certain species consist of populations of very few individuals — which makes them genetically vulnerable and even more exposed to the danger of extinction, for example, through individual loss during the reproductive period. Then there are the cases in which changes in the environment cause an erosion of biotopes, which gradually shrink until suddenly they effectively become ecological islands, with consequent adaptation problems that are difficult to overcome. The danger of extinction varies for different geographical areas, biotopes, and groups; butterflies seem to be threatened to a greater extent than are the other Lepidoptera (perhaps only because their ecology and distribution are better known), and the tropical and forest species are in greater danger than are those in temperate zones. Most Lepidoptera species are concentrated in the tropics, where an area of forest as large as forty or fifty football fields disappears every minute. It is therefore not surprising that the tropical species are the most threatened. There are plenty of reports on the decline of Lepidoptera in the scientific literature, but only a few countries have the data about their fauna that will be indispensable if they wish to take measures for their protection. In Great Britain, where there is a traditional awareness of this kind of problem, there are faunal maps showing the precise geographical distribution of the country's Lepidoptera; a catalogue exists of species and biotopes in danger; the demographic progress of certain species is charted over the years on a

Collection and Conservation of Lepidoptera

It is often said that the enthusiasm of collectors does much harm to natural populations of Lepidoptera, but in practice there is no way of estimating the damage actually done by collecting since it would be difficult to carry out the large-scale, direct observations needed to produce reliable data. The effects seem to vary enormously from species to species, from biotope to biotope, from one period to another. The Lepidoptera most in demand by collectors belong mainly to the tropical populations of Papilionoidea (examples include *Papilio*, *Parides*, *Graphium*, *Ornithoptera*, *Delias*, *Appias*, *Heliconius*, *Prepona*, *Agrias*, *Charaxes*, *Morpho*, and *Euploea*), Uraniidae (*Chrysiridia* and *Urania*), Saturniidae (*Attacus*), Sphingidae, and some of the Noctuidae. In the temperate zones the following groups are the most sought after: Papilionidae (*Papilio*, *Parnassius*, *Zerinthia*, and so on), Nymphalidae (*Apatura*, *Erebia*, *Pseudochazara*, *Melitaea*), Lycaenidae (*Lycaena*, *Lysandra*, *Maculinea*), and Pieridae (*Colias*) as well as Zygaenidae and Noctuidae. Species that collectors do not catch for themselves (not everyone can go to South America or Africa) can be bought. There is a flourishing trade in butterflies and moths — which is illegal when it infringes the rules that forbid collection of protected species — and there are retailers to be found in every city in the world; wholesalers operate in strategic places (Taipei, Hong Kong, and Singapore, for instance, are the main clearing houses for the Indo-Malayan region) and are able to satisfy the most exacting requirements. The price of specimens varies according to their rarity, beauty, size, and state of preservation. A butterfly that might earn a Malay collector, say, $0.50 might be sold in Singapore for $12 to $20 and in a European shop for twice as much. Some species are actually quoted on the Insects Exchange in Frankfort and may cost over $400. A very considerable number of butterflies is put on the market: 15 million specimens advertised in Taiwan and over 50 million butterflies from Brazil.
How should we regard this phenomenon in relation to the conservation of Lepidoptera and their environment? It is a rather delicate problem, not least because of the emotional reaction it arouses in some conservationist circles, which amounts to a bitter condemnation of all lepidopterological collecting as "collectionism." However, the less thoughtful and more dogmatic conservationists overlook one important point: The enormous majority of butterflies are caught after the fertilized eggs have been laid. Unless this fact is taken into account in discussion of the ecological problems, the correctness of the information is compromised and attempts to educate the general public are vitiated. We may well be shocked at the number of specimens caught and displeased by some of the uses made of them (like the notorious frame of exotic butterflies bought and hung up in drawing rooms); yet just

local, regional, and national scale; cases of dispersal, immigration, or colonization by species foreign to the resident fauna are well documented; detailed studies of single species and biotopes are made; and protected species are put under ecological surveillance. In some other European countries (Sweden, Germany, Switzerland, and France) and elsewhere (notably in Canada, the United States, Australia, and New Zealand) a wealth of data on the lepidopteran fauna is available, and there are many measures in force to protect threatened species and biotopes.

In the United States the Endangered Species Act of 1973 provides for the listing, protection, and recovery of all species threatened with extinction ("Endangered" category) or liable to become so ("Threatened" category). The law, administered by the United States Fish and Wildlife Service, provides for the protection of Lepidoptera. As of 1984, nine North American Lepidoptera were listed, all as Endangered:
Schaus's swallowtail (*Papilio aristodemus ponceanus*)
Lange's metalmark (*Apodemia mormo langei*)
San Bruno elfin (*Incisalia mossii bayensis*)
Palos Verdes blue (*Glaucopoyche lygdamus palosverdesensis*)
mission blue (*Acaricia icareoides missionensis*)
lotis blue (*Lycaeides idas lotis*)
Smith's blue (*Euplilotes enoptes smithi*)
El Segundo blue (*E. battoides allyni*)
Kern primrose sphinx moth (*Euproserpinus euterpe*)

More than 100 other North American Lepidoptera are considered strong candidates for future listing and protection. Listed species are protected from "taking" by law, and commerce in them and export are prohibited. Land may be acquired for their protection. A good example is the Antioch Dunes National Wildlife Refuge, acquired on behalf of Lange's metalmark. A conservation and recovery plan for the species is presently being carried out. R. A. Arnold and J. A. Powell have documented the butterfly's declining population status; recovery of its host buckwheat populations may reverse the trend. In Great Britain the Nature Conservancy Council and the Institute of Terrestrial Ecology work hand in hand to carry out research and conservation actions for rare and declining species. Several preserves, such as Monk's Wood, carry out management plans for Lepidoptera enhancement, while more specific actions have been taken for a few species in serious trouble, such as the black hairstreak (*Strymonidia pruni*), the large copper (*Lycaena dispar*), and the large blue (*Maculinae arion*) — now extirpated.

Progress in other parts of the world on behalf of Lepidoptera is generally not so advanced. A notable exception is the conservation and farming of the birdwing butterflies (*Ornithoptera*) of New Guinea.

Any measures for the protection of Lepidoptera must be seen as part of the general problem of environmental conservation.

Why, What and How to Conserve

1. The conservation of nature has three main aims, which may be divided into aesthetic, moral, and scientific and practical aspects. With regard to the first two considerations, let us at once say that they are relevant but that they cannot or indeed should not necessarily meet with unanimous agreement. We shall not, therefore, deal with aesthetic questions, such as whether or not diversity and variety are preferable in natural environments to poverty and monotony, and ethical questions, for example, whether or not anyone has a right to destroy deliberately a heritage that also belongs to future generations. We shall focus our attention on the third point: The preservation of natural environments guarantees that ecological processes can continue to function. Obvious

because collecting Lepidoptera involves removing a number of individuals from a population, this does not automatically result in permanent damage to the species in its wild state. In fact we know that the adults, once they have fulfilled their reproductive function, make no difference in the survival of the species. After mating the fate of the species corresponds for practical purposes with the fate of the fertilized eggs; for it no longer matters whether one butterfly among the thousands that will die of old age falls prey to a spider or a net. The decline in the numbers of Lepidoptera in the world is virtually always due to other, much more damaging, human activities, such as the use of chemicals that wipe out insects at every stage of their development, the terrible deforestation of great areas of the earth's surface, and indiscriminate settlement by man. In any case, although there have been rare examples in which collectors have been shown to be responsible for the local extinction of moths or butterflies [as with the British populations of the large copper (*Lycaena dispar*)], collectors realize that what was permissible in the past is not generally acceptable today. Knowing that they need to explore the countryside without leaving ruin in their wake, they are aware that the greatest care is necessary on coming across small populations or those that are threatened in any way; protected species must obviously be left alone. It is good practice to collect only a few adult specimens and limit as much as possible the taking of preimaginal stages, even of common and abundant species. Finally, light traps should be used with extreme caution since they are not selective and may also catch protected species and those which are rare or in decline.

biological limitations make us incapable of creating an absolutely artificial environment in which we can live; consequently our species has to adapt itself to the biosphere and submit to the obligations it imposes on us (by obligations we mean here the maintenance of the regular functioning of ecological processes and of the environmental complex that is the result of millions of years of evolution). Considering our absolute ignorance of the functioning of ecosystems, to modify them with the hope that we should derive therefrom some advantage is at present inconceivable. At this point the scientific aspect becomes a corollary of the practical considerations that compel us to conserve our environment. Only knowledge of all forms of life and of the ecological relationships among them can make the rational control of organic resources possible and hence their balanced economic exploitation. To achieve such an aim, the full range of different environmental situations must be available for the purpose of carrying out research and experiments. What is more, maintenance of the variety and diversity of natural environments also guarantees the preservation of a gene bank of plant and animal species that, as our knowledge progresses, may one day be converted into useful new resources.

2. With the disastrous increase in damage to the environment the need to apply ecology to the conservation of nature has shifted from the level of single populations (conservation of a species) to that of communities and ecosystems (conservation of the natural environment). The principles emerging from scientific research into ecology have led to the view that it is impossible to conserve a species in its wild state without conserving the whole environment. Together with the current state of the environment (the damage being done is not just local and circumscribed but generalized, affecting whole ecosystems) this forces ecologists to adopt a systematic, global approach, the only approach that can provide a positive solution to the problem of the conservation of nature. For the reasons just mentioned current conservation strategy gives priority to protecting entire ecological communities rather than single species. For single species in danger of extinction, of course, the approach is different. In such cases direct, urgent action is required to identify the causes of the threat and modify the conditions that endanger the species. Once this critical phase is over, the species can be considered to be out of danger and its ecological control will once more come within the sphere of broader conservationist measures.

3. Although it is important to take into account regional differences in terms of the ecological and socioeconomic situation that require specific solutions, there exists policy — for intervention to conserve the environment — that is shared by many ecologists today. According to this view, conservation of the environment should aim at "compartmentalized planning" of a region. In practice this would involve moving from the present state of affairs, in which compromise measures are taken, to a phase in which a given area is divided — not of course rigidly — into a certain number of the environmental compartments required by man (for example, urban-industrial complexes, land for agricultural production, a series of natural zones with different levels of protection, and so on). By taking into account the connection that links the various subsystems and by overseeing the transfer of energy and materials among them, this type of planning would make it possible to identify by appropriate analytical techniques (such as computer simulation) the amount of space that should be allocated to each territorial division in order to keep the whole ecosystem in a relatively stable condition of equilibrium.

Bibliography

Barlow, H. S. (1982), *An Introduction to the Moths of Southeast Asia*, Malayan Nature Society.

Brock, J. P. (1971), "A Contribution towards an Understanding of the Morphology and Phylogeny of the Ditrysian Lepidoptera," *J. Nat. Hist.*, 5: 29–102.

Carcasson, R. H. (1981), *Butterflies of Africa*, Collins, London.

Common, I. F. B. (1970), "Lepidoptera," in *Insects of Australia*, Melbourne University Press.

Corbet, A. S., and H. M. Pendlebury (1978), *The Butterflies of the Malay Peninsula*, 3rd ed., Malayan Nature Society.

Covell, C. V., Jr. (1984), *A Field Guide to the Moths of North America East of the Plains*, Houghton Mifflin, Boston.

D'Abrera, B. (1971–), *Butterflies of the World: Australian Region, Afrotropical Region, Neotropical Region, Oriental Region*, Lansdowne, Lansdowne & Classey, Hill House.

Ford, E. B. (1955), *Moths*, Collins, London.

Ford, E. B. (1957), *Butterflies*, Collins, London.

Gómez-Bustillo, M. R., and F. Fernández-Rubio (1974–1979), *Mariposas de la Peninsula Ibérica* (4 vols.), Publicaciones del Ministerio de Agricultura, Madrid.

Heath, J. (1976–), *The Moths and Butterflies of Great Britain and Ireland* (3 vols. pub. of 11 planned), Blackwells & Curwen, Harley.

Higgins, L. G., and N. D. Riley (1970), *A Field Guide to the Butterflies of Britain and Europe*, Houghton Mifflin, Boston.

Hodges, R. W. (ed.) (1971–), *The Moths of America North of Mexico* (12 volumes published to date), Classey & R.B.D. Publications, Washington.

Howe, W. H. (ed.) (1975), *The Butterflies of North America*, Doubleday, Garden City.

Klots, A. B. (1951), *A Field Guide to the Butterflies of Eastern North America*, Houghton Mifflin, Boston.

Lewis, H. L. (1973), *Butterflies of the World*, Albertelli-Leventhal.

Nye, W. B. (1975–), *The Generic Names of the Moths of the World*, British Museum of Natural History, London.

Opler, P. A., and G. O. Krizek (1984), *Butterflies East of the Plains: An Illustrated Natural History*, Johns Hopkins University, Baltimore.

Owen, D. F. (1971), *Tropical Butterflies*, Oxford, London.

Pinhey, E. C. G. (1975), *Moths of Southern Africa*, Tafelberg, Cape Town.

Pyle, R. M. (1981), *The Audubon Society Field Guide to North American Butterflies*, Knopf, New York.

Riley, N. D. (1975), *A Field Guide to the Butterflies of the West Indies*, Collins, London.

Scott, J. A. (1985), *The Butterflies of North America*, Stanford University, Palo Alto.

Watson, A., and P. E. S. Whalley (1975), *The Dictionary of Moths and Butterflies in Color*, McGraw-Hill, New York.

Whalley, P. (1981), *The Mitchell Beazley Pocket Guide to Butterflies*, Beazley, London.

Wickler, W. (1968), *Mimicry in Plants and Animals*, McGraw-Hill, New York.

Index